Agates: Types, Mineralogy, Deposits, Host Rocks, Ages and Genesis

Agates: Types, Mineralogy, Deposits, Host Rocks, Ages and Genesis

Editor

Galina Palyanova

MDPI • Basel • Beijing • Wuhan • Barcelona • Belgrade • Manchester • Tokyo • Cluj • Tianjin

Editor
Galina Palyanova
Siberian Branch of the Russian
Academy of Sciences
Russia
Novosibirsk State University
Russia

Editorial Office
MDPI
St. Alban-Anlage 66
4052 Basel, Switzerland

This is a reprint of articles from the Special Issue published online in the open access journal *Minerals* (ISSN 2075-163X) (available at: https://www.mdpi.com/journal/minerals/special_issues/agates).

For citation purposes, cite each article independently as indicated on the article page online and as indicated below:

LastName, A.A.; LastName, B.B.; LastName, C.C. Article Title. *Journal Name* **Year**, *Volume Number*, Page Range.

ISBN 978-3-0365-2183-1 (Hbk)
ISBN 978-3-0365-2184-8 (PDF)

Cover image courtesy of Galina Palyanova

Contents

About the Editor

Galina Palyanova Ph.D, is a leading scientist at V.S. Sobolev Institute of Geology and Mineralogy of the Siberian Branch of the Russian Academy of Sciences. She is currently a lecturer at the Department of Mineralogy and Geochemistry of Novosibirsk State University. She graduated from the Novosibirsk State University with a degree in geochemistry. Her main area of research is ore mineralogy, as well as thermodynamic and experimental modeling of ore-forming processes. She has published over 90 research articles in international journals. Her monograph Physical and Chemical Features of the Behavior of Gold and Silver in the Processes of Hydrothermal Ore Formation was awarded a medal and a diploma by the Russian Mineralogical Society. She has also been awarded medals by the Russian Ministry of Higher Education and the Russian Academy of Sciences, and an honorary diploma from the Ministry of Natural Resources and Energy Resources for a great contribution to the development of the mineral resource base of Russia. One of the latest directions of her research is the study of ore minerals in agates and their genesis.

Editorial

Editorial for Special Issue "Agates: Types, Mineralogy, Deposits, Host Rocks, Ages and Genesis"

Galina Palyanova

Sobolev Institute of Geology and Mineralogy, Siberian Branch of Russian Academy of Sciences, 630090 Novosibirsk, Russia; palyan@igm.nsc.ru

Citation: Palyanova, G. Editorial for Special Issue "Agates: Types, Mineralogy, Deposits, Host Rocks, Ages and Genesis". *Minerals* **2021**, *11*, 1035. https://doi.org/10.3390/min11101035

Received: 6 September 2021
Accepted: 23 September 2021
Published: 24 September 2021

Publisher's Note: MDPI stays neutral with regard to jurisdictional claims in published maps and institutional affiliations.

Agates are famous, beautiful, and fascinating stones found all around the world. They are classically associated with volcanic rocks but can be found in sedimentary, metamorphic, and igneous environments. Although agates are composed almost entirely of SiO_2, it is the trace quantities of various other elements that give agates their color and lead to their characteristic banding. Despite many in-depth studies and improving analytical and investigative methods, the source of the silica required for the growth of agate remains largely unknown. The first known descriptive reports about agates are from the 18th century. Serious scientific investigations concerning the chemical composition properties and genesis of agates started in the middle of the 19th century. As a result of these numerous studies, a wide spectrum of theories about the formation of agates exists today. The complex, multi-step process of agate formation is not yet completely understood. The aim of this Special Issue is to bring together researchers from different countries to further our understanding of the genesis of agates.

The Special Issue "Agates: Types, Mineralogy, Deposits, Host Rocks, Ages and Genesis" contains two reviews [1,2] and six research articles devoted to detailed studies of agates from Morocco [3], China [4], Russia [5–7], and Cuba [8].

This Special Issue was prepared during the 2020 and 2021 COVID-19 pandemic. Terry Moxon [2] died on 5 January 2021, from COVID-19. He was one of the leading scientists in the research of agates, an excellent reviewer, and a responsible coauthor. Evgeny Sidorov [7] died on 20 March 2021, of complications from COVID-19. He was a very bright, reliable, positive, creative, extraordinary, kind, generous, and wise person as well as an exceptional organizer. I dedicate this Special Issue to Terry Moxon and Evgeny Sidorov.

The review by Götze et al. [1] is an outstanding contribution to this Special Issue. Their paper provides a comprehensive compilation of mineralogical and geochemical data of agates from worldwide occurrences in different host rocks and summarizes the results of extensive studies on agate samples of different origin and type. This article is an excellent introduction to agates for potential researchers who want to study them. Their paper covers almost the entire spectrum of agate problems. Götze et al. [1] provide a preliminary model of agate formation. According to their model, agates begin as silica sols and amorphous silica from silicic acid. Crystallization begins with non-equilibrium spherulitic growth of chalcedony. The banding and colors in agate are "governed by processes of self-organization" with differences due to factors such as porosity, crystallite size, kind of silica phase and incorporated color pigments.

Moxon and Palyanova [2] present a thorough review of the genesis of agates in basic volcanic rocks including relevant literature covering the last 250 years. This review covers research on agate from 1770 to the present, which during this time evolved from physical observations of host rocks and agate thick sections to the sophisticated modern analytical determination of physical and chemical properties. They conclude that agates in mafic igneous rocks form at <100 °C. The precursor to agate is amorphous silica, and agates continue to change over long periods of time, ending with the formation of microcrystalline quartz. They recommend studies of agates hosted by very young rocks.

Pršek et al. [3] report new data on agates from Asni and Agouim (Western Atlas Morocco) and compare their results with previous studies of nearby localities Sidi Rahal and Kerrouchen (Atlas Mountains, Morocco). The authors identified the similarities and differences between agates from these four localities in Morocco. They concluded that the possible sources of silica-bearing fluids were related to the syn- and post-volcanic alteration of the host rocks.

Zhang et al. [4] studied Zhanguohong agates from Beipiao, Liaoning Province (China), hosted by intermediate-felsic volcanic breccia of the Early Cretaceous Yixian Formation. The authors document that the color of the agate changes from yellow, to orange, and then to red when the hematite content increases and the goethite content decreases. They provide a model of the formation process of the rhythmic banding in Zhanguohong agates connected with the fluctuations in the moganite and α-quartz contents and the zonation of the Fe-bearing particles (mixtures of hematite, goethite, and silica phases).

Svetova and Svetov [5] investigated Onega agates from the Paleoproterozoic volcanic complex on the Southeast Fennoscandia (Russia) within the Onega Basin. Agate mineralization is widespread in amygdules and in inter-pillow void spaces of basalts, picrobasalts, basaltic andesites, and agglomeratic tuffs. The authors reconstruct the possible conditions for the formation of Onega agates. They propose that the agates formed during the Paleoproterozoic from the hydrothermal fluid of magmatic and/or meteoric waters are associated with gabbro-dolerite sills.

Svetova et al. [6] provide the first detailed investigations of black agates hosted by volcanic rocks of the Zaonega Formation within the Onega Basin (Karelian Craton, Fennoscandian Shield, Russia). Black agates are rare in nature. The authors describe three main texture types of black agates: concentrically zoned, spherulitic, and mossy. They demonstrate that disseminated carbonaceous matter associated with (fibrous) chalcedony or fine-grained quartz is the major coloring agent. They propose that the source of carbonaceous matter in agates is from underlying carbon-bearing shungite rock, which was redistributed by the Paleoproterozoic hydrothermal system.

Palyanova et al. [7] present the results of studies of the copper-containing agates from the Avacha Bay (Eastern Kamchatka, Russia). They determine that the copper minerals in agates from this location are native copper, cuprite, and various copper sulfides (chalcocite, djurleite, digenite, anilite, yarrowite, and rarely chalcopyrite). Sphalerite, native silver, and barite are also found in these agates. The native copper crystallized simultaneously with early silica. Copper sulfides in the pore and interstitial space of spheroid-layered silica aggregates and cuprite and barite microveins indicate their later deposition. Fluid inclusions suggest that the agates formed at temperatures of <110 °C from very low salinity (<0.3 wt.% NaCl equivalent) solutions.

Götze et al. [8] investigated agates hosted by Paleocene/Eocene tuffs from El Picado/Los Indios, Moa region (the Eastern part of Cuba). The agates consist of high amounts of opal and moganite. Surficial alteration of basic volcanic rocks provides abundant silica for the crystallization of agates in fissures and cavities of the volcanic host rocks from heated meteoric waters.

I hope all the articles in this Special Issue will be a helpful and valuable resource for anyone who is interested in the genesis of agates and will provide a foundation for further research.

Funding: Work is done on state assignment of Sobolev Institute of Geology and Mineralogy of Siberian Branch of Russian Academy of Sciences financed by Ministry of Science and Higher Education of the Russian Federation.

Acknowledgments: I thank the authors of all articles and the organizations that have financially supported research in the areas related to agates. I would like to thank the editor-in-chief of the mineral deposits section, the editorial board, and reviewers for their work on this Special Issue. I am grateful to the late Terry Moxon for his invaluable help. He helped with English proofreading and reviewing of some articles for this Special Issue.

Conflicts of Interest: The author declares no conflict of interest.

References

1. Götze, J.; Möckel, R.; Pan, Y. Mineralogy, Geochemistry and Genesis of Agate—A Review. *Minerals* **2020**, *10*, 1037. [CrossRef]
2. Moxon, T.; Palyanova, G. Agate Genesis: A Continuing Enigma. *Minerals* **2020**, *10*, 953. [CrossRef]
3. Pršek, J.; Dumańska-Słowik, M.; Powolny, T.; Natkaniec-Nowak, L.; Toboła, T.; Zych, D.; Skrepnicka, D. Agates from WesternAtlas (Morocco)—Constraints from Mineralogical and Microtextural Characteristics. *Minerals* **2020**, *10*, 198. [CrossRef]
4. Zhang, X.; Ji, L.; He, X. Gemological Characteristics and Origin of the Zhanguohong Agate from Beipiao, Liaoning Province, China: A Combined Microscopic, X-ray Diffraction, and Raman Spectroscopic Study. *Minerals* **2020**, *10*, 401. [CrossRef]
5. Svetova, E.N.; Svetov, S.A. Mineralogy and Geochemistry of Agates from Paleoproterozoic Volcanic Rocks of the Karelian Craton, Southeast Fennoscandia (Russia). *Minerals* **2020**, *10*, 1106. [CrossRef]
6. Svetova, E.N.; Chazhengina, S.Y.; Stepanova, A.V.; Svetov, S.A. Black Agates from Paleoproterozoic Pillow Lavas (Onega Ba-sin, Karelian Craton, NE Russia): Mineralogy and Proposed Origin. *Minerals* **2021**, *11*, 918. [CrossRef]
7. Palyanova, G.; Sidorov, E.; Borovikov, A.; Seryotkin, Y. Copper-Containing Agates of the Avacha Bay (Eastern Kam-chatka, Russia). *Minerals* **2020**, *10*, 1124. [CrossRef]
8. Götze, J.; Stanek, K.; Orozco, G.; Liesegang, M.; Mohr-Westheide, T. Occurrence and Distribution of Moganite and Opal-CT in Agates from Paleocene/Eocene Tuffs, El Picado (Cuba). *Minerals* **2021**, *11*, 531. [CrossRef]

Review

Mineralogy, Geochemistry and Genesis of Agate—A Review

Jens Götze [1],*, Robert Möckel [2] and Yuanming Pan [3]

[1]   Institute of Mineralogy, TU Bergakademie Freiberg, Brennhausgasse 14, 09599 Freiberg, Germany

[2]   Helmholtz-Zentrum Dresden-Rossendorf, Hemholtz Institute Freiberg for Resource Technology, Chemnitzer Str. 40, 09599 Freiberg, Germany; r.moeckel@hzdr.de

[3]   Department of Geological Sciences, University of Saskatchewan, Saskatoon, SK S7N 5E2, Canada; yuanming.pan@usask.ca

*   Correspondence: jens.goetze@mineral.tu-freiberg.de; Tel.: +49-3731-39-2638

Received: 18 September 2020; Accepted: 17 November 2020; Published: 20 November 2020

Abstract: Agate—a spectacular form of SiO_2 and a famous gemstone—is commonly characterized as banded chalcedony. In detail, chalcedony layers in agates can be intergrown or intercalated with macrocrystalline quartz, quartzine, opal-A, opal-CT, cristobalite and/or moganite. In addition, agates often contain considerable amounts of mineral inclusions and water as both interstitial molecular H_2O and silanol groups. Most agate occurrences worldwide are related to SiO_2-rich (rhyolites, rhyodacites) and SiO_2-poor (andesites, basalts) volcanic rocks, but can also be formed as hydrothermal vein varieties or as silica accumulation during diagenesis in sedimentary rocks. It is assumed that the supply of silica for agate formation is often associated with late- or post-volcanic alteration of the volcanic host rocks. Evidence can be found in association with typical secondary minerals such as clay minerals, zeolites or iron oxides/hydroxides, frequent pseudomorphs (e.g., after carbonates or sulfates) as well as the chemical composition of the agates. For instance, elements of the volcanic rock matrix (Al, Ca, Fe, Na, K) are enriched, but extraordinary high contents of Ge (>90 ppm), B (>40 ppm) and U (>20 ppm) have also been detected. Calculations based on fluid inclusion and oxygen isotope studies point to a range between 20 and 230 °C for agate formation temperatures. The accumulation and condensation of silicic acid result in the formation of silica sols and proposed amorphous silica as precursors for the development of the typical agate micro-structure. The process of crystallisation often starts with spherulitic growth of chalcedony continuing into chalcedony fibers. High concentrations of lattice defects (oxygen and silicon vacancies, silanol groups) detected by cathodoluminescence (CL) and electron paramagnetic resonance (EPR) spectroscopy indicate a rapid crystallisation via an amorphous silica precursor under non-equilibrium conditions. It is assumed that the formation of the typical agate microstructure is governed by processes of self-organization. The resulting differences in crystallite size, porosity, kind of silica phase and incorporated color pigments finally cause the characteristic agate banding and colors.

Keywords: agate; quartz; chalcedony; silica minerals; micro-structure; trace elements; O-isotopes; paragenetic minerals

---

1. Introduction

Agates belong to the most fascinating mineral objects in nature because of their wide spectrum of colors and spectacular morphologies. Therefore, they play a dominant role as gemstones and cut stone since antiquity. The name *"Agate"* can be dated back to ca. 350 B.C. (Theophrast) and was probably related to the discovery of agates in the river *Achates* (recently *Drillo*) in Sicily. Today, agate deposits and agate treatments are known from historical and recent sites all over the world [1–4].

Both in historic and recent times, many speculations have been made about the processes leading to the formation of agates. First descriptive reports and discussions about agates are known from the 18th century [5,6]. Serious scientific investigations concerning chemical composition, properties and genesis of agates started in the middle of the 19th century [7]. As a result of these numerous studies, a wide spectrum of theories, both serious and speculative, about the formation of agates exists today (see compilations in e.g., [1,8]).

Early theories hypothesizing a formation of agates from circulating SiO_2-rich solutions in volcanic rocks, assume that the characteristic agate banding is the result of rhythmic silica supply to natural voids and cavities ("outer rhythm" [7]). Several years later, Daubrée [9] published interesting experiments about the alteration of silicate glasses by alkaline solutions (>400 °C, 1000 atm). These reactions resulted in concentric chalcedony layers, similar to observed structures in natural agates. Comparable results were later published by Nacken [10]. Liesegang [11,12] presented his fundamental results about the formation of rhythmic color bands due to diffusion processes of metal ions in silica gels. He concluded that a zonation and banding can be created by "inner processes" such as simple chemical reactions. However, later studies revealed that banding and color distribution in agates are not always identical [8]. The characteristic color banding in agates seems to be more likely related to the micro-structure and the distribution of finely dispersed color pigments within the SiO_2 matrix.

Recent mineralogical and geochemical investigations during the last decades significantly increased our knowledge about the mineralogy and genesis of agates. Numerous publications about certain agate occurrences or specific aspects of agate formation have been published (e.g., [1,8,13–60]).

According to these recent studies, agates can be regarded as banded chalcedony, which is intergrown or intercalated with other silica phases. Moreover, agates often contain considerable amounts of mineral inclusions and water (1–2 wt%) as both interstitial molecular H_2O and silanol groups [13,21,40]. These impurities may form spectacular internal structures or may be responsible for the different coloration of agates [13,29,40,61].

The chemical and mineralogical complexity of agate is a result of its complex formation history. In principle, agates occur in practically all rock types however, their formation in volcanic rocks is the most abundant. Therefore, the majority of agate deposits worldwide is related to SiO_2-poor (andesites, basalts) and SiO_2-rich (rhyolites, rhyodacites) volcanic rocks. Other types of agate may occur as hydrothermal vein agate or due to silica accumulation in sediments and sedimentary rocks [1,8,15,49]. The chemical composition of agates as well as the presence of specific paragenetic minerals (carbonates, clay minerals, zeolites, iron compounds) indicate that the formation of volcanic agates seems to be associated with late- or post-volcanic alteration processes within the host rocks (e.g., [24,37,62,63]).

Although general ideas concerning the formation of agates exist, the origin of agate remains incompletely understood. The discussions are controversial especially due to the fact that no one has unambiguously documented agate formation in real time and agates have never been successfully synthesized in the laboratory. Only certain silica mineralization in active geothermal regions or at the sea-floor (white smokers) is comparable to agates [64].

The present paper provides a global review of what is known on agate properties, based on the literature and completed by numerous additional data. On this basis, an attempt is made to establish a preliminary model of agate formation.

2. Materials and Methods

In the present paper results of extensive studies on agate samples of different origin and type are summarized and compared with data from literature. The analyzed material includes more than 300 agate samples from all over the world and from different parent rocks (basic and acidic volcanic rocks, hydrothermal vein agates, agates from sedimentary host rocks—see Table S1). The agate samples as well as the surrounding host rocks were visually described and documented, and selected aliquots prepared for further investigations.

The mineral and chemical composition of the agate host rocks was generally analyzed by a combination of *X*-ray diffraction (XRD) and *X*-ray fluorescence (XRF) measurements on selected and prepared sample material. The qualitative and quantitative phase compositions were analyzed using an URD 6 (Seifert/Freiberger Präzisionsmechanik, Freiberg, Germany) with Co Kα-radiation in the range 5–80° (2θ). Data evaluation was realized using Analyse RayfleX v.2.352 software (GE Sensing & Inspection Technologies GmbH, Ahrensburg, Germany) and subsequent Rietveld refinement with Autoquan v.2.7.00 (GE Sensing & Inspection Technologies GmbH, Ahrensburg, Germany) [65]. XRF measurements on powdered samples mixed with Li-tetraborate were carried out with a PANalytical Axios Minerals spectrometer and WROXI package (PANalytical, Almelo, The Netherlands) [37].

The identification of different SiO_2 phases in the agate samples was done by a combination of *X*-ray diffraction and Raman spectroscopy [29]. Additional results were obtained from microscopic investigations by polarizing, cathodoluminescence (CL) and scanning electron microscopy (SEM) on polished thin sections (30 µm). Conventional polarizing microscopy was made with a Zeiss Axio Imager A1m microscope (ZEISS, Thornwood, NY, USA). These investigations were completed by SEM studies using a JEOL 6400 SEM with EDX detector (JEOL Ltd., Akishima, Japan). CL measurements were made with an optical CL microscope HC1-LM (LUMIC, Bochum, Germany) on carbon-coated, polished thin sections [66]. CL microscopy and spectroscopy was performed with a Peltier cooled digital video-camera (OLYMPUS DP72, OLYMPUS Deutschland GmbH, Hamburg, Germany) and an Acton Research SP-2356 digital triple-grating spectrograph with a Princeton Spec-10 CCD detector (OLYMPUS Deutschland GmbH, Hamburg, Germany), respectively [67].

The paramagnetic centers of powder agate samples were analyzed by electron paramagnetic resonance (EPR) spectroscopy using a Bruker EMX spectrometer (Saskatchewan Structural Science Centre, Saskatoon, Canada). The equipment operated with microwave frequencies of ~9.63 GHz and 9.39 GHz at 295 K and 85 K, respectively. Experimental conditions included modulation frequency of 100 kHz, modulation amplitude of 0.1 mT, and microwave powers from 0.2 to 20 mW to obtain optimal conditions for different center types. The spectral resolutions were ~0.146 mT for wide scans from 50 to 650 mT and 0.024 mT for narrow scans from 300 to 350 mT.

Trace-elements in agates were analyzed by instrumental neutron activation analysis (INAA—XRAL Laboratories, Mississauga, ON, Canada), inductively coupled plasma mass spectrometry (ICP-MS) and laser ablation ICP-MS (LA-ICP-MS), respectively [30,68]. Additional sections perpendicular to the agate banding were prepared for selected agate samples to analyze the spatial distribution of trace elements. The thick sections were analyzed using a Thermo Finnigan Element 2 mass spectrometer with a DUV 193 laser ablation system [69] and alternatively with a double-focusing sector field mass spectrometer ELEMENT XR with a NewWave 193 nm excimer laser probe (Thermo Scientific, Waltham, MA, USA) [70]. Additional measurements were made with conventional ICP-MS using a Perkin Elmer Sciex Elan 5000 quadrupole instrument with a cross-flow nebulizer and a rhyton spray chamber (Perkin Elmer Inc., Baesweiler, Germany) [71] to facilitate the quantification of the very low contents of REE and some other elements in the agate samples.

Isotope and fluid inclusion studies on quartz and chalcedony samples were carried out to get information about the formation temperatures of the agates. The oxygen isotope composition of SiO_2 phases was examined by a laser-based micro-analytical method in the Stable Isotope Laboratory of the University Lausanne, Lausanne, Switzerland. Separate chalcedony and macrocrystalline quartz pieces were prepared, chemically treated and analyzed with a Thermo Fisher Scientific MAT 253 mass spectrometer [72]. Fluid inclusions in macrocrystalline quartz of the agates (polished 100 µm thick sections) were investigated using a Linkam THMS 600 heating-freezing stage. For the calibration of homogenization and freezing measurements, two synthetic fluid inclusion standards (SYN FLINC; pure H_2O, mixed H_2O-CO_2) were used [37]. In addition, the composition of gaseous and liquid fluids in the SiO_2 matrix of the agates was analyzed using a thermogravimetry-mass-spectrometry system (NETZSCH STA 409; Netsch, Selb, Germany), which was coupled to a QMS 403/5 quadrupole mass spectrometer (Pfeiffer Vacuum, Aßlar, Germany) [73].

Last but not least, the investigations included mineralogical and geochemical studies on paragenetic minerals and organic materials in agates to get more detailed information about alteration, accumulation and crystallisation processes. In general, associated minerals were identified by a combination of polarizing and scanning electron microscopy (SEM), X-ray diffraction and Raman spectroscopy. Raman measurements were carried out using a confocal Jobin Yvon (Horiba) LabRam-HR spectrometer (Horiba, Bensheim, Germany) on an Olympus BX41 optical microscope, with an 1800-lines-per-millimeter diffraction grating and a Peltier-cooled, Si-based CCD detector. The analyses were complemented by Infrared (IR) absorption spectra acquired using a Bruker Hyperion microscope interfaced to a Bruker Tensor 27 Fourier transform infrared spectrometer [35].

The trace-element and isotopic composition of paragenetic calcite in certain agates was analyzed because of its ubiquitous occurrence with agates in the form of intergrowths and pseudomorphs. Trace elements were measured by INAA (XRAL Laboratories, Toronto, ON Canada). The C and O isotope composition of paragenetic calcite was analyzed with the method of acid digestion using a Finnigan MAT 252 mass spectrometer in the Stable Isotope Laboratory of the University of Lausanne (Lausanne, Switzerland) [74].

3. Geological Occurrences and Types of Agates

In general, agate occurrences are distributed around the world on all continents, and agates have already been formed very early in the Earth's history. The oldest known occurrence, the Warrawoona agate in Western Australia, was found in 3.48 Ga old metamorphosed rhyolitic tuffs [75]. More than one billion years old agates are also known from the basalts of the Lake Superior region in the USA and Canada.

A closer view reveals that agate occurrences are in particular connected with geological periods of strong volcanic activities such as huge basaltic lava flows or eruptions of acidic lava from Permian to Tertiary. Chemical and mineralogical analyses of the host rocks show that most global agate occurrences are related to both SiO_2-poor (andesites, basalts) and SiO_2-rich (rhyolites, rhyodacites) volcanic rocks (Figure 1).

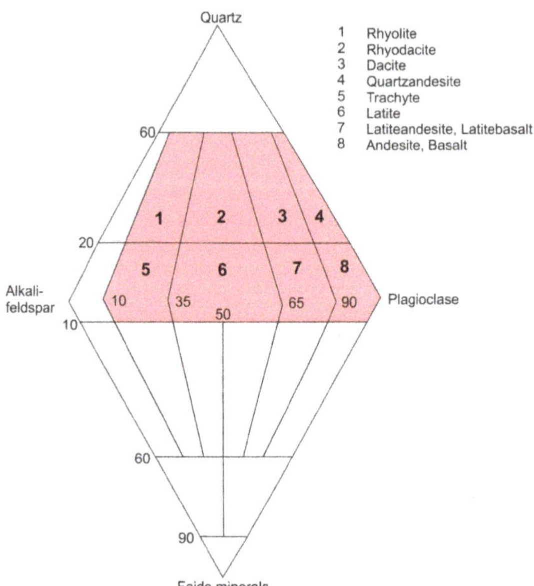

Figure 1. Streckeisen diagram showing the field (**red**) of agate-bearing volcanic rocks.

In acidic volcanics, agates originate from the infill of silica into cavities of spheroidal aggregates, so called lithophysae (thundereggs), whereas agate formation in basic volcanic rocks happens in former vesicular cavities (Figure 2a–d). The occurrences of *volcanic agates* are in particular a result of alteration processes of the volcanic host rocks. Therefore, they are preferentially formed in the marginal parts of volcanic bodies and ignimbrite layers, which are enriched in water and other volatiles [22,37].

Figure 2. Compilation showing the different main agate types; (**a**) lithophysa (thunderegg) within the flow texture of the rhyolitic host rock from Nesselhof, Thuringia (Germany); (**b**) large thunderegg with agate from rhyolitic ignimbrite of St. Egidien (Saxony, Germany); (**c**) andesitic host-rock with both empty and agate-filled vesicles at the Rancho Coyamito (Chihuahua, Mexico); (**d**) finely banded agate from a trachyandesitic host-rock of Xuanhua (Hebei province, China); (**e**) hydrothermal vein agate in metamorphic host rocks (gneisses) from Röthenbach (Saxony, Germany); (**f**) sedimentary banded microcrystalline granular quartz from a clay seam in carbonate host-rocks from Dulcote (Mendip, UK); (**g**) clastic agate pebble from river sediments of the Yangtze river near Nanjing (Jiangsu province, China); Scale bar is 5 cm.

It is very likely that the process of agate formation in many cases already started during the volcanic activities. Several investigations indicated the participation of residual magmatic fluids in the agate formation and the heating of meteoric fluids, which are necessary for the alteration of the rocks and the release of silica and other chemical elements forming secondary minerals (e.g., [22,24,37]). However, in certain occurrences such as the Paraná Continental Flood Basalt Province, Brazil or the Deccan Trap Province, India strong indications were found that the agate formation took place remarkably later than the emplacement of the surrounding volcanic host rocks [76–78].

Besides the preferred formation in volcanic rocks, agates can also be formed by other processes and in other rock types. In different host rocks, vein agates and agate-like structures occur in breccia, cracks and fissures, which are often mineralized by hydrothermal solutions. *Hydrothermal vein agates* can be found in mm- and cm-sized veins (Figure 2e) but can also reach thicknesses of several dm and lengths of several hundreds of meters. The vein agates are often related to hydrothermal ore deposits. The formation of vein agates is not restricted to volcanic host rocks but can also be found in igneous and metamorphic host rocks.

The formation of agates in sedimentary environments is relatively rare compared to that in volcanic host rocks. *Sedimentary agates* have been found especially as irregular forms in stratigraphic sequences of carbonate rocks and clastic sediments (Figure 2f). In addition, silicification of residues of animals and plants is common in the surface region of certain sediments and volcano-sedimentary units. Agate structures are well known from silicified wood remains worldwide and silicified dinosaur bones from Utah, USA or silicified corals from Florida, USA [79–81]. Another common feature is the secondary deposition and redistribution of agates from primary deposits. Agates in clastic sediments (river gravel and marine sediments) are known from the surroundings of many agate occurrences worldwide (Figure 2g).

4. Agate Properties

4.1. Mineralogy and Micro-Structure of Agates

4.1.1. SiO$_2$ Phases in Agate

Numerous mineralogical investigations have shown that most agates consist of more than one silica phase [23,29,40,82]. The most common mineral in agate is trigonal α-quartz. It is present both as macrocrystalline and granular micro-crystalline variety, as well as in form of the disordered fibrous quartz varieties chalcedony (length-fast) and quartzine (length-slow).

Chalcedony is the common and dominant form of silica in agates. One of the most conspicuous features under the polarizing microscope is the presence of characteristic chalcedony "fibers" (Figure 3a), which was first reported by Brewster [83]. Investigations by X-ray diffraction confirmed the assumption that these structures consist of α-quartz, and the dimensions of the "fibers" were measured with diameters of 0.1–1 µm and lengths of up to several mm [84].

However, electron microscopic studies could not confirm such a fiber structure of chalcedony at high magnification [8,18,85]. Instead, the chalcedony "fibers" represent an intergrowth of quartz microcrystals of ca. 0.1 up to 3 µm in size (Figure 3b), which are oriented with the c-axis perpendicular to the "fiber" direction (length-fast), whereas their orientations are parallel to the c-axis in quartzine (length-slow). The fibrous "chalcedony crystals" are polysynthetically twinned according to the Brazil-twin law, with the c-axis slightly twisted around the fiber axis [23,86].

Heaney [23] developed a model of "dislocation growth ", which can explain the formation of the fibrous structure of chalcedony and the high frequency of defects, Brazil twinning and twisting in the chalcedony "fibers". Accordingly, the formation of dislocations in the crystal structure during crystal growth promotes the arrangement of atoms/molecules into spiral layers, which is energetically favoured compared to the incorporation into plane crystal faces. Such dislocations can be initiated by point defects or incorporated impurity ions in the crystal structure. The varying torsion of the c-axis (the optical axis) of the quartz crystallites causes different optical orientation and results in the formation of so called "Runzelbänderung" in polarized light (Figure 3a).

Quartzine is rare in agates and occurs only on the outer edge of the geodes or as interlayers between chalcedony and macrocrystalline quartz (Figure 4). Folk and Pittmann [87] concluded that the presence of quartzine may point to a deposition in sulphate-rich solutions or evaporitic conditions. However, up to now the specific conditions leading to the formation of quartzine are not completely understood.

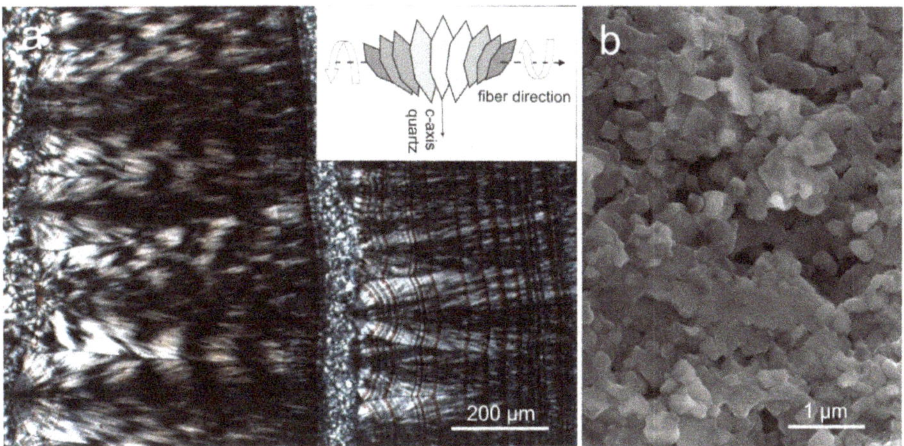

Figure 3. (**a**) Micrograph of an agate from Schlottwitz, Germany in transmitted light (crossed polars) showing chalcedony "fibers" with the characteristic "Runzelbänderung" caused by twisting of the optical axis of the quartz crystallites oriented with their c-axes perpendicular to the "fiber" axis; the inset schematically shows the formation of chalcedony "fibers"; (**b**) SEM micrograph of the same area revealing numerous quartz microcrystals <1 µm in size.

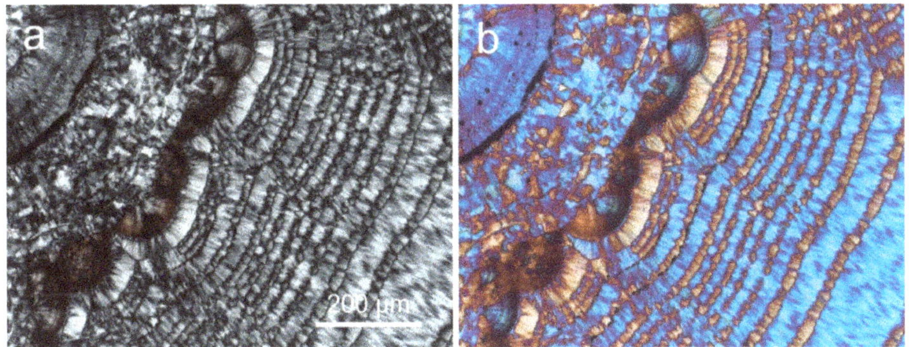

Figure 4. Agate from Xuanhua (Hebei province, China) in transmitted light (crossed polars) (**a**) and with λ compensator (**b**) showing alternating bands of fibrous chalcedony (blue) and quartzine (yellow).

Recent studies have shown that opal-CT, opal-C, cristobalite and moganite are also common in many agates worldwide [29,40,45,54]. Opal-CT represents an intermediate stage between pure amorphous and crystalline silica. It often occurs as tiny spherical aggregates (lepispheres) formed by thin crystal blades [82]. The crystal structure consists of fine interlayers of crystalline cristobalite and tridymite sometimes intergrown with amorphous opal-A (Figure 5). Opal-CT was detected e.g., in agates from the Richardson Ranch, Oregon (USA), Los Indios (Cuba), Sidi Rahal (Morocco) and Jalgaon/Savda, India [45,49].

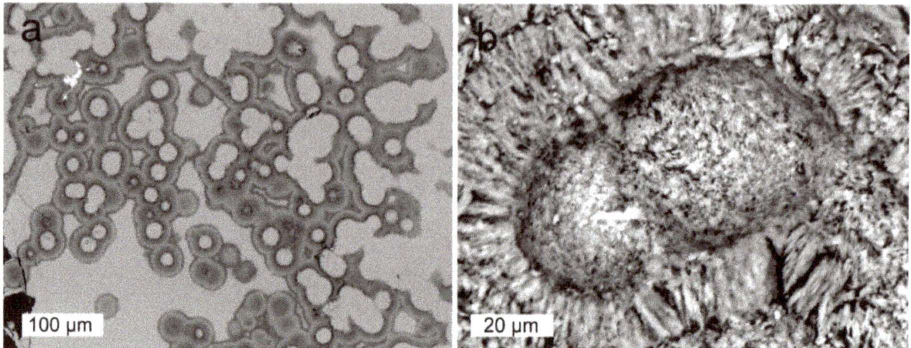

Figure 5. (**a**) SEM migrographs of spherical opal-CT aggregates within a chalcedony matrix in an agate from El Picado, Cuba; (**b**) close-up of (**a**) showing lepispheres of opal-CT consisting of radially arranged crystal blades.

Moganite was first described by Flörke et al. [88] from ignimbrites of the Mogan formation, Gran Canaria. The crystal structure of this monoclinic SiO_2 modification is characterized by alternate stacking of layers of right- and left-handed quartz, with a periodic twinning according to the Brazil-twin law at the unit-cell scale [86]. Although the occurrence of pure moganite is rare, the intergrowth of chalcedony with moganite is frequent in agates, chert and flint [29,54,59,82,89,90].

Because of the similar optical properties and the narrow intergrowth with chalcedony it is difficult to reveal the presence of moganite in agate. However, Raman spectroscopic studies showed that moganite and quartz can be distinguished based on their different spectral characteristics [91]. Using these results, Götze et al. [29] first documented the variations in moganite content in different parts of agates and even within the chalcedony banding. Accordingly, Raman profiles or Raman mapping [92] can provide detailed information about the quantitative spatial distribution of moganite in agates (Figure 6).

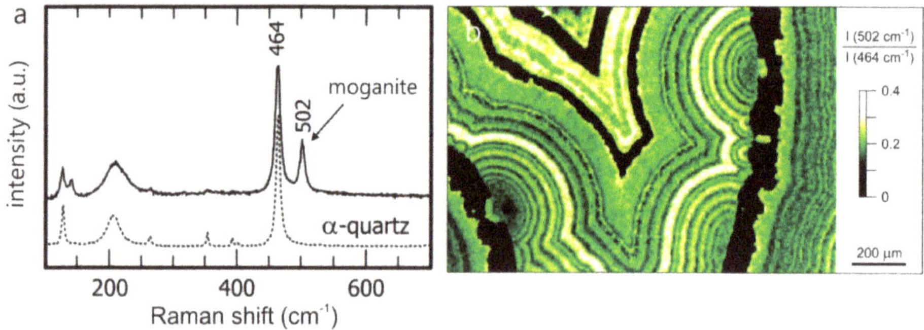

Figure 6. (**a**) Raman spectra of α-quartz (dotted line) and moganite-bearing chalcedony showing the characteristic peaks at 464 cm^{-1} and 502 cm^{-1}, respectively; (**b**) Raman mapping revealing the moganite distribution in an agate from St. Egidien (Saxony, Germany) based on the intensity ratio of the two characteristic Raman bands (modified after Nasdala et al. [92]—courtesy Lutz Nasdala).

In contrast to the microcrystalline SiO_2 species, amorphous opal-A is rare in agate. Previously it was concluded that agates contain about 10% of opal-A [93,94]. However, Flörke [95] first suggested that most agates do not contain any opal-A. It can be assumed that most of the identified amorphous silica reported in the older literature is likely caused by the powder preparation process for XRD measurements. Amorphous opal-A is mainly restricted to recent siliceous sinters, where it relatively rapidly converts to opal-CT [96].

The presence and spatial distribution of different silica phases is not only a result of the primary crystallisation processes but can also be influenced by secondary processes such as ageing, metamorphosis and/or temperature. These secondary processes can significantly change the primary phase composition, especially due to the conversion of unstable or metastable silica phases (opal-A, opal-CT, moganite) into quartz. The comparison of natural agate samples with synthetically treated agates showed that the conversion of metastable silica phases into quartz is accompanied by coarsening of crystallite size and loss of structural water [19,21,75,97,98]. Moxon and Carpenter [61] reported that in 300 to 1100 Ma old agate samples from 9 regions six had moganite contents below 6 wt% and three contained 1 wt% or less. The systematic change in phase composition and structural parameters may have potential to estimate the age of agates.

4.1.2. Microstructure of Agates and Agate Banding

One of the most conspicuous features of almost all agates is the typical agate banding. For the characterization of the general microstructure of agates, at least two types of agate banding have to be distinguished: (I) wall-lining agates ("fortification agates") with agate banding that is more or less conforming the outline of the host cavity (Figure 7a), and (II) horizontally layered agates ("Uruguay agate", water line agate) with silica layers parallel to the previous horizontal orientation (Figure 7b). There are certain intermediate states and combinations of these two main agate types (Figure 7c).

Figure 7. Examples for different types of banding in agates; (**a**) wall-lining ("fortification") agate from Ojo Laguna (Chihuahua, Mexico); (**b**) horizontal banding ("Uruguay agate") in an agate from Bightbridge, Simbabwe; (**c**) mixed type with both wall-lining and horizontal banding as well as macrocrystalline amethyst in the center from Tres Pinheiros-Fontoura Xavier (Rio Grande do Sul, Brazil); Scale bar is 2 cm.

The general structure of wall-lining agates often consists of three main parts/zones: 1. outermost layer of either microcrystalline, granular quartz or spherulithic chalcedony (or sometimes quartzine) in many cases including other minerals like e.g., calcite, hematite and/or pseudomorphs, 2. main zone of fibrous chalcedony, which develops from the spherulitic chalcedony, and 3. central part, where chalcedony passed over to macrocrystalline quartz, sometimes containing a hollow centre (Figure 8). Repeated layers with varying crystallite size and/or intercalations of fibrous chalcedony and granular quartz (or quartzine) are possible.

Figure 8. Cross section of a wall-lining agate from Wiederau (Saxony, Germany) from margin (left) to the center (right) in transmitted light (crossed polars) representing the general textural succession of spherulitic and micro-granular chalcedony—fibrous chalcedony—macrocrystalline quartz (width of the micrograph is 10 mm).

The formation of horizontally layered "Uruguay agates" can be explained by the gravitational sedimentation of sols with flocculated, large SiO_2- particles. Under conditions of fast coagulation (flocculation) of the monomeric silicic acid (H_4SiO_4) horizontal layers are formed [99]. The formation of sol particles requires a high silica concentration and therefore, horizontally layered agates are often characteristic features for specific agate occurrences (Figure 9a). Walger [62] reported that horizontal layers in agates from Permian volcanic rocks in Germany are strictly parallel to the surface and therefore, can be used as geological level. The sporadic presence of discordant silica layers in agates (Figure 9b,c) indicates that the orientation of the nodule has been changed during the formation of the agate bands.

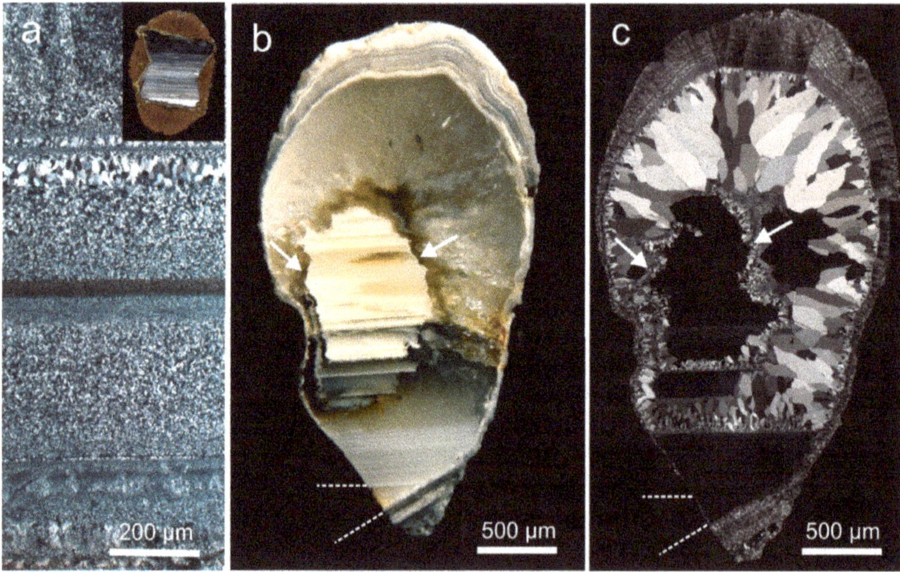

Figure 9. (**a**) Transmitted light (crossed polars) micrograph of an "Uruguay agate" from the Richardson Ranch (Oregon, USA, see inset); the different horizontal layers show strong variations in crystallite size (**b**,**c**) agate from Gröppendorf (Saxony, Germany) as original sample (**b**) and in transmitted light (crossed polars, (**c**)) with tilted horizontal layers and ongoing re-crystallisation of quartz replacing former horizontal bands (see arrows).

Microscopic investigations with high magnification revealed that visible macroscopic bands may consist of up to 1000 micro-bands per mm. A closer look reveals that agate banding appears due to periodic change in the characteristics of the microscopic sub-particles. Accordingly, the visible banding is caused by variations in the type of silica phases, their crystal size and habit, as well as porosity and the distribution of micro-inclusions of minerals (e.g., Fe-oxides/-hydroxides) within the SiO_2 matrix (Figure 10, compare Figures 4 and 6). Holzhey [100] published diameters of the submicroscopic SiO_2 particles between 0.07 to 0.7 µm for wall-lining agates and 0.2 to 2.5 µm in horizontally banded agates, which is similar to the diameters of microscopic globules in opal (0.25 to 0.4 µm; [101]).

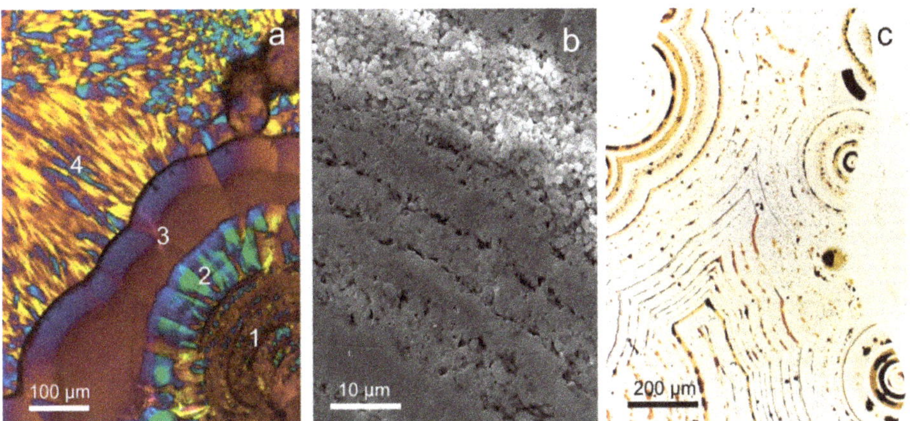

Figure 10. (**a**) Micrograph in transmitted light (crossed polars, λ-compensator) of an agate from the Whale Bay, Iceland showing a sequence of microcrystalline, granular quartz (1) quartzine (2) opal-C (3) fibrous chalcedony (4); (**b**) SEM micrograph of an agate from St. Egidien (Saxony, Germany); the high magnification reveals significant variations in crystal size and porosity in the different chalcedony layers; (**c**) transmitted light micrograph of an agate from Wiederau (Saxony, Germany); fine-dispersed Fe-oxides/-hydroxides are aligned along the micro-banding and cause the visible color banding.

Beside the characteristic agate banding, additional characteristic microtextures can provide information concerning the physico-chemical conditions of formation. For instance, primary growth textures such as colloform or botryoidal textures indicate precipitation of silica gel in free space, whereas comb-like, crustiform or zonal textures in quartz crystals point to a direct crystallisation from hydrothermal fluids [102]. Growth lines in euhedral quartz crystals, so called "Bambauer lamellae", are preferentially formed under strongly varying physico-chemical conditions during agate formation [46,52,55,60]. Most of these features can be related to epithermal conditions of formation and frequently occur in hydrothermal vein agates.

Other microtextures (feathery/flamboyant, mosaic/jigsaw-puzzle) are significant indications for their generation during recrystallisation of metastable silica phases such as amorphous silica precursors or chalcedony [46,49,55,60,102]. Replacement textures (lattice-bladed, pseudo-acicular) result from the secondary replacement of soluble phases (e.g., calcite, barite) by quartz. In general, such quartz textures due to the recrystallisation of amorphous silica or replacement textures are good indications of boiling conditions in epithermal environments. An excellent compilation about these microtextures in quartz is given by Dong [102].

4.1.3. Point Defects in Agate

Chalcedony, quartzine and macrocrystalline quartz are the most frequent SiO_2-forms in agates. Despite their different structural specifics, all three species show the same basic crystal structure of trigonal, low-temperature alpha-quartz that is composed exclusively of $[SiO_4]^{4-}$ tetrahedra with all oxygens joined together in a three-dimensional network. Different factors (e.g., concentration of SiO_2 and impurities in the mineralizing solution, temperature, growth velocity, etc.) can result in the formation of defects in the quartz lattice during crystallisation. Besides extended defects such as micro-inclusions of minerals and fluids or twinning and dislocations, defects at the atomic scale of the crystal lattice (zero-dimensional point defects) are present [103–106].

First investigations of the defect structure of agates from different parent rocks of certain world-wide localities by a combination of electron paramagnetic resonance (EPR) spectroscopy, cathodoluminescence (CL) and trace element analysis showed that defects related to silicon and/or oxygen vacancies are the most frequent point defects in agate [107]. In addition, the substitution of Si atoms by Al, Ge and/or Fe is common. In agate samples from different localities the following paramagnetic centers were detected: O_2^-, O_2^{3-}, O_3^-, E'_1, $[AlO_4]^0$, $[FeO_4/M^+]^0$, and $[GeO_4/M^+]^0$. In contrast, defect centers of the type $[TiO_4/Li^+]^0$ or $[TiO_4/H^+]^0$, which were common in quartz of the parent volcanic rocks, were not detected in agate [107]. In general, the abundance of the O_2^-, O_2^{3-}, O_3^-, centers (related to silicon vacancy) and E'_1 centers (oxygen vacancy) in chalcedony is remarkably higher than in macrocrystalline quartz. This high defect density indicates rapid growth of silica under non-equilibrium conditions probably with a non-crystalline precursor.

Figure 11a shows the spectra of three selected agate samples from Germany in a wide scan measured at a microwave power of 2 mW. The EPR spectra contain a characteristic oxygen-vacancy electron center E'_1 [108] and an orthorhombic Fe^{3+} center at the effective g value of 4.28 [103,109]. A closer look at the central magnetic region (Figure 11b) shows further multiple species of silicon-vacancy hole centers (e.g., superoxide and ozonide radicals such as B, B′, C, and C′; [110–115]). The agate from St. Egidien, Saxony (Germany) is particularly interesting due to abundant B/B′ and C/C′ centers, even more than quartz samples from high-grade uranium deposits [109,115]. Weak signals at g = 1.997 and 1.994 belong to the Ge (B) and Ge E'_1 centers [108]. The simulated g values of the $[AlO_4]^0$ center are shown in the experimental spectrum in Figure 11c compared with the analyzed spectrum of an agate from Hausdorf (Saxony) containing 709 ppm Al.

The comparison of the EPR spectra of worldwide agate samples shows, that the oxygen-vacancy electron center E'_1 is the dominant defect center but in varying concentrations (Figure 12). In particular chalcedony in agates from acidic volcanic rocks is characterized by elevated concentrations of the E'_1 center, much more than chalcedony in agates from basic volcanic rocks and macrocrystalline quartz within the agates. In general, quartz in agates has much more abundant oxygen-vacancy electron centers (by several orders of magnitude) than hydrothermal and pegmatite quartz that were measured for comparison (e.g., from Minas Gerais, Brazil, and Sichuan, China).

In addition, EPR spectra of the chalcedony parts of the agates often show high intensities of silicon-vacancy hole centers (Figures 11 and 12), whereas the number of these centers is lower in the macrocrystalline quartz part of the agates (Figure 11d). Nevertheless, the EPR analysis of clear macrocrystalline quartz crystals in agates provided powder-like spectra indicating that the quartz crystals are not really homogeneous but consist of numerous "micro-crystals".

Luminescence methods are suitable analytical tools for the visualization of point defects and therefore, provide useful information about the defects in the agate structure and their spatial distribution [117]. The common use of UV light (photoluminescence—PL) produces luminescence images, which reveal the heterogeneous microtexture of agates (Figure 13). The frequently observed bright green luminescence is caused by activation due to uranyl $((UO_2)^{2+})$ ions [36] and therefore, clearly indicates trace amounts of uranium in the SiO_2 network (Figure 13b).

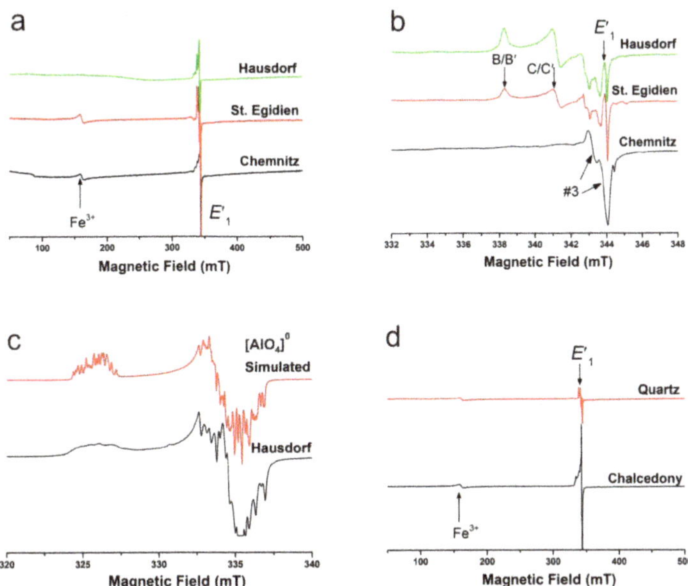

Figure 11. Powder EPR spectra of agate samples from Hausdorf, St. Egidien and Chemnitz (Saxony, Germany); (**a**) wide scans (microwave power of 2 mW and 295 K) showing a rhombic Fe^{3+} center as well as a pronounced E'_1 center in all samples; (**b**) narrow scan for the central magnetic field region (microwave power of 20 mW and 295 K) showing multiple species of silicon-vacancy hole centers (B/B′, C/C′, and #3); (**c**) narrow scan of the central magnetic field region (microwave power of 20 mW and 85 K) showing the $[AlO_4]^0$ center and silicon-vacancy hole centers in the agate from Hausdorf; also shown for comparison is a simulated spectrum of the $[AlO_4]^0$ center (data from Walsby et al. [116]); (**d**) wide scans (microwave power of 2 mW and 295 K) showing that the E'_1 center in chalcedony part of the agate sample from St. Edigien has a markedly higher intensity than its counterpart in macrocrystalline quartz from the center of the agate geode.

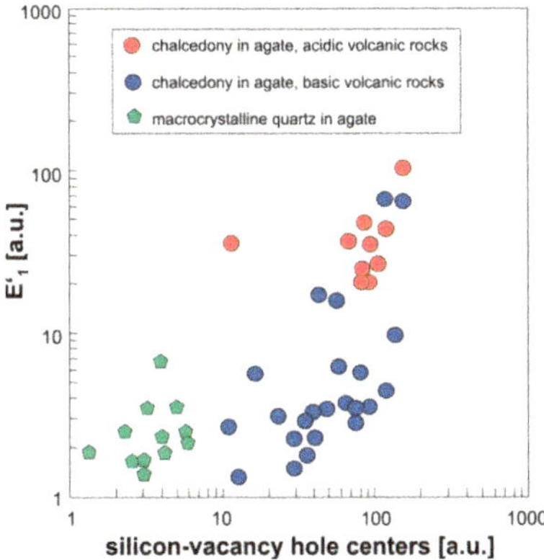

Figure 12. Relative abundances of silicon-vacancy hole centers (O_2^-, O_2^{3-}, O_3^-) and the oxygen-vacancy electron center E'_1 in chalcedony of agates from acidic and basic volcanic rocks, respectively, as well as macrocrystalline quartz in agates (data from Götze et al. [107]).

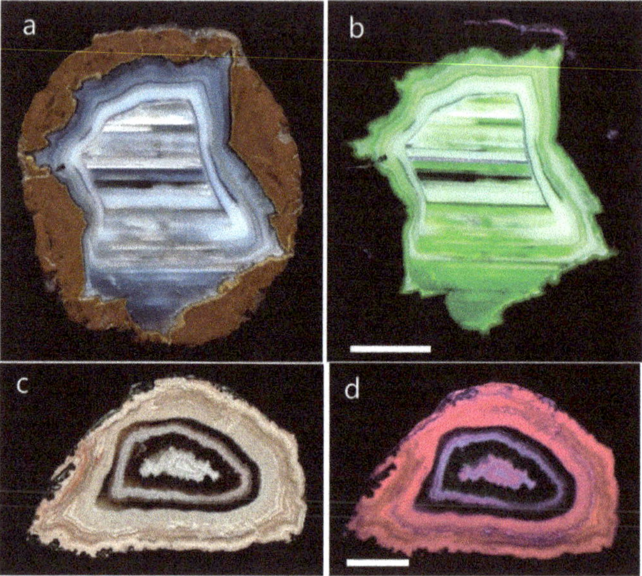

Figure 13. Agates from the Richardson Ranch, Oregon (USA, (**a**,**b**)) and from Idar-Oberstein (Germany, (**c**,**d**)) in day light (left) and under short-wave UV excitation (Photoluminescence, right); scale bar is 2 cm; the green luminescence is caused by uranyl ions (UO_2^{2+}), whereas the dominant reddish PL can be related to structural defects in the SiO_4-network.

Cathodoluminescence microscopy and spectroscopy can also be used to reveal differences in the microstructure of macrocrystalline quartz and chalcedony in the agates. Cathodoluminescence microscopy enables the visualization of zoning and other internal structures in agates and quartz incrustations at the micro-scale, which often markedly different to observations with conventional polarizing microscopy (Figure 14). For instance, most of the apparent homogeneous quartz crystals show a wide variety of internal textures under CL including oscillatory zoning, sector zoning, or skeletal growth (see Figure 14b,f). These irregular internal textures and sector zoning point to silica crystallisation under non-equilibrium conditions.

The visible CL of agates can vary drastically (Figure 14). On one hand, agate samples can exhibit more or less one dominant CL colour, whereas in other agates SiO_2 phases exhibit multiple CL colors with different shades of blue, violet, green, yellow, and red. The visible luminescence colours are caused by different CL emission bands from the blue to the red spectral region with variable intensities, which are associated to various structural defects (e.g., oxygen/silicon vacancies or broken bonds) and/or trace elements (e.g., Al, Li, Fe) in the lattice [117]. Moreover, CL can help to visualize different SiO_2 generations (Figure 14g) or secondary alteration effects such as radiation damage by the detection of radiation halos (Figure 14h). These characteristic CL properties can often be used to reconstruct specific geological processes.

Information concerning the lattice defects causing the different luminescence colors can be obtained by spectral measurements. Based on the fact that the defects causing the different CL emissions reflect the variable physico-chemical conditions of formation, CL spectra of agates often differ from those of quartz from crystalline rocks [107,118,119]. In the CL spectra of agates at least three broad emission bands can be detected: a dominating red band at 650 nm, a yellow band at about 570 nm and a blue band of mostly low intensity (Figure 15).

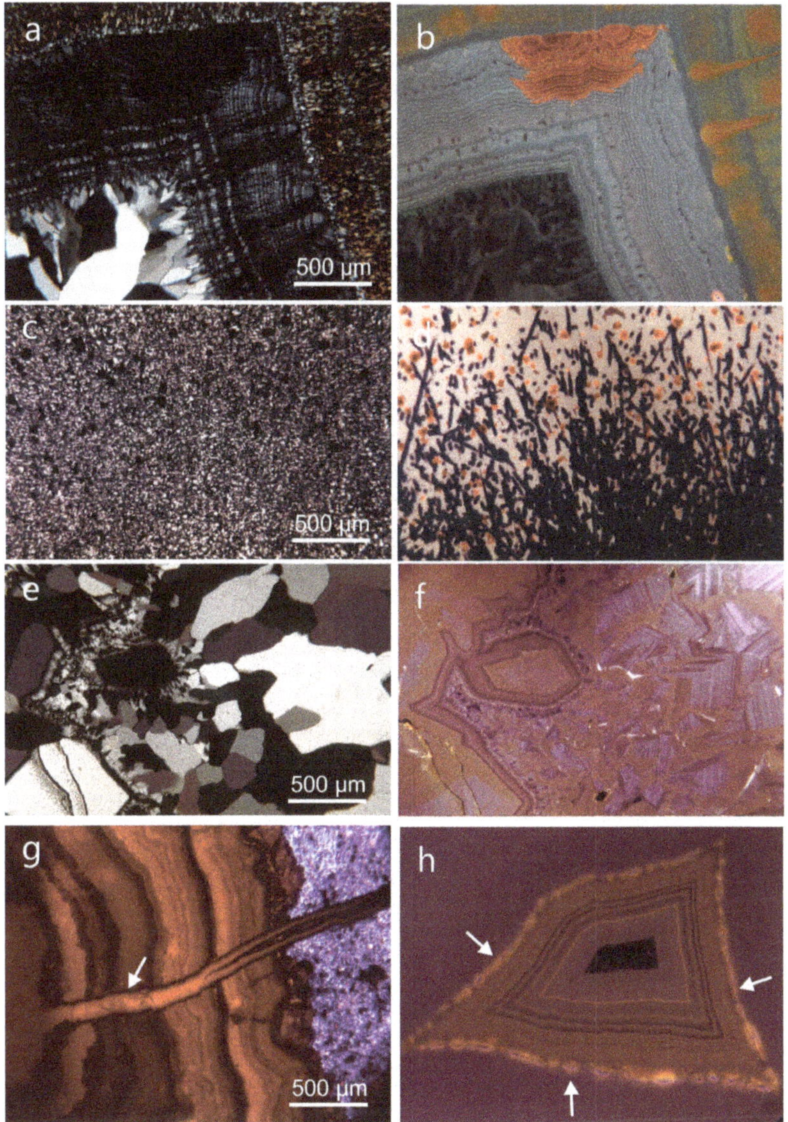

Figure 14. Micrographs of various agates in transmitted light (crossed polars) and cathodoluminescence (CL); (**a**,**b**) agate in liparite from Beipiao (Liaoning province, China) with multiple luminescence colors; note the heterogeneous internal CL textures of quartz crystals (**b**) appearing homogeneous in polarized light (**a**); (**c**,**d**) hydrothermal vein agate from Halsbach (Saxony, Germany) exhibiting spectacular microtextures in CL (**d**) not visible in polarized light (**c**); (**e**,**f**) agate from Gröppendorf (Saxony, Germany); the macrocrystalline quartz crystals appear homogeneous in polarized light (**e**) and show remarkable oscillatory as well as sector zoning in CL (**f**); (**g**) agate in rhyolite from Gehlberg (Thuringia, Germany); CL reveals a flow channel crosscutting the agate banding (arrow); (**h**) CL micrograph of an agate from Chemnitz (Saxony, Germany); the arrows point to a growth zone with numerous radiation halos due to radioactive micro-inclusions.

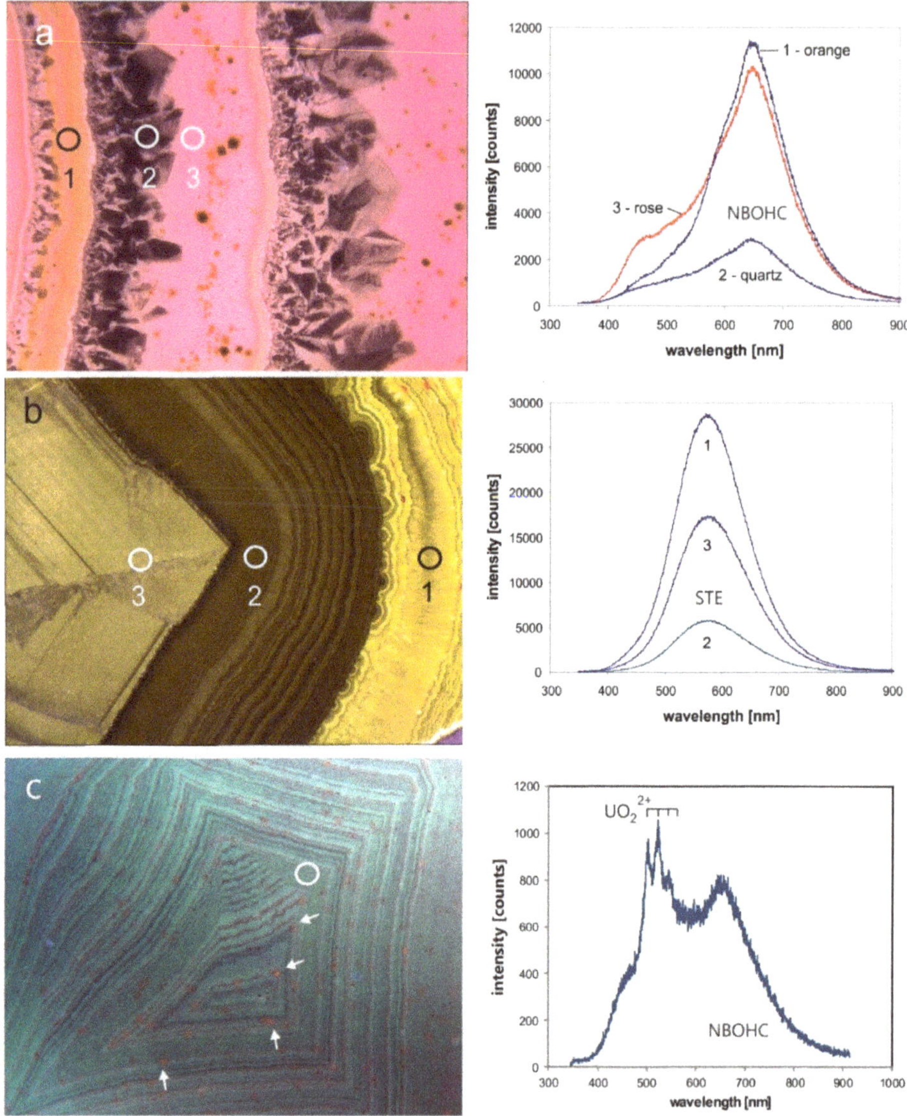

Figure 15. CL micrographs and related CL spectra of agates from different locations and host rocks; the circles mark the analytical spots of spectral measurements; (**a**) sedimentary "Dryhead agate" from the Pryor mountains, Montana (USA) showing growth zones with variable CL dominated by the 650 nm emission (non-bridging oxygen hole center, NBOHC); (**b**) agate in ignimbrite from St. Egidien (Saxony, Germany) with yellow CL; the spectra exhibit a broad emission band around 570 nm (self-trapped exciton, STE) with varying intensities; (**c**) agate in basalt from Soledade, Rio Grande do Sul (Brazil) showing greenish CL; the visible CL color results from the overlapping of a broad band at 650 nm (NBOHC) and multiple peaks between 500 and 600 nm of the uranyl ion (UO$_2$$^{2+}$); note the radiation halos around U-bearing inclusions (arrows).

The emission bands at 450 nm (2.69 eV) and 650 nm (1.91 eV) are caused by lattice defects [105,118,120]. The 650 nm band (Figure 15a) is the most common CL emission in chalcedony and can be related to the non-bridging oxygen hole center (NBOHC; [121]). The 650 nm emission is very sensitive to electron irradiation resulting in an increase of the band intensity during electron bombardment due to the conversion of different precursors (e.g., silanol groups: Si-O-H, Na impurities: Si-O-Na) into hole centers [105]. The specific CL behavior in chalcedony can be explained by elevated values of silanol groups [13,118].

Another defect induced luminescence is yellow CL (emission band at 570 nm—2.17 eV), which preferentially appears in macrocrystalline quartz and chalcedony of agates in acidic volcanic rocks (Figure 15b). The 570 nm emission band is related to high oxygen deficiency and local structural disorder in quartz. Götze et al. [68] related this specific CL emission to fast crystallisation from a non-crystalline precursor at low-temperature (mostly <250 °C).

The other type of luminescence active defects in agates is attributed to trace elements. Aluminium is the most frequent trace element in quartz and chalcedony (up to a few 1000 ppm). Several studies showed that the alkali (or hydrogen) compensated $[AlO_4/M^+]$ center is responsible for a transient blue emission band at ~390 nm (3.26 eV) [122–124]. The 390 nm emission is very sensitive to electron irradiation and the strong decrease under an electron beam is caused by the dissociation and electromigration of the charge compensating cations out of the interaction volume under the influence of the irradiation induced electrical field.

A striking greenish luminescence in agate is related the uranyl ion (UO_2^{2+}) and can be excited both by short-wave UV light (<300 nm) or an electron beam [36]. The visible greenish luminescence in agates is characterized by a typical emission line at ~500 nm accompanied by several equidistant lines due to the harmonic vibrations of oxygen atoms in the uranyl complex (Figure 15c). The mechanism of the uranyl luminescence is very effective and can be detected in agates with U contents as low as 1 ppm.

4.2. Geochemistry of Agates

4.2.1. Trace Elements

Trace elements represent impurities in the SiO_2 matrix of agates and may be structurally incorporated into the regular lattice, in inter-lattice positions or can be related to micro-inclusions of fluids and minerals [106,125]. The chemical elements were mostly incorporated during agate formation and therefore, are important geochemical indicators of geological processes reflecting the origin of mineral-forming fluids and/or the specific conditions of crystallisation.

Table 1 presents a compilation of trace-element data from the literature representing different kinds of agates from worldwide occurrences. The results show a broad range of trace-element concentrations reflecting the influence of the geological background of the specific locations and the agate type. In principle, trace element concentrations are low and the content in macrocrystalline quartz is mostly lower than that in the associated chalcedony. Only certain elements are relatively enriched in agates. These are in particular Al, Ca, Fe, K and Na, which are probably mobilized and transported together with SiO_2 from the host rocks during agate formation. Moreover, elevated concentrations of halogens (up to 380 ppm F and up to 154 ppm Cl; [63]) indicate that these two elements play an important role in fluids of the alteration and transport processes. In contrast, Sc, Nb, Ta, Th, Ti and other immobile elements are present in very low concentrations, often below the detection limit of trace-element analysis; only Powolny et al. [55] reported significant amounts of Zr in agates from Borówno, Poland.

Table 1. Variations of selected trace-element concentrations (in ppm) of chalcedony and macrocrystalline quartz in agates of different origin (data from [1,26,27,30,31,36,37,58,68,69,126]).

Element	Agates in Acidic Volcanic Rocks		Agates in Basic Volcanic Rocks		Vein Agates		Agates in Sedimentary Rock	
	Chalcedony	Quartz	Chalcedony	Quartz	Chalcedony	Quartz	Chalcedony	Quartz
Al	109 … 7300	34 … 1380	15 … 1170	30 … 700	640 … 7000	75 … 546	103 … 809	1 … 13
Ca	<10 … 540	<10 … 27	<10 … 710	<10 … 120	20 … 400	<16 … 45	350 … 1400	630 … 1200
Fe	30 … 4500	<30 … 978	7 … 3900	<30 … 670	150 … 3000	150 … 1700	26 … 1600	<1 … 30
Mn	<1 … 120	<1 … 10	<1 … 90	<1 … 46	1 … 27	3 … 6	<1	<1
K	40 … 3640	30 … 600	20 … 580	66 … 420	150 … 1360	25 … 480	na	na
Na	90 … 850	22 … 600	85 … 750	20 … 140	50 … 350	100 … 240	48 … 323	45 … 158
Mg	5 … 556	3 … 70	1 … 364	4 … 287	16 … 220	46 … 105	na	na
F	<25 … 380	<25 … 52	<25 … 228	<25 … 270	79 … 156	31 … 200	na	na
Ge	3.2 … 22	0.9 … 30	0.5 … 8.5	0.5 … 15	1.3 … 4.8	0.6 … 4.5	<1 … 2	<1
B	<10 … 135	<2 … 155	3 … 146	<2 … 120	<2 … 18	6 … 46	na	na
U	<1 … 21	<1 … 14	<1 … 21	<1 … 4	<1 … 15	<1 … 4	<1 … 72	5 … 18

na = not analyzed.

Chemical analyses of separate agate zones or element profiles perpendicular to the banding provided more detailed information concerning the trace-element distribution within the microstructure of the agates [1,26,27,30,31,36,37,58,68,69,126]. The concentration profiles show variations for most elements and only some elements behave relatively constant (compare Figure 16, Table 2).

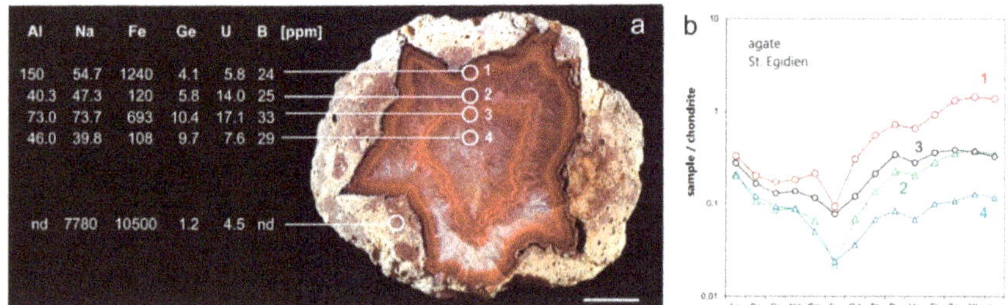

Figure 16. (**a**) Trace-element data from ICP-MS analyses of different zones (1, 3 = chalcedony; 2, 4 = macrocrystalline quartz) in an agate from St. Egidien, Saxony (Germany) compared to the composition of the rhyolitic host rock (the circles mark the areas of chemical analyses; nd—not detected); (**b**) chondrite-normalized REE distribution patterns (normalization according to data of Mason [127]) of the same agate zones shown in (**a**).

Table 2. Concentrations of trace elements (in ppm) from LA-ICP-MS local chemical analyses of a profile across an agate from St. Egidien, Saxony (Germany) (data from Götze et al. [68]); the numbers relate to the analytical points indicated in Figure 17.

Spot	Ge	Al	Fe	B	Ga	Ca	Li	Na	K	Rb
1	77.4	-	11.5	36	0.07	-	0.98	77.4	41.1	0.45
2	16.4	366	46.9	34.2	0.71	62.4	1.66	101	162	1.86
3	22.2	329	48.2	33.3	0.81	58.3	1.89	67.4	168	1.5
4	94.9	-	11.3	33.4	0.16	-	0.89	-	61.8	0.5
5	53.5	-	12.4	23.5	0.04	-	0.86	-	49.2	0.32
6	93.6	-	14.5	31.3	0.12	-	0.51	-	37.4	0.44
7	26.8	-	18.7	8	0.14	-	0.59	-	32	0.15
8	68.9	-	12.8	19.2	0.12	-	0.47	-	33	0.24
9	31.6	103	154	28.7	0.38	9.7	0.49	37.9	89.2	0.58
10	36.3	<6.0	47.7	29.2	0.17	9.4	0.39	-	51.9	0.33
11	28.9	-	47.8	41	0.11	-	0.49	-	55.1	0.43
12	20.4	-	61.9	35.7	0.12	-	0.59	-	60.7	0.33
13	13.2	<6.0	55.4	45.6	0.22	-	0.35	-	60.2	0.31
14	21.2	162	66.9	25.8	0.98	-	0.30	-	149	0.97
15	0.5	-	1.9	<1.0	0.18	53.3	0.30	-	1.2	-

(-) = not detectable; in general Ti < 1.2 ppm, Ba < 0.1 ppm, Sr < 0.1 ppm, Mn < 0.6 ppm.

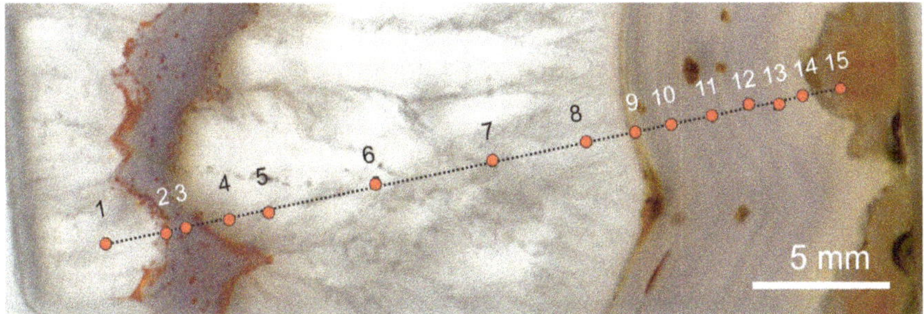

Figure 17. Micrograph of an agate from St. Egidien, Saxony (Germany) with a profile showing the analytical spots (spot numbers relate to analytical data in Table 2) of trace element analysis by LA-ICP-MS.

Results of these studies revealed a remarkable behavior for specific chemical elements. Remarkable high concentrations of Ge (>90 ppm), B (46 ppm) and U (>20 ppm) were detected in agates, which often exceed the average concentration of the Earth's crust (for Ge = 1.4 ppm, B = 12 ppm) and sometimes the element concentration in the surrounding host rocks (Figure 16). Interestingly, the concentrations of Ge, B and U in macrocrystalline quartz are sometimes higher than in associated chalcedony (Figure 16). High Ge contents can be explained by the similar geochemical character of Si and Ge, which causes the common transport and incorporation of Ge into the quartz structure [125,128]. The occurrence of Ge and B defect centers in quartz of agates was confirmed by recent EPR measurements [68,107].

The observed high concentrations of U within some of the agates are noteworthy. In the literature, U contents of up to 1200 ppm are published for chalcedony in agates from acidic volcanic rocks [129]. The U contents of sedimentary agates from the Dryhead area (Montana, USA) are amongst the highest values (compare Table 1). The location is close to a Uranium deposit, and therefore there is high availability of uranium in the vicinity [31]. The mobility of uranium during the alteration of volcanic rocks was investigated by Zielinski [130], who observed a parallel accumulation of Si and U. Recent investigation showed that a substitution of Si ions in the quartz lattice by U is unlikely due to the different crystal-chemical properties, and instead the incorporation as uranyl silicate complex is favoured [36,131]. First, uranium can be adsorbed from solution by natural silica compounds (silica colloids) and then trapped as uranyl silicate complex in a stable silica matrix [131]. Luminescence studies confirmed the existence of uranyl compounds in quartz and chalcedony of agates [36] (compare Section 4.1.3).

First results were also published for rare earth elements (REE) in agates [30,37,55], which are preferentially bound to fluid inclusions and therefore, reflect the composition of the agate mineralizing fluids. The chondrite normalized REE patterns of different agate zones in Figure 16 are characterized by pronounced negative Eu anomalies and a slight slope of the light REE (La—Sm), whereas the shape of the heavy REE patterns (Gd—Yb) is increasing. Moreover, agates may show slight "tetrad effects", i.e., a subdivisions of the REE series into four concave-upward groups [132,133]. Such REE distribution patterns can be explained by a primary crustal signature that is overprinted by an enrichment of heavy REE due to the preferred complexation of released HREE by carbonate- and F-complexes during the alteration and transport processes [134]. However, various scenarios for mineralizing SiO_2-bearing fluids for agate formation are possible, sometimes also resulting in positive Eu-anomalies [55]. Indications for the mixing of extremely different compounds (e.g., volatiles, meteoric water) are also provided by the presence of Ce anomalies in some agates [34,37,135].

The measured trace-element profiles illustrate, that there can be a wide scatter of concentrations for certain trace elements, whereas others are relatively constant in their spatial distribution. For instance, balanced concentrations at a low level were recorded for the elements Li, Rb, Ba, Sr, Ga or Ti (Table 2). Contrast variations can be detected for Fe concentrations, which are mostly related to the color of the agates (Figure 16). Strongly colored (especially reddish and yellow) chalcedony bands have the highest Fe contents due to micro-inclusions of certain Fe-compounds (e.g., hematite, goethite). These zones can also contain elevated concentrations of other transition elements such as Mn, Ni or Cr [1,30]. Concentrations of up to 400 ppm Ni and 500 ppm Cr are reported by Barsanov et al. [136] for agates from Timan and Caucasus. The elements are probably incorporated in fine-grained iron oxides, where concentrations of up to 14 wt% Mn, >1000 ppm Ni and >550 ppm Cr were measured [1].

The general trend of the lateral trace element distribution from the agate host rock to the center shows a slight decrease in element concentrations within chalcedony without a direct relation between trace-element contents and banding, and with lowest contents of most elements in macrocrystalline quartz. These variations can probably be explained by a "self-purification" process during agate formation. Certain chemical elements are firstly precipitated together with the primary amorphous silica and accordingly are finely dispersed in the silica matrix. The following chalcedony growth may initially incorporate high contents of trace elements. During further crystallisation, the concentration of impurities decreases. Heaney [23] and Merino et al. [27] stated that the formation of frequent defects, Brazil twinning and the typical twisting in chalcedony are caused by the high trace-element contents in silica. Chalcedony fibers in agates may therefore be the result of dislocation growth. The continuous lowering of trace elements in the crystallisation medium may then finally cause the formation of coarse-grained quartz.

4.2.2. Isotope Composition

Isotope data of the SiO_2 phases in agates may provide information about the isotopic composition of fluids from which the agates were formed and allow a first estimation of formation temperatures. The composition of the oxygen isotopes ($\delta^{18}O$) in agates from various locations and of different types reveals a rather wide range from 13 to 33‰ (Figure 18). In most cases macrocrystalline quartz from the respective specimens has slightly lower $\delta^{18}O$ values than chalcedony of the same sample and more closely matches the values of hydrothermal quartz precipitated by late magmatic fluids [137]. The difference between the $\delta^{18}O$ values of chalcedony and quartz can possibly be explained by the lower equilibrium constants in the system quartz–water compared to the systems tridymite–water and cristobalite–water [138,139]. Accordingly, amorphous SiO_2, chalcedony, cristobalite and tridymite may have higher $\delta^{18}O$ values than quartz formed under the same temperature conditions.

During crystallisation of quartz from a mineral-forming fluid the initial $^{18}O/^{16}O$ ratio changes due to the preferred incorporation of the heavy ^{18}O-isotope into the quartz lattice [137]. Because of the temperature dependence of this fractionation the measured isotope ratios can be used to calculate formation temperatures on the basis of the fluid oxygen isotopic composition [139]. However, no data exist regarding their oxygen isotope composition, since a direct analysis is mostly not possible because of the extreme small amount of fluid inclusions. Therefore, the temperatures of isotope fractionation were calculated assuming different fluid compositions of meteoric water (−10‰), oceanic water (0‰) and magmatic water (+8‰).

The calculation of agate formation temperatures based on these data provided rather low temperatures between 20 °C and 230 °C [30,31,37,140,141], depending on the isotopic composition of the involved waters. These temperatures cover a broad range between sedimentary origin and hydrothermal formation and have to be discussed in combination with other geochemical data. Fluid inclusion studies provided similar temperatures and confirm the results from the oxygen isotope investigations (see Section 4.2.3).

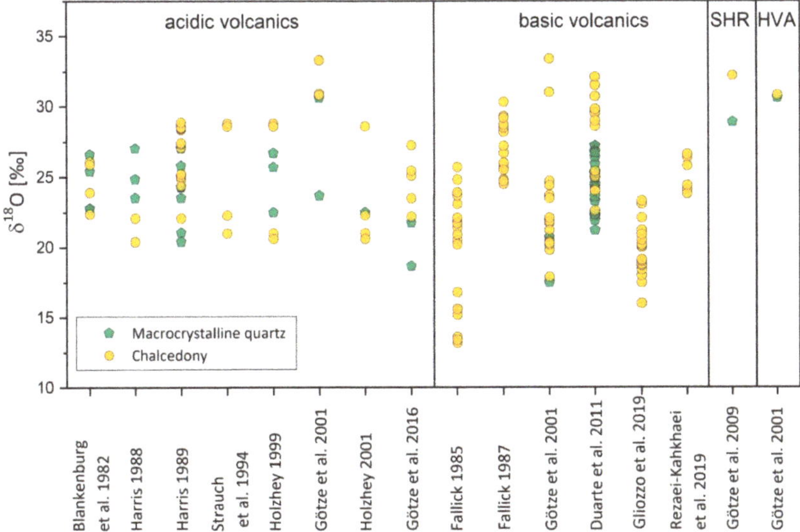

Figure 18. Compilation of oxygen isotope data ($\delta^{18}O$) of agates from different origins (SHR = sedimentary host rock; HVA = hydrothermal vein agate); data are compiled from [14,30,31,37,58,101,140–147].

Results of spatially resolved oxygen isotope analyses of different agate bands reveal strong variations of the $\delta^{18}O$ values (partially >10‰) over small scales in some agates, whereas others show a homogeneous isotopic composition (Figure 19). Assuming that the analyzed isotopic compositions are of a primary nature, the variations could be explained either by fluctuations of temperatures resulting in changes of the conditions of isotope fractionation, or variations in the primary isotope composition of the fluids, i.e., participation of fluids of different geochemical composition. Multiple temperature variations are not really likely, whereas the supply of SiO_2-bearing solutions by mixing of surface water, heated meteoric water and/or magmatic fluids was discussed by several authors [13,30,37,41]. Different ratios of the participating fluids could then cause variations in the isotope composition.

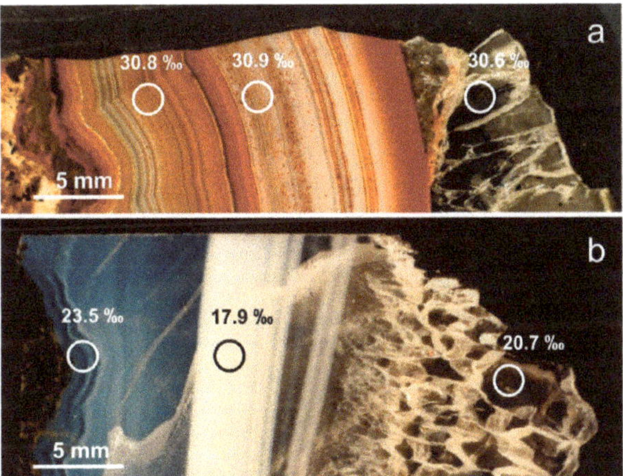

Figure 19. Hydrothermal vein agate from Schlottwitz, Germany (**a**) and agate from mafic lavas of the Ardownie quarry, Scotland (**b**) with results of spatially resolved oxygen-isotope analyses (circles).

Alternatively, isotope fractionation under non-equilibrium conditions could explain the observed isotopic variations due to various kinetic processes such as diffusion, adsorption, solution-precipitation, phase transformations or boiling. Kita and Taguchi [148], for instance, found kinetic effects in the fractionation between SiO_2 and geothermal waters during precipitation of SiO_2 that resulted in variations of the $\delta^{18}O$ values of some ‰. Accordingly, the assumption of agate formation from silica-bearing geothermal waters via an amorphous precursor could also explain heterogeneities in the isotopic composition.

Strong variations were also reported for δD values in chalcedony (−130‰ to −40‰) and macrocrystalline quartz (−90‰ to −50‰) in agates from different localities including variations between different agate bands within single agate samples [30,145]. These data support the idea of fluid mixing and non-equilibrium conditions during agate formation. Therefore, the isotope data of agates have to be interpreted with care.

4.2.3. Fluid Inclusions and Water in Agate

Several studies have shown that agates may contain considerable amounts of water and other fluid components [1,13,15,21,34,37,40,73,149]. The H_2O content in the agate matrix varies between 0.5 and 3 wt% [13,21,40] and is preferentially present as OH (Si-OH silianol groups). This is in contrast to opal, which may contain up to 10 wt% H_2O, predominantly as molecular water. Flörke et al. [13] assumed that the incorporation of OH and H_2O in chalcedony is especially related to structurally damaged crystal boundaries.

Frondel [150] first stated that the content of structural water in different agate zones can differ due to variations in the amount of different SiO_2 phases and their microstructure within the agates. Graetsch et al. [151] investigated separate parts of chalcedony from Brazilian agates and determined water contents between 0.65 and 1.42 wt% in the wall-lining agate and 0.56 up to 0.92 wt% in horizontal agate layers. The ratio between silanol water and molecular water varied between 0.4 and 1.0. Based on these data they postulated temperatures of agate formation below 250 °C and pressures lower than 50 MPa.

Moxon and Rios [40] and Moxon [21] found a decreasing amount of silanol water with increasing geological age during their investigations of agates from worldwide occurrences in 23 to 2717 million years old parent rocks. The total water content varied in the agates between 0.23 and 1.78 wt% and the average amount of structural silanol water was found to be 76% of the total water content. They concluded that silanol groups were released during the recrystallisation of moganite to chalcedony and therefore, an estimation of the age of agates based on the amount of structural water should be possible. These conclusions supported previous assumptions that agates from Brazil and New Zealand were formed long after their host rocks [21].

Small amounts of gases and liquids from mineral-forming fluids can be incorporated into growing crystals during their formation. The type, frequency and chemical composition of such fluid inclusions can provide valuable information for the reconstruction of the mineral-forming processes. Several fluid inclusion studies in agates have shown that more or less no visible fluid inclusions exist in cryptocrystalline quartz (chalcedony) because of the extreme small crystal size. Therefore, analyses are only possible in macrocrystalline quartz crystals from the agate centers. The frequency of measurable primary inclusions in the quartz crystals of agates is in general very low (Figure 20) and the inclusions are extremely small (<20 μm).

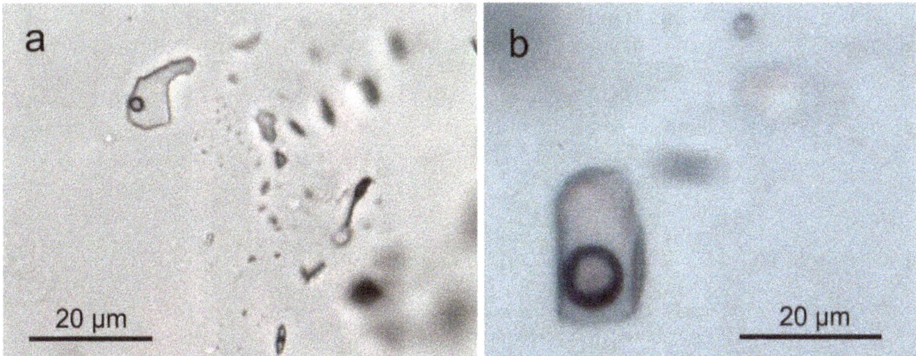

Figure 20. Primary two-phase (gas-liquid) inclusions in macrocrystalline quartz of agates from Saxony (Germany): (**a**) Burgstall, (**b**) St. Egidien.

Published homogenization temperatures of quartz from agates scatter in a very broad temperature range between <100 °C and >500 °C [15,37,149,152,153]. However, the interpretation of the measured homogenization temperatures has to be done very carefully. Different degrees of filling in the inclusions (i.e., ratio of gas to liquid phase) provide indications for fluid trapping under heterogeneous conditions. This conclusion is supported by a strong variation of the measured homogenization temperatures. Such heterogenization can be caused by splitting of fluids into an aqueous phase with high salinity and a gaseous phase with low salinity at high temperatures (>375 °C) due to drop of pressure [154,155].

Another possibility is phase separation in the subcritical temperature range (<375 °C) at low pressure due to boiling in an open system (aqueous phase with saturated water vapor). These conditions result in the occlusion of gas/liquid mixtures of varying ratios. Therefore, homogenization temperatures cannot directly be used as trapping temperatures of the fluids [154]. The minimum temperature of trapping is then represented by the lowest measured temperatures of fluid homogenization. Taking into account all these facts, a heterogeneous trapping of inclusions has to be assumed for most of the investigated agates from acidic and basic volcanic rocks. Table 3 summarizes minimum homogenization temperatures for agates of some German occurrences. These data confirm the results concerning agate formation temperatures obtained from oxygen isotope studies (compare Section 4.2.2).

Table 3. Minimum homogenization temperatures (T_{hom}) obtained from fluid inclusion studies of agates from different occurrences in Germany.

Agate Location	Host Rock/Agate Type	T_{hom}
Gehlberg (Thuringia)	rhyolite/lithophysa	95 °C
Chemnitz (Saxony)	ignimbrite/lithophysa	176 °C
St. Egidien (Saxony)	ignimbrite/lithophysa	186 °C
Mügeln (Saxony)	rhyolite/lithophysa	177 °C
Burgstall (Saxony)	rhyolite/lithophysa	172 °C
Gröppendorf (Saxony)	melaphyre/amygdale	134 °C
Schlottwitz (Saxony)	hydrothermal vein agate	80 °C

In addition, cryometric measurements provided information concerning the chemical composition of the investigated inclusions. In agates from basic and intermediate rocks, quartz mostly contains inclusions having low salinity (<4 eq% NaCl). In contrast, agates from acidic volcanic rocks sometimes contain inclusions with up to 30 eq% NaCl indicating the involvement of magmatic solutions. Besides NaCl, especially KCl and $CaCl_2$ have been detected as salt phases [37,149]. For instance, eutectic temperatures of −10.5 °C and the absence of salt hydrates in fluid inclusions of agates from Saxonian agates favored KCl as the main salt component in the fluid system [37]. Measurements of the freezing point depression according to Hall et al. [156] revealed salt concentrations of 0.65–2.4 wt% KCl.

To get more information about the geochemistry of the gaseous and liquid fluids involved in the transport of silica, systematic studies of agates were performed by evolved gas analysis (thermogravimetry-mass-spectrometry; [34,73]). The volatile components were released and transferred into a vacuum during heating at various temperatures in the temperature range between 25 and 1450 °C and identified according to their mass-charge-ratio and the corresponding intensity ratios of the detected volatile fragments.

The results provided information regarding the participation of different fluids during agate formation. The detected volatiles consisted of compounds of C, N, S, F and Cl with oxygen and/or hydrogen (e.g., H_2O, HF, NO, S, SO, CO, CO_3^{2-}). In particular in agates from volcanic rocks, several hydrocarbon compounds as well as carbonic acids could be identified [34,44,73]. The incorporation of hydro-carbon compounds into the SiO_2 matrix points to co-precipitation of both compounds, probably originating from the same source. Hydrothermal methane and/or other hydro-carbon compounds in the volatiles participating in the silica accumulation could have served as precursors for the detected organic fluids in the agates.

The data emphasize that a transport of elements and chemical compounds in aqueous solutions cannot be the only process governing agate formation. Therefore, the discussion of the agate genesis has to be focused not exclusively on the system SiO_2–H_2O but has to consider a complex multi-component system including both meteoric water and magmatic fluids (CO_2, HF, HCl, hydrocarbons, etc.).

4.3. Paragenetic Minerals in Agates

Many field observations and analytical results document that besides the different silica minerals numerous other mineral phases are formed pre-, syn- and post-genetically with the agates. Depending on the type of host rocks and the time of formation these minerals can occur as mineral inclusions in quartz, paragenetic minerals intergrown with SiO_2, or pseudomorphs, and their size can range from microscopic inclusions up to several cm in size. These differences can be related to the different geological background and differences in the conditions of mineralization of the agates.

The compilation of detected minerals in agates in Table 4 illustrates that all mineral classes occur. Furthermore, several organic phases were detected in agates such as carbonaceous material, solid bitumen or disordered graphite-like substances [35,44–46]. Carbonates (especially calcite), iron oxides/hydroxides and clay minerals are the most frequent and present in more or less all types of agates, whereas minerals of the zeolite group are typical phases in volcanic agates [1,15,49,78]. It is noteworthy, that certain chemical elements may be present in different chemical compounds in the same agate. For instance, native copper can occur together with copper oxide and copper sulfide, or pyrite/marcasite in association with iron oxides/hydroxides in agate geodes. This is a line of compelling evidence for highly variable physicochemical conditions (especially redox conditions) during mineral formation.

Table 4. Compilation of detected minerals in agates (unpublished data and data from [1,15,18,22,24,28,30,31,34,35,38,42,44–47,49,52–55,59,60,76,78,157–165]); minerals and mineral groups in bold are the most frequent phases in agates.

Mineral Group	Minerals
Elements	copper, lead, sulphur, graphite
Sulfides	pyrite, marcasite, sphalerite, galenite, chalkopyrite, covelline
Oxides/Hydroxides	**hematite, goethite**
	todorocite, ramsdellite, birnessite, pyrolusite, rancieite, hollandite, cryptomelane/psilomelane, manganite
	cuprite, rutile, anatase, spinel (magnesio-chromite)
Carbonates	**calcite**, aragonite, dolomite, siderite, ankerite, rodochrosite, strontianite, magnesite, smithsonite, malachite, azurite, bastnesite-(Ce)
Sulfates	barite, celestine, anhydrite, gypsum
Phosphates	apatite, monazite-(Ce), rabdophane-(Ce), xenotime-(Y)
Halides	fluorite
Silicates	**clay minerals** (kaolinite, illite, montmorillonite, beidellite, saponite, nontronite)
	zeolites (mordenite, heulandite/clinoptilolite, harmotome, chabasite, natrolite, analcime, scolezite, mesolite, stilbite, thomsonite, laumontite, brewsterite, philipsite)
	glauconite–celadonite series, chlorite, serpentine, talc, prehnite, datolithe, epidote, apophyllite, chrysocolle, orthoclase, albite, plagioclase

The observed mineral associations (and their stability fields) in volcanic agates indicate a preferred formation under hydrothermal and low temperature conditions. The most frequent mineral phases such as clay minerals, zeolites, iron oxides/hydroxides and carbonates can be related to the alteration processes of the volcanic host rocks and emphasize that besides enormous amounts of SiO_2 also Al, Fe, Ca, Na and K are released during these processes. The presence of water- and hydroxyl-bearing minerals, carbonates and fluorite in volcanic agates highlights the role of H_2O, CO_2 and F-complexes as main volatile compounds for the transport of substances [34].

In contrast to the agates from volcanic rocks, hydrothermal vein agates can be formed within veins and fissures of the crystalline host rocks directly from circulating hydrothermal solutions. Therefore, SiO_2 minerals can precipitate and crystallize together with a number of other hydrothermal minerals such as carbonates, sulphates, fluorite and/or certain ore minerals (oxides, sulphides).

In sedimentary rocks, where agates are formed by filling of cavities and empty pore space with SiO_2 or the silicification of concretions and pseudomorphism, these processes often take place during sedimentation or early diagenesis, when the sediments contain enough moisture. Characteristic mineral associations commonly include carbonates, sulphates, clay minerals, sulphides and oxides/hydroxides [15,31,42,157,158]. In addition, large crystals of quartz, calcite, celestine or barite can form in the center of the geodes.

Carbonates, especially calcite ($CaCO_3$), belong to the most frequent minerals in volcanic agates [30,38,76,78,159,160]. Coarse-grained calcite has often been formed as overgrowth on quartz or chalcedony in the central part of the agate geodes, but calcite layers intergrown with chalcedony are also possible (Figure 21). A special case is the so called "calcite agate" [38,161,162], an "agate" completely consisting of calcite (Figure 21d). The multiphase character of the calcite formation and several replacement processes can be proven by microscopic investigations (Figure 21b,c).

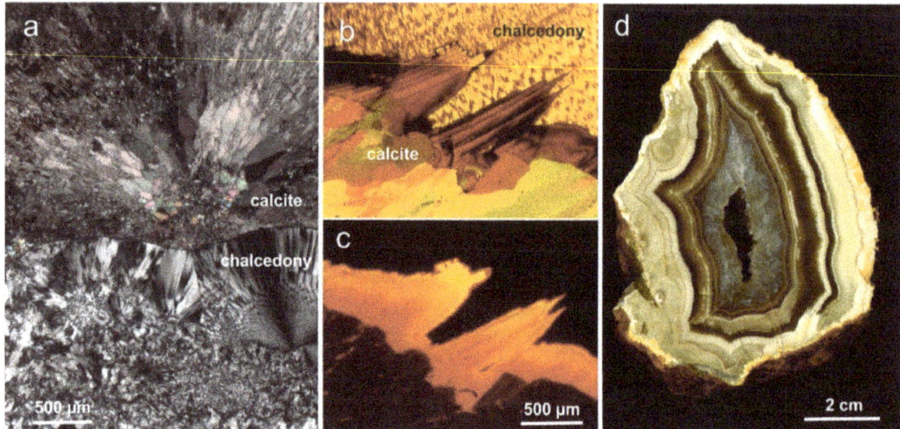

Figure 21. Appearance of calcite in agates: (**a**) micrograph in transmitted light (crossed polars) of the transition zone between chalcedony and calcite in an agate from Xuanhua, province Hebei (China); note the sharp boundary between the two phases and the spherulitic growth of calcite; (**b**,**c**) micrographs in transmitted light (crossed polars, (**b**)) and CL (**c**) of an agate from Agate Point, Canada showing two calcite generations (dark brown, orange CL) in chalcedony matrix (dark); the second calcite generation replaces the banded chalcedony and forms pseudomorphs; (**d**) "calcite agate" from Krásný Dvoreček (Czech Republic) with the typical banded appearance of agate but consisting completely of carbonate.

Geochemical investigations of calcite in agates sometimes showed several carbonate generations and proved a temperature of crystallisation in the range between ca. 20 and 230 °C [1,30,76,77,160], which confirms the estimated temperatures of agate formation. In addition, oxygen and carbon isotope studies of calcite in agates provided indications concerning participation of at least two different fluids during formation. Relatively high $\delta^{18}O$-values (up to +25‰) point to a sedimentary source, whereas $\delta^{13}C$-values between −5 and −15‰ can be related to a primary magmatic origin [1,30,78,159,160]. The isotope signature of calcite can probably be explained by processes of fluid mixing during formation, where carbon inherits its primary magmatic signature (volcanic H_2CO_3 as the dominant carbon species), whereas the signature of oxygen isotopes is overprinted by the secondary influence of meteoric water (H_2O).

Another remarkable feature in worldwide agate occurrences is the appearance of iron-bearing phases in agates, indicating high concentrations of iron in the mineralizing fluids [1,15,30,31,37,47,59,159,163]. Different types of iron oxides and hydroxides in the marginal parts of the agates as well as within the chalcedony matrix cause the typical colorations in red, brown and yellow (Figure 22). In addition, enrichment of iron compounds in the form of crusts or earthy masses in the central parts of agates can be observed.

Microscopic investigations revealed that at least three different types of iron oxides (hydroxides) can be distinguished, which are closely connected with the agate formation (Figure 22). Iron oxides (e.g., hematite) of probably pre- or syngenetic formation often appear at the interface between host rock and agate or in the outermost chalcedony layer. The crystals show spherulitic as well as acicular or dentaloid forms, the latter probably originating from the replacement of former existing carbonates (pseudomorphs).

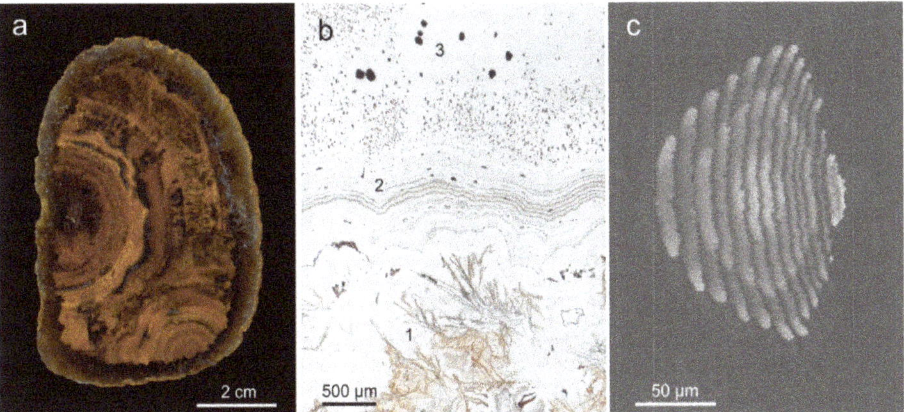

Figure 22. Fe-oxides/hydroxides in agates: (**a**) "agate" from Xuanhua, province Hebei (China) almost completely consisting of Fe-oxides/-hydroxides (hematite, goethite) showing a spherulitic growth; (**b**) micrograph in transmitted light of an agate from Xuanhua, province Hebei (China) showing the silica matrix with the distribution of Fe-oxides/-hydroxides: dentritic (1), aligned along the chalcedony banding (2) and large, irregular particles (3); (**c**) SEM-BSE image showing details of the microtexture of Fe-oxides/-hydroxides in an agate from St. Egidien, Saxony (Germany), which follows the microtexture of the calcedony banding indicating simultaneous crystallisation.

Other types of iron oxide inclusions have been observed within the agate matrix (Figure 22b). On one hand, laminated and lenticular agglomerates of very small hematite particles (<10 µm) occur. In contrast, the other type is represented by larger, sometimes globular or irregular aggregates of up to several 100 µm in size. The fine-grained particles were probably released from the silica matrix during crystallisation, moved together with the growth front and accumulated along the banding. In contrast, larger iron oxide inclusions show interesting structures consisting of several rounded disks. These structures indicate that the iron oxides neither existed before the crystallisation of the chalcedony, nor have been incorporated later. Most likely, iron oxides and silica precipitated simultaneously from a precursor sol.

The complex processes during agate formation and the accompanied variations of the physico-chemical conditions often result in a wide variety of inclusions and replacements of earlier existing phases in agates (e.g., pseudomorphism and perimorphism). Pseudomorphism of quartz is frequently detectable after carbonates (calcite, aragonite) and sulfates (barite, anhydrite/gypsum), but also after zeolites and pyrite [1,15,28]. The earlier existing minerals have been replaced by SiO_2-rich fluids (Figure 23). In most cases the original minerals are completely replaced and the former mineral is at most recognizable on basis of its preserved crystal shape. Therefore, relics of the pre-existing minerals are only detectable using highly advanced analytical methods. Figure 23 shows a sample of agate pseudomorphs from Liebgensmühle near Leisnig, (Saxony, Germany) with pseudomorphs of an elongated shape. Microscopic CL studies under high magnification revealed relics of primary dolomite and inclusions of kaolinite probably from alteration processes (Figure 23c). However, the agate almost completely consists of SiO_2. In other cases, interstices between pre-existing minerals can be filled by agate such as in the case of the well-known Paraiba agates in Brazil. In result, the agates develop polygonal outer shapes resembling crystal faces (so called "polygonal agate"; Figure 23d).

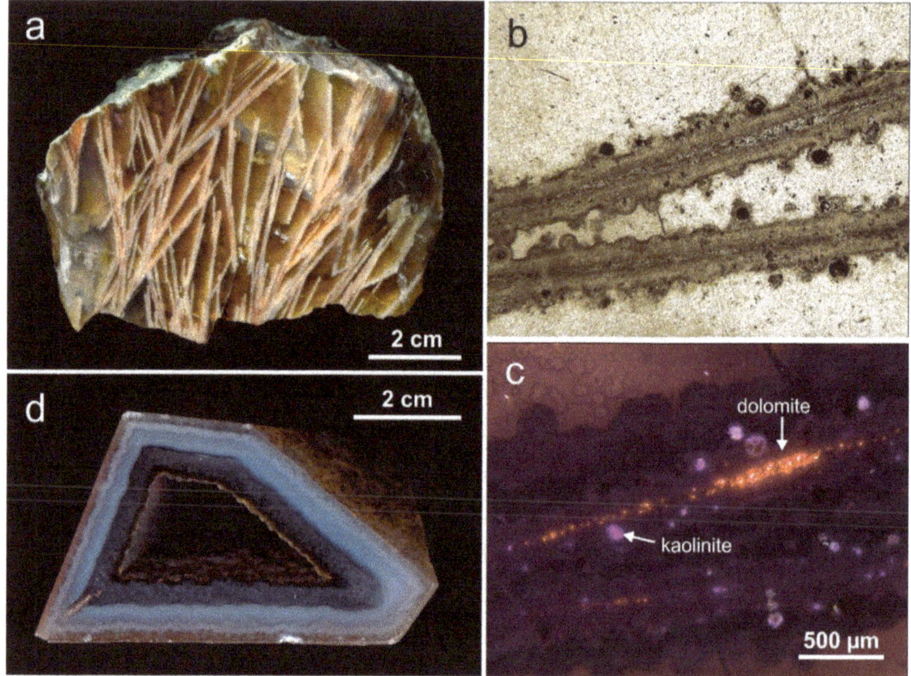

Figure 23. (**a**) Pseudomorphic mineral formation in an agate from Liebgensmühle (Saxony, Germany); (**b**) micrograph in transmitted light showing details of the lath-shaped crystals; (**c**) CL micrograph of the same area, which reveals relics of dolomite in the core of the lath-shaped crystals as well as tiny kaolinite from alteration processes; (**d**) agate from Cachoeira dos Índios, Paraiba (Brazil); the polygonal shape developed due to infill of silica into interstices between laminated calcite (or barite) crystals.

5. Discussion

5.1. Origin of Cavities for Agate Formation

A precondition for the formation of agates is the availability of cavities or open spaces, in which silica can be accumulated. Depending on the geological environment, different types of holes and fissures may exist, which significantly influence the appearance and characteristics of the associated agates. The outer shape of an agate is more or less a replica of the former existing cavity (compare Figure 2). Therefore, in different types of rocks the kind of cavities is variable.

In basic volcanic rocks (basalts, andesites), the accumulation of silica and the formation of agates take place in former gas/liquid bubbles of the solidified lava (amygdales). Because of the high temperatures and low viscosities of these basaltic melts, the lava is flowing over large areas and many gas bubbles are formed in the internal, coalescing during ascent and got stuck during cooling of the solidifying basalt (Figure 24a). These gas bubbles often exhibit rounded or drop-like shapes and form vesicular textures. The primary relations are often visible in the outcrop showing vesicles that are arranged in the flow structure of the former lava (Figure 24b).

Figure 24. (**a**) Vesicular trachyandesite in the agate quarry of Dishuya (Hebei province, China); the numerous cavities formed by fluid bubbles that got stuck in the cooling lava; the inset shows an agate-filled cavity (amygdale); (**b**) agate-filled former bubbles arranged in the flow texture of the solidified andesitic lava from the Steinkaulenberg, Idar-Oberstein (Germany).

The formation of cavities in acidic volcanic rocks occurs via several stages. The process starts with the formation of spherical forms of crystallisation, so called high-temperature crystallisation domains (HTCD) [166]. HTCD with one or more cavities are called lithophysae (thundereggs), which exclusively occur in SiO_2-rich lavas, ignimbrites or subvolcanic bodies and crystallize above the glass transition temperature T_g [167,168]. The position of the lithophysae within the flow texture of the volcanic rocks often provides evidence concerning their early crystallisation before the final solidification of the melt (Figure 2a). Cavity formation is assumed to be the result of transient tensional stress in the melt close to the HTCD crystallisation front (Figure 25a; [166]).

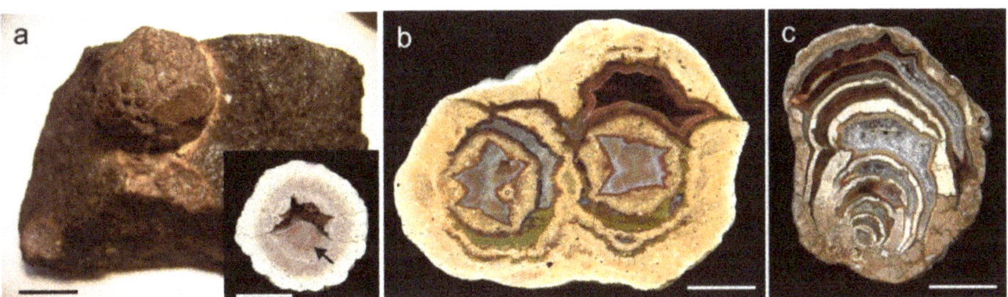

Figure 25. Different kinds of lithophysae with agates in silica-rich volcanic rocks: (**a**) lithophysa in an ignimbritic pitchstone from Zwickau, Saxony (Germany); the inset shows a lithophysa from Chemnitz, Saxony (Germany) with crystallisation center (arrow) and empty cavity; (**b**) double-lithophysa with multiple agate cavities from St. Egidien, Saxony, Germany; (**c**) thunderegg from the Baker Egg Mine, Deming (New Mexico, USA) with multiple crescent-shaped agate cavities; scale bar is 2 cm.

The complexity of the processes can result in the development of single, double and multiple lithophysae (Figure 25b) as well as in complex structures due to the twofold (or multiple) opening and filling with silica or alteration products such as clay minerals or iron compounds (Figure 25c). Lithophysae normally appear in sizes of several cm or dm, but examples with diameters of more than 4 m ("megaspherulites") have also been reported from several localities in the USA, Argentina and China [169].

Hydrothermal vein agates represent a completely different genetic type of agates. These agates form directly within veins and fissures in different crystalline host rocks, which have been formed by tectonic processes (Figure 2e). The vein-like structures are often connected with deep-seated faults with ascending hydrothermal solutions. Vein agates and veinlike silica mineralization often occur together with ore mineralization in hydrothermal systems, e.g., with carbonates, sulfates, sulphides or certain gold deposits worldwide (e.g., [102,170]). In the case of sufficient silica supply, the vein structures can be completely filled with agate and quartz. The sealing of the veins during continuous hydrothermal activity can cause increasing pressure in deeper levels and the fracturing of the primary vein structure [171]. These fractures are often filled with a secondary generation of agate or quartz, which is visible by healed brecciation ("brecciated agate" or "Trümmerachat" Figure 26a).

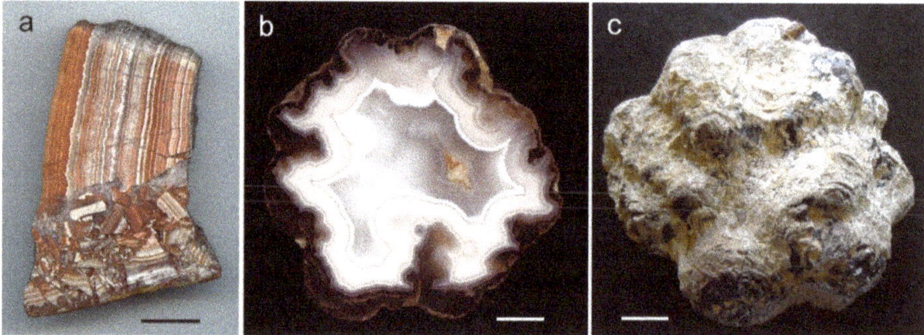

Figure 26. (**a**) Hydrothermal vein agate ("Trümmerachat") from Schlottwitz, Saxony (Germany) showing brecciation and healing by secondary silicification in the lower part; primary agate formation took place in Permian time (263.7 ± 2.2 Ma), whereas brecciation and secondary cementation by silica happened much later during tectonic activities in the Upper Oligocene/Miocene [172]; (**b**) agate of sedimentary origin from Rutba, Western desert (Iraque); (**c**) back side of the agate in (**b**) illustrating the typical cauliflower-like surface; scale bar is 2 cm.

In sedimentary rocks, agates are predominantly formed by silicification of concretions or pseudomorphism due to the replacement of soluble mineral species such as sulphates (e.g., anhydrite) or carbonates (calcite, dolomite) by silica [31,42,157,158,165,173,174]. Residues of plants and animals can also be silicified [79,80,175,176]. The silica originates from circulating pore water, which can precipitate under specific conditions. Geochemical data indicate a preferred silicification in shallow sediments, especially in areas with increased salinity (lagoons, marine shelf) under near-surface conditions and temperatures below 40 °C [177]. Sedimentary agates are often characterized by typical features such as a cauliflower-like surface, the frequent presence of quartzine besides chalcedony or the lack of horizontal banding (Figure 26b,c) [15].

5.2. Temperature of Agate Formation

One key point of scientific discussions concerning agate genesis is the temperature of formation. There is no doubt concerning the hydrothermal nature of most vein agates, and agate formation in sediments is running at relatively low temperatures under surface or diagenetic conditions. However, the arguments about the formation of agates in volcanic rocks are extremely widespread. There exist theories about an agate formation both at high temperatures (>400 °C) [1,15,56,178] and ideas about agate genesis at low temperatures (20–100 °C) in a sedimentary-diagenetic environment (e.g., [14,140,179,180]).

Investigations of active thermal vents such as in the Yellowstone Park or on Iceland, or the recent formation of agate-like structures on the seafloor [64] in so called "white smokers" point to a temperature range below 100 °C. However, recent observations concerning the formation of colloidal silicic acid in high-temperature hydrothermal fluids (300 up to >700 °C) indicate that agate formation at high temperatures in volcanic rocks seems to be possible [181]. Interestingly, experiments concerning the synthesis of chalcedony successfully run under elevated temperatures: White and Corwin [182] produced chalcedony during hydrothermal treatment of silica glass at 400 °C/340 bar, Flörke [183] synthesized chalcedony at temperatures above 400 °C, and Oehler [184] succeeded with syntheses at 100–300 °C and 3 kbar.

Modern analytical methods such as fluid inclusion studies, isotope or trace-element analyses provide useful information for the estimation of the temperatures of mineral-forming processes. In the case of fluid inclusion studies in agates, several limitations have to be taken into account. A direct determination of the temperature in agate (chalcedony) is mostly not possible because of the lack of measurable fluid inclusions. Data are only provided by fluid inclusion studies of the macrocrystalline quartz layers. Moreover, a heterogeneous trapping of inclusions has to be assumed for most of the investigated agates from acidic and basic volcanic rocks.

Taking into account all these facts, minimum homogenization temperatures for agates from volcanic rocks of some German occurrences scatter between ca. 95 °C and 186 °C (compare Table 3). The hydrothermal vein agate from Schlottwitz, Saxony (Germany) provided a temperature of 80 °C. These temperatures are in the same range as crystallisation temperatures calculated from inclusion studies of associated minerals in agates (e.g., calcite) [160].

An estimation of the temperature of formation of agates is also possible by using oxygen isotopes. Oxygen isotope data of agates from worldwide occurrences (Figure 18) are mainly in the range between +20‰ and +30‰ but show a scatter of $\delta^{18}O$ values over a broad range between +13‰ and +33‰. The estimated temperatures of agate formation using the fractionation curve of Matsuhisa et al. [139] result in a range between surface temperatures and ca. 230 °C. An exact calculation of temperatures is mostly not possible because of the unknown origin of the primary fluids, but the temperatures are in accordance with the temperatures obtained from fluid inclusion studies.

The results of both methods show relatively high formation temperatures for volcanic agates, which indicate that the mobilization and accumulation of silica started already during a late phase of or soon after the volcanic activity [37]. Considering temperatures obtained from the fluid inclusion homogenization, temperature calculations for different fluid compositions clearly indicate the mixing of dominant amounts of late magmatic (hydrothermal) fluids with smaller amounts of meteoric water [37]. The spatially resolved isotope analyses revealed that the scatter in $\delta^{18}O$-values may be as high as 10‰ even in adjacent chalcedony layers (compare Figure 19) confirming the idea of fluid mixing and/or isotope fractionation under non-equilibrium conditions.

Götze et al. [185] first tried to estimate the temperature of agate formation using the Al concentration in quartz ("Al-geothermometer" of Dennen et al. [186]). The temperatures were calculated from the amount of $[AlO_4]^0$ centers determined by EPR measurements [187]. The $[AlO_4]^0$ concentrations in macrocrystalline quartz of the agates provide reliable temperatures of formation between 65 and 200 °C, which are very close to the estimated temperatures using fluid inclusions or oxygen isotopes (Figure 27). In contrast, calculated temperatures for chalcedony of up to 655 °C as well as strong variations between different chalcedony bands within one agate sample indicate that these temperatures are invalid. It is assumed that the high temperatures are artifacts, which result from non-equilibrium crystallisation processes.

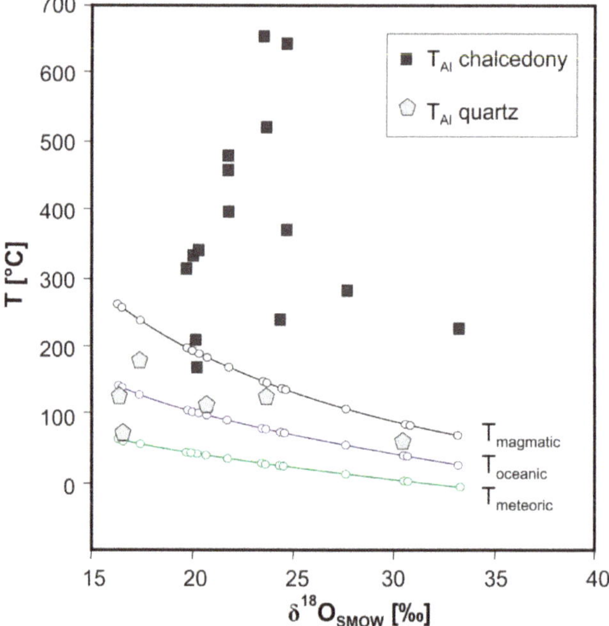

Figure 27. Temperatures of agate formation calculated from $[AlO_4]^0$ concentrations in chalcedony (T_{Al} chalcedony) and macrocrystalline quartz (T_{Al} quartz) compared to temperatures obtained for the same agate samples assuming oxygen isotope equilibration with meteoric ($T_{meteoric}$), oceanic ($T_{oceanic}$) or magmatic water ($T_{magmatic}$); data from Götze et al. [185].

The evaluation of the different methods applied for a calculation of the temperature of agate genesis illustrates that reliable results in most cases are only possible via the analysis of macrocrystalline quartz from the agates. Direct measurements of chalcedony bands provided only reasonable results for oxygen isotopes. In summary, it can be concluded that the temperature interval of agate formation ranges from surface conditions up to temperatures of ca. 230 °C. These temperatures are confirmed by the stability fields of associated paragenetic minerals, homogenization temperatures of fluid inclusions in paragenetic calcite, the low degree of maturity of carbonaceous matter found as cogenetic inclusions within some agates (e.g., [35,45,46,60]) as well as specific microstructural properties of agates. Results of Götze et al. [68] emphasized that the yellow CL, which can often be detected in agates, is a typical indication for low-temperature hydrothermal environment (mostly <250 °C) and is related to fast crystallisation processes from non-crystalline silica precursors.

The geochemical data indicate a mixing of heated meteoric water with magmatic fluids, which alter the volcanic host rocks and release SiO_2. There is no indication for a participation of supercritical fluids, although it cannot be excluded. Additionally, a formation of agate directly from an igneous melt is unlikely.

5.3. Origin and Supply of Silica

The SiO_2 sources for agate formation can clearly be related to the geological environment of the agates. Silica in hydrothermal vein agates and sedimentary agates originates from ascending hydrothermal fluids and silica-rich pore solutions, respectively. In contrast, SiO_2 in volcanic agates represents a product of late- or post-volcanic alteration/weathering of the volcanic host rocks [24,28,37,41,62,63,180].

The formation of agates in volcanic rocks especially occurs in marginal parts of the volcanic bodies and zones with high activities of fluids and water. Geochemical data indicate that heated meteoric water in interaction with magmatic fluids intensively reacts with the host rocks resulting in certain alteration processes ("autometasomatosis"). These alteration processes are reflected in the chemical composition of the agates showing elevated concentrations of elements from the rock matrix (e.g., Al, Fe, K, Na, Ca; compare Table 1) and the occurrence of characteristic secondary minerals in agates (see Table 4). In particular unstable volcanic glass and minerals (e.g., pyroxene, feldspar) are converted into clay minerals, zeolites, and Fe-oxides/-hydroxides, accompanied by the release of enormous amounts of SiO_2 and alkali elements (Table 5). Although the principle processes in basic and acidic volcanics are similar, the specific rock composition and the chemical environment (T, Eh, pH, CO_2, SO_2, water, etc.) strongly influence the alteration processes [78,188–195].

Table 5. Mineral composition (in wt%) of a relatively fresh agate-bearing pitchstone from Chemnitz, Saxony (Germany) in comparison to the surrounding altered rock.

Phase	Pitchstone	Altered Pitchstone
Amorphous (glass)	63.7 ± 2.2	-
Plagioclase	16.1 ± 0.9	14.3 ± 1.0
K-feldspar	5.5 ± 0.5	8.7 ± 0.9
Quartz	1.7 ± 0.2	9.6 ± 0.7
Biotite	1.1 ± 0.9	-
Magnetite	0.8 ± 0.2	-
Ilmenite	0.5 ± 0.2	-
Montmorillonite	10.6 ± 1.9	62.0 ± 1.4
Klinoptilolite	-	5.4 ± 1.0

In mafic volcanic rocks, a greenish layer of sheet silicates (celadonite) can often be observed at the interface between agate geode and host rock. In addition, several minerals of clay and zeolite groups occur as secondary minerals in association with agates [78,196,197]. The process of the rock decomposition and silica release in acidic volcanic rocks is mainly characterized by the formation of sheet silicates such as kaolinite, illite or montmorillonite. Holzhey [63] measured up to 18 wt% illite in the agate-bearing altered rhyolites of the Thuringian forest (Germany). Kryza [198] calculated that the metasomatic processes in 1kg rhyolite can mobilize up to 235 g SiO_2. Therefore, these processes significantly contribute to the release of silica for the agate formation.

Sufficient amounts of fluids (H_2O, but also CO_2 or F) have to be available to enable alteration and transport processes of released silica to the agate cavities. The silica transport in aqueous solution is mainly limited by the extreme low solubility of SiO_2 in water. Several factors control silica solubility such as the pH (the solubility strongly increases at pH > 9), the temperature (100–140 mg/L at 20 °C; 300–380 mg/l at 90 °C) [199], and the silica phase which is in equilibrium with the solution. According to Landmesser [8], for instance, a silica concentration of 250 mg/L SiO_2 in the solution would require up to 0.5 m^3 water to flow in and out a cavity to form an agate of ca. 100 g. Since sufficient pathways for free liquids are lacking in most rocks, the silica supply and accumulation by diffusion processes via intergranular pore space to the cavities is favored.

Due to the larger size (1–100 nm) and rather low diffusion rates of colloidal SiO_2 particles, the proposed main process of silica transport silica is via diffusion of monomeric silicic acid (H_4SiO_4) [8,200]. Recent investigations confirm this hypothesis, showing the release of polymeric and monomeric silicic acid during weathering processes of various silicate minerals. While monomeric silicic acid is the dominating species in a broad range of pH 1–9, polymeric forms including more than ten silicon atoms require higher pH values of above pH 10 [201]. Taking this into account and the fact that most weathering solutions show a pH value <9, the assumed transport mechanism for dissolved silica is via monomeric $Si(OH)_4$ (or oligomer).

The continuous supply of silicic acid results in the accumulation of silica and the filling of the cavities. Subsequent condensation processes lead to the formation of SiO_2 sols and amorphous silica. In a first step, condensation processes cause the bridging of molecules of monomeric silicic acid and the development of Si–O–Si siloxane bonds [100].

$$\equiv Si\text{-}OH + HO\text{-}Si\equiv \quad \rightarrow \quad \equiv Si\text{-}O\text{-}Si\equiv + H_2O$$

As a result of this process, polymeric silicic acid is formed consisting of up to 10 Si atoms, which are linked via Si–O–Si bonds [201]. Further polycondensation results in the formation of discrete particles (sol). During succeeding reactions, the particles grow, form chains, and finally a three-dimensional network (gel; Figure 28). Even in the case of supersaturation, SiO_2 does not precipitate immediately because of the initial formation of di- and polysilicic acid (stable negative sol). These sols are hydrophilic and hardly flocculate due to the water film. Factors which promote the flocculation and the transition from gel to a SiO_2 precipitate are the lowering of temperature and pH (e.g., mixing of fluids, contact with ground water, etc.), the presence of positively charged Al- and Fe-hydroxides from the residual solutions of the clay mineral formation, or electrolytic reactions [179]. The formation of agates from such polydisperse colloidal systems can also explain the manifold and partially very complex agate structures.

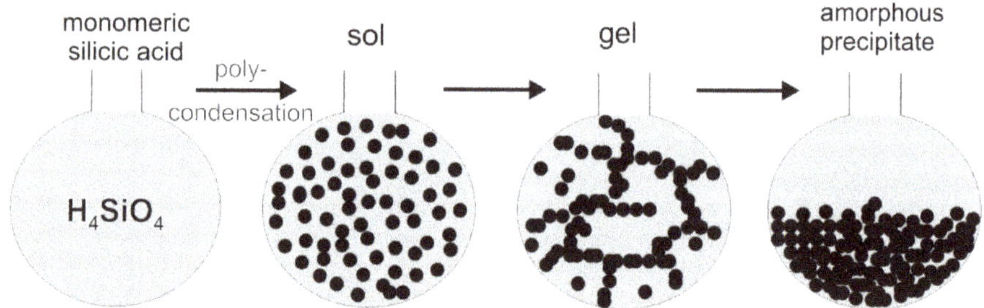

Figure 28. Schematic development of silica from monomeric silicic acid to an amorphous precipitate during the transport and accumulation processes.

Paragenetic minerals like e.g., calcite and fluorite and significant concentrations of elements (e.g., B, Ge, U) in agates indicate that the composition of fluids which altering the volcanic rocks and also mobilize and transport the dissolved chemical compounds can vary significantly from "pure" water. CO_2, for instance, is essential for the formation of calcite from a fluid phase. In addition, chemical transport reactions (CTR) of gases and liquids are realized most probably by stable fluoride compounds like for example SiF_4, BF_3, GeF_4 or UO_2F_2 [34]. Si, B, Ge and U transport can be much better explained by CTR than exclusive transport by H_4SiO_4 in water. Actually, F concentrations of 25 up to 360 ppm have been found in agates of various occurrences [126]. In addition, REE distribution patterns measured in agates indicate that released HREE are preferably incorporated in carbonate- and F-complexes during alteration and transport processes [134]. Pronounced Ce anomalies in certain agates also provide indications for a possible mixing of different fluids from contrasting sources (e.g., magmatic volatiles, meteoric water) [55,135].

5.4. Formation of Agate Banding and Color

The results of the microstructural and geochemical investigations show that visible agate banding appears due to variations of silica phases, their respective crystal size and habit, and porosity. The formation of spherulites from an amorphous (non-crystalline) silica precursor [202] and the following crystallisation characteristics of SiO_2 and self-organization of silica and impurities are the most important processes in the development of the characteristic agate microstructure [25,203,204].

Assuming that the initial cavity was filled with a silica sol and/or amorphous silica, the silica's flocculation and crystallisation process must have been initiated by either changes in the physico-chemical conditions (e.g., T, pH) or reactions of positively charged compounds (e.g., Fe- or Al-hydroxides) with the negatively charged colloidal silica. The crystallisation mostly starts from the wall of the agate chamber with the formation of micro-granular or spherulitic quartz proceeding with fibrous chalcedony inwards to the center of the nodule and often ends up with macrocrystalline quartz (Figure 29). Indications for crystallisation processes of amorphous precursors under non-equilibrium conditions are provided by the results of CL and EPR studies, and are also documented by heterogeneous internal textures of macrocrystalline quartz crystals, which appear homogeneous in the optical microscope (Figure 29b,c).

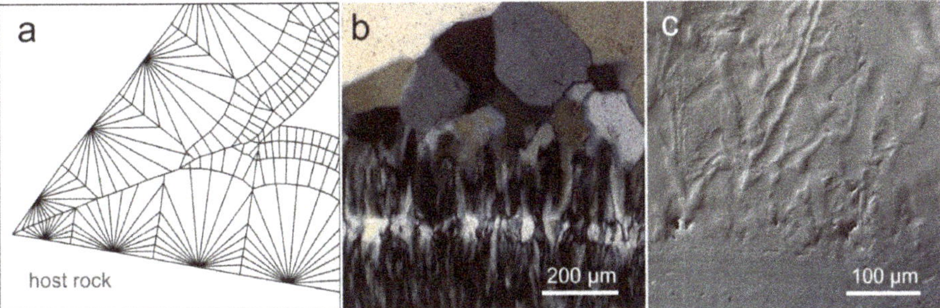

Figure 29. (**a**) Scheme of the formation of agate banding starting with separate spherulites and continuing inwards with fibrous chalcedony layers; (**b**) micrograph in transmitted light (crossed polars) of an agate from Los Indios (Cuba) showing a smooth transition from fibrous chalcedony to macrocrystalline quartz; (**c**) Nomarski DIC micrograph of the interface chalcedony/quartz revealing skeletal growth and irregular internal textures of macrocrystalline quartz, which point to rapid crystallisation under non-equilibrium conditions.

Wang and Merino [205], Heaney [23] and Merino et al. [27] published crystallisation models describing the formation of agate banding in the light of self-organizing processes. Therefore, the fibrous texture of chalcedony could be explained by the dynamics of silica and impurity (i.e., trace elements) transport in the surrounding environment. It is assumed that the impurities are initially precipitated together with the primary amorphous silica leading to a fine dissemination in the respective matrix. The first emerging chalcedony spherulites therefore may contain relatively high concentrations of trace elements. Because of the interaction of morphologically rather unstable crystallisation fronts of quartz and cation-enhanced quartz growth, a first generation of twisted chalcedony with high impurity content is formed. The ongoing silica release and lowering of trace elements during further crystallisation result in a residual medium of aqueous character with low silica concentration. Consequently, the advancing crystallisation front becomes more and more morphologically stable and macrocrystalline quartz evolves. The interface between chalcedony and quartz is often accompanied by a characteristic drop of both impurities and structural defects. This idea is supported by frequent inclusions of iron compounds and elevated contents of Al in the agates. For instance, Götze et al. [206] measured up to 0.55 wt% Al in early opal-CT of agates from Los Indios (Cuba) in contrast to 8–13 ppm Al in macrocrystalline quartz of the same sample.

Self-organization and Ostwald-ripening result in appropriate crystallisation sequences and the crystallinity of SiO_2 increases with proceeding maturity: amorphous $SiO_2 \rightarrow$ opal-CT/cristobalite $\rightarrow$ moganite $\rightarrow$ chalcedony (microcrystalline SiO_2) $\rightarrow$ macrocrystalline quartz (Figure 30). The succession of maturation can be explained by the decrease of the chemical potential and the Gibbs free energy ΔG according to the OSTWALD rule. During this process, the initial water content of up to 12% in amorphous silica decreases down to almost 0% in crystalline quartz [21]. The decrease in volume during ageing and dehydration of silica is compensated by fine dispersed pore space. The formation of completely filled nodules requires a continuous supply of silicic acid during the crystallisation and "self-organization" process, which may diffuse/penetrate through the existing layers of the agate. Alternating layers of chalcedony and macrocrystalline quartz indicate possible multi-step processes of silica supply and agate formation [51].

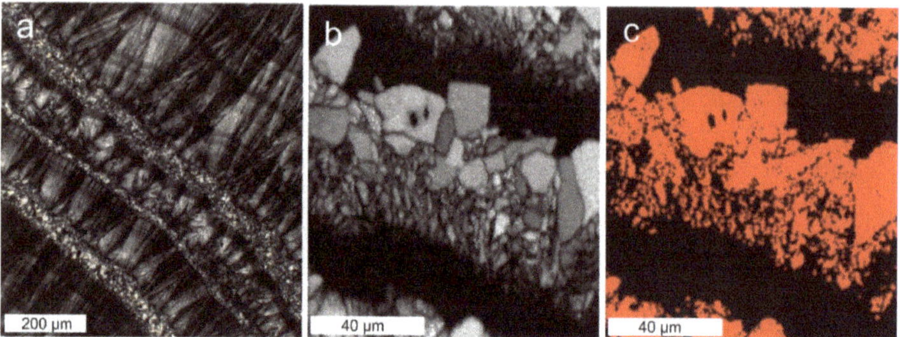

Figure 30. Micrographs showing the development of the band micro-structure in agates: (**a**) rhythmic banding of granular quartz and fibrous chalcedony in an agate from Los Indios (Cuba); (**b**) electron backscatter diffraction (EBSD) band contrast image revealing an increase in brightness with increasing SiO_2 crystallinity; (**c**) EBSD micrograph showing the distribution of α-quartz (red colour) in the agate bands; (**b,c**) demonstrate that the dark band areas consist of amorphous or strongly disordered silica.

It cannot be excluded that during ripening of amorphous silica certain layers/zones of different constitution (silica and/or water content, accumulation of trace elements, etc.) had already formed as precursors for the formation of a variable agate microstructure due to processes of diffusion and chemical differentiation (segregation; [57,207]). The formation of agate bands could then be related to a nonlinear reaction–diffusion–advection system, where fractal patterns are formed due to Belousov-Zhabotinsky (B-Z) chemically-oscillating reactions, similar to those in other minerals and materials [208–210]. Pabian and Zarins [24] proposed the B-Z reaction as a mechanism for agate genesis with an initial silica gel and cyclic chemical effects caused by later contact with alkaline water. However, apart from the continual need of circulating alkaline water, the mechanism fails as any gel would eventually collapse. Nevertheless, the agate formation from polydisperse colloidal systems allows the explanation of the wide spectrum of agate structures.

Up to now, only limited information concerning the timescale of the filling and crystallisation processes are available from technical processes or mathematical calculations. Oehler [184] crystallized silica gel to quartz in the laboratory in between 12 h and 180 days. However, these experiments run under high-pressure/high-temperature conditions, which might not apply for agate formation. Ortoleva et al. [25] calculated a duration of about 5000 a for the crystallisation and growth process of a spherical agate with a diameter of 3 cm. Other calculations indicate that the duration of the processes may range from some hundreds to some ten thousands of years [211]. Of course, the time will strongly be influenced by the specific conditions of formation.

The origin of color(s) and color banding in agates is closely connected with the formation of the agate microstructure. The most common colours in agate are white, grey and blue (Figure 31). These colours are caused by variations in crystallite size, microstructure (including water content) and porosity. SEM studies showed that white agate bands are not fibers but a relatively thick plate edge-like structure, whereas the clear areas are globular and mostly have high porosity [19,21]. These plate edges are able to totally reflect incident light while the non-white area allows some white light to be transmitted. The blue color is caused by the dispersion of light on the micro-particles (Tyndall effect). In particular, blue agate bands are relatively sensible for aging processes. Sometimes they lose their concise banding during several years (Figure 31b). This might be related to the loss of water, since storage in water can at least partly reactivate the blue color (Figure 31c).

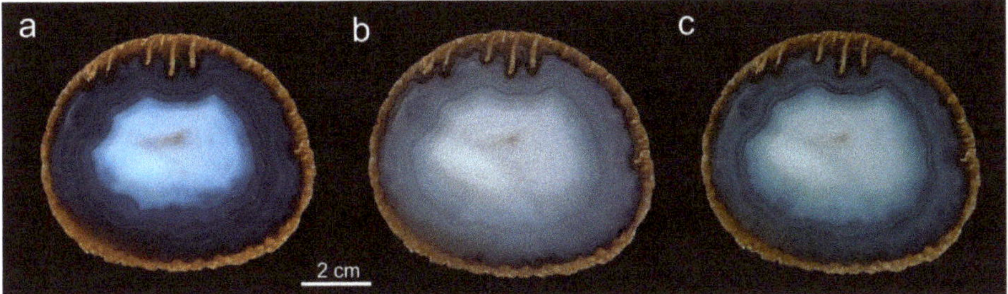

Figure 31. Aging of agate accompanied by loss of distinct banding and blue color demonstrated by an agate sample from Soledade, Rio Grande do Sul (Brazil): (**a**) freshly cut and polished agate with deep blue color and distinct banding; (**b**) the same agate after 3 years at room temperature and under natural light; (**c**) the agate after 6 weeks storage in water.

Although amethyst and smoky quartz can frequently be observed as macrocrystalline incrustations in the agate center, color centers related to point defects in chalcedony are rare. Up to now there is only evidence for the origin of violet color of chalcedony in "Grape agate" from Pantai Ngalo, Sulawesi (Indonesia) due to the amethyst color center [212]. In most cases, the colorful bands are correlated to the distribution and fixation of impurities in the silica matrix during the crystallisation process and often show conformity to the structural banding. Liesegang [11] tried to explain the formation of color banding by diffusion processes of metal ions in silica gels (Liesegang rings), and he suggested that also the rhythmic crystallisation of the agates is caused and influenced by these impurities. The appearance of Liesegang rings in a gel is indeed very similar to that what we see in a natural agate. However, the formation of differently colored bands and the distribution of color pigments in agates are mainly governed by the self-organizational processes of the silica matrix during crystallisation.

The most frequent color pigments are iron phases; different chemical compounds of manganese were also detected. Certain amorphous and crystalline Fe-oxides and Fe-hydroxides (in particular, hematite and goethite) are mostly responsible for the frequent red and also yellow coloration of agates. Different color shades are caused by different kinds of iron compounds, varying amounts or grain size of the color pigments. The mixture of different compounds also results in varying shades of colors.

The presence of iron oxides within the SiO_2 matrix is a result of high iron activity in the initial silica-bearing, mineralizing medium. The accumulation of fine-dispersed pigments along the banding of the chalcedony layers, parallel to the spherulitic structures, is probably a result of "self-purification" of the crystallizing SiO_2 phases. The incorporation of iron into the quartz structure is limited and therefore iron is rejected from the emerging quartz crystals. Consequently, it is transported with the crystallisation front and accumulated/adjusted along the banding (Figure 32).

If distinct banding in agates is lacking, color pigments can be distributed randomly in the silica matrix resulting in irregular color distribution (Figures 32 and 33a). Secondary effects due to alteration, weathering or infiltration of impurities along cracks can also change the primary colour banding in agates. In these cases, leaching or oxidation processes result in the formation of coloration or decoloration zones along the rims of the nodules or along secondary cracks (Figure 33b,c). Such secondary color distribution discordant to the agate banding is often called "chromatography" effect.

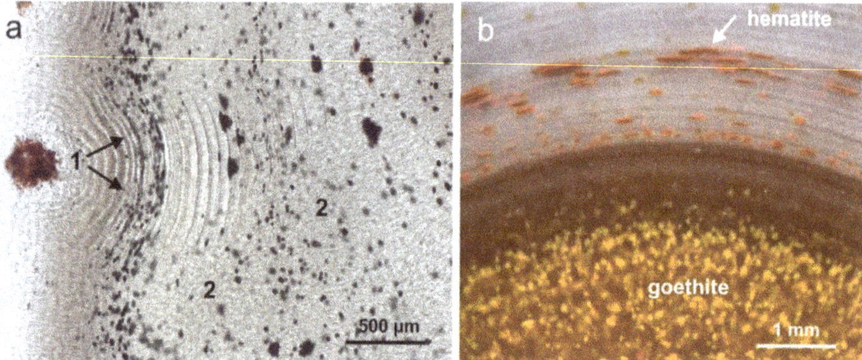

Figure 32. Incorporation of iron compounds as color pigments in agate: (**a**) micrograph in transmitted light of an agate from Hausdorf, Saxony (Germany) with iron oxides aligned along the banding (1) and random distribution in areas without banding (2); (**b**) agate from Pujiang, province Zhejiang (China) showing yellow color due to incorporation of goethite and red bands caused by hematite inclusions; note that hematite particles are flattened and aligned along the banding.

Figure 33. (**a**) Agate from Sierra del Chachahuen, Mendoza (Argentina) without distinct banding and irregular color distribution; (**b**) secondary discoloration (left side and along cracks, see arrows) in an agate from San Rafael, Mendoza (Argentina); (**c**) agate from Xuanhua, province Hebei (China) with effects of secondary bleaching and discoloration; scale bar is 2 cm.

5.5. Microstructural Peculiarities and Possible Biosignatures in Agates

Some agates may show structural peculiarities, which deviate from the common microstructures. One of the widely discussed aspects is the presence of so called "infiltration channels" ("escape tubes" or "flow channel"; Figure 34a), which predominantly occur in agates of mafic volcanic rocks. The shape and name of these structural features suggest that they represent channels of silica supply during agate formation [211]. Although in some specific cases the real impression of silica inflow exists (Figure 34b), a closer examination shows that in most agates these structures may have formed by deformation of non-solidified silica in a plastic state. Microscopic studies revealed that the primary orientation of chalcedony "fibers" changes at the interface and the banding is deformed. It is possible that this deformation is caused by increasing pressure during the formation process and central parts of the banding are squeezed outwards. Sometimes, agates even show multiple deformation features (Figure 34c).

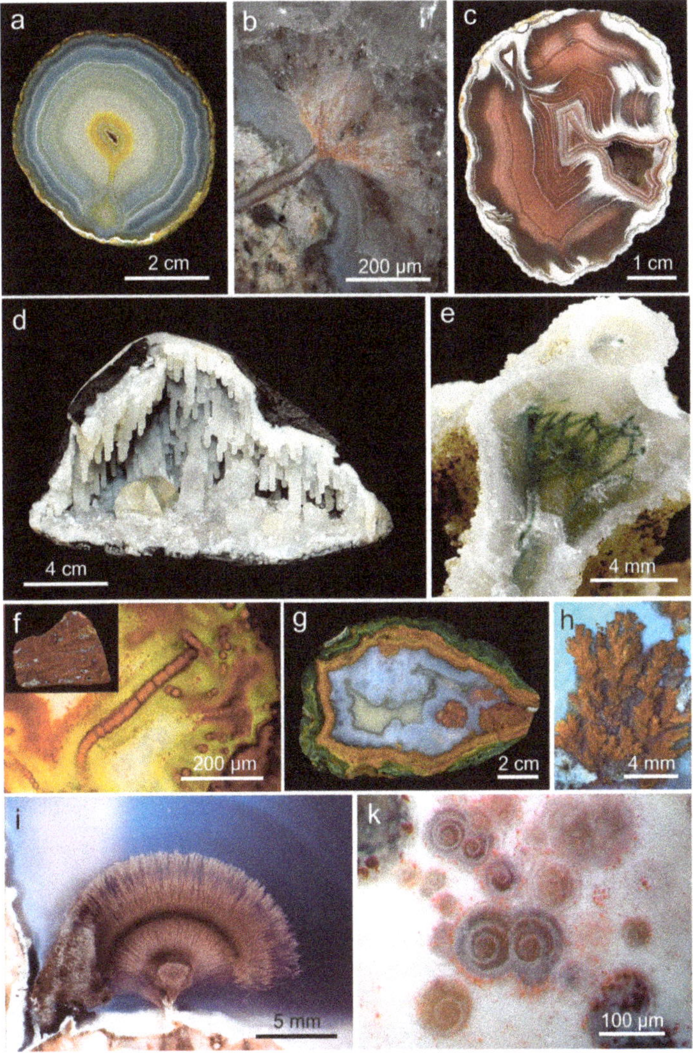

Figure 34. Structural peculiarities in agates: (**a**) agate from Moctezuma, Chihuahua (Mexico) showing a so called "infiltration channel" ("escape tube"); (**b**) micrograph of the marginal part of an agate from Zschaitz, Saxony (Germany) resembling the inflow channel of silica into the agate, which cross-cuts the outermost chalcedony layer; (**c**) agate from Agate Creek, Queensland (Australia) with multiple deformation features; (**d**) cavity from the Dekkan Trap basalt, India completely filled with quartz stalactites (SFF); (**e**) green filamentous structure in a moss agate from Jalna, India; (**f**) Fe-rich agate of probably biogenic origin from the Mátra Mts., Hungary (inset) with filamentous pseudomorphs in the microcrystalline silica matrix; (**g**) agate from Kardzali (Bulgaria) with plume structures; (**h**) details of the reddish-brown plume structures in chalcedony matrix; (**i**) bundle-like structures resembling the shape of natural sheaf of wheat in an agate from the Darrell Friend Ranch near Ashwood, Oregon (USA) (sample courtesy G. Schiecke, photo A. Gabert); the unusual shape is probably caused by calcite traced in silica gel; (**k**) agate from Berschweiler near Idar-Oberstein (Germany) with inclusions of spiral forms probably developed from an amorphous silica precursor by self-organisation growth.

The formation of stalactitic fabrics in agates is also a rare anomaly in agates (Figure 34d). It is known only from a few deposits worldwide such as in the USA, Brazil, Uruguay, Germany, Poland, Russia or India. The Baker Egg Mine in New Mexico (USA) is famous for the high amount (up to 10%) of geodes with stalactites [43]. It is assumed that the stalactites have been formed simultaneously with the wall layers from a liquid that had completely filled the cavity. Recent studies on such pseudostalactitic fabrics (sub-surface filamentous fabrics—SFF; [213]) in cavities of the Deccan Trap basalts in India (Figure 34d) revealed that the growth of the SFF is not limited to vertical directions and might be initiated by microbial activities [195]. Residues of organic material (fungal chitin) in the innermost filamentous core as well as the morphometric characteristics of certain filamentous fabrics indicate a biological contribution during an early state of SFF formation. The internal structure of SFF is mostly made up of a core of clay minerals (in particular smectite), celadonite and zeolites, followed by concentric layers of chalcedony. Similar results were obtained for moss agates (Figure 34e), which also show clear indications of biosignatures [33,195,213,214].

Other examples worldwide show that biosignatures and indications of biomineralization processes can be detected in agates and agate-like silica mineralization [33]. Several studies on ancient and modern hydrothermal systems revealed that microorganisms are closely connected with both terrestrial and deep-sea hydrothermal vents [215]. Almost all of the microfossils associated with hydrothermal systems are filamentous. These microorganisms are thermophilic (i.e., the organisms grow best at temperatures > 45 °C, sometimes >100 °C) and often obtain their energy from the oxidation of inorganic compounds ("chemolithoautotroph") (e.g., [216–219].

Müller et al. [220] found Fe-rich cryptocrystalline silica masses with agate-like structures in the southern part of the Mátra Mts. (Hungary), in the neighbourhood of the Gyöngyöstarján and Gyönggyösoroszi area (Figure 34f). These agate-like aggregates in fractures and cavities of an andesite consist mainly of quartz, moganite, opal-CT, hematite, goethite (together with calcite, nontronite, celadonite) and show characteristic signatures of microbial activities (Figure 34g).

Because of the general lack of residues of organic materials within agates, the detection of biosignatures is mostly based on morphological criteria indicating former activities of living organisms. However, recent studies have proved that certain chemical pathways exist, that produce inorganic mineral matter resembling living forms [221–225]. The processes resulting in such crystal architectures include a couple of different chemical interactions including the complexation of ions, selective adsorption onto growing crystal surfaces, or temporary stabilization of an amorphous precursor phase [226]. These results emphasized that morphological criteria alone cannot evidence the biogenic origin of mineral matter.

In conclusion, unusual mineral forms and structures in agates can also be related to such inorganic chemical processes, as in the case of "plume agates" (Figure 34g,h). Such plume structures represent the well-known phenomenon of "silica gardens"—i.e., plant-like structures that formed due to the addition of certain metal salts to sodium silicate solutions. Such aggregations of inorganic self-assembly originate from a combination of osmosis, chemical reactions and buoyancy [222,224]. Moreover, carbonates of certain alkaline earth elements can develop highly unusual shapes during crystallisation in a silica-rich environment such as in the case of calcite in silica gel [226]. Laboratory experiments resulted in finger-like structures or bundles resembling the shape of natural sheaf of wheat (Figure 34i). Further studies revealed that these spectacular aggregations consist of numerous calcite rhombohedra that were self-assembled along their c-axis and form laminar aggregates with rhythmically banded patterns [227].

A new kind of agates with unusual features was recently found in volcanic rocks of certain occurrences in the Saar-Nahe region (Germany). These agates include spirals of several tens of μm up to several hundreds of μm in size within the silica matrix (Figure 34k). Because of the specific shape, the conspicuous creations first led to a controversial discussion concerning a possible biogenic origin (microfossils such as foraminifera or snails) or an inorganic formation [228]. However, detailed investigations revealed that the spirals probably developed from an amorphous silica precursor by self-organization growth and polymerization initiated by a screw dislocation (dislocation growth; [39]). The data indicate that agate formation runs through several structural states of SiO_2 with amorphous silica as the first solid phase, and the specific conditions can produce manifold spectacular forms making agates a unique product of nature.

6. Conclusions

The present paper provides a comprehensive compilation of mineralogical and geochemical data of agates from worldwide occurrences in different host rocks. Based on these data in combination with field observations and results of laboratory experiments an attempt is made to develop a model concerning the formation of agates. The presented results illustrate that the processes of agate formation are complex and often involve multiple steps. Moreover, agate formation in volcanic rocks, in sedimentary host rocks or the formation of hydrothermal vein agate can run under different conditions and involve divergent processes.

The majority of agates can be related to mafic (basalts, andesites) and acidic (rhyolites, rhyodacites) volcanic host rocks. Many field observations and analytical results indicate that volcanic agates form in a late stage of volcanism or after solidification of the volcanic host rocks. Most of the silica necessary for agate formation is supplied by alteration processes of the surrounding volcanic rocks by heated meteoric water, hydrothermal/magmatic fluids and other volatiles such as CO_2 or F-complexes. In vein agates, SiO_2 derives from ascending hydrothermal-magmatic solutions, whereas silica in sedimentary agates is mostly accumulated by SiO_2-rich pore solutions.

The transport of silica in the form of monomeric silicic acid H_4SiO_4 or oligomers is predominantly realized by diffusion processes to cavities in the host rocks. In basic volcanic rocks, vesicular cavities form from included gases and liquids during the solidification of the lava, whereas in acidic volcanics lithophysae (high-temperature crystallisation domains) are formed above T_g in the cooling melt as host for the agates. Vein agates can be formed in fissures and veins within different types of host rocks, where a free movement of silica-bearing mineralizing fluids through a system of cracks is possible. In sedimentary rocks, formation of agates is often accompanied by replacement of pre-existing soluble minerals (sulfates, carbonates) by silica.

The accumulation and condensation of silicic acid result in the formation of silica sols and amorphous silica as precursors for the development of the typical agate structures. The process of crystallisation often starts with the spherulitic growth of chalcedony or micro-granular silica continuing into chalcedony fibres. Macrocrystalline quartz crystallizes when the SiO_2 concentration in the mineralizing fluid is low. The abundance of lattice defects and elevated concentrations of certain trace elements in agates as well as heterogeneous internal textures and sector zoning in macrocrystalline quartz crystals provide clear indications for crystallisation under non-equilibrium conditions. The formation of the typical agate microstructure is most likely governed by processes of self-organization. Depending on the specific physico-chemical conditions (e.g., T, pH, SiO_2 concentration, trace-element contents), these crystallisation processes may proceed in a very complex way and result in characteristic structures, SiO_2 phases, colors and inclusions. Oxygen isotopes, Al concentrations and fluid inclusion studies provided a temperature range for agate formation between ca. 20 and 230 °C.

During agate genesis not only silica minerals are formed, but also numerous other mineral phases. The spectrum of paragenically formed minerals can significantly vary in dependence on the genetic type of agate. These differences can be related to the different types of host rocks and differences in the conditions of mineralization. The most common minerals associated with agates are clay minerals, Fe-oxides/-hydroxides, minerals of the zeolite group and carbonates (especially calcite).

The complex processes of agate formation, often running under non-equilibrium conditions, result in a wide variety of peculiarities such as spectacular internal textures, "moss agates", "plume agates" or stalactitic aggregations. Moreover, there are compelling indications for microbial activities that are sometimes involved in agate creation, at least in a very early state of formation. These specific features make agates a unique and spectacular product of nature. Although a lot of basic data concerning the formation of agates are now established, several open questions remain.

Supplementary Materials: The following are available online at http://www.mdpi.com/2075-163X/10/11/1037/s1, Table S1: Compilation of investigated agate occurrences with references to presented data and relevant published results, respectively.

Author Contributions: J.G., R.M., and Y.P. conducted different analytical measurements, evaluated the mineralogical and geochemical data, and provided appropriate parts of the manuscript. J.G. compiled and wrote the final version of the manuscript. All authors have read and agreed to the published version of the manuscript.

Funding: This research received no external funding.

Acknowledgments: The present publication would not have been possible without the help and promotion of numerous colleagues, students and friends over the last decades. Collaboration during sampling and scientific discussions with Terry Moxon (Doncaster, UK), Gerhard Holzhey (Erfurt, Germany), Johann Zenz (Gloggnitz, Austria), Klaus Schäfer (Vollmersdorf, Germany) successfully supported the preparation of the manuscript. Ulf Kempe (Freiberg, Germany), Reinhard Kleeberg (Freiberg, Germany), Harry Berek (Freiberg, Germany), Lutz Nasdala (Vienna, Austria) and Axel Müller (Oslo, Norway) are gratefully acknowledged for their analytical suppport and the fruitful discussions. We thank Galina Palyanova (Novosibirsk, Russia) for the invitation to write this review article for the special issue "Agates: Types, Mineralogy, Deposits, Host Rocks, Ages and Genesis" and the helpful comments. Reviews of Magdalena Dumańska-Słowik (Kraków, Poland), Terry Moxon (Auckley, Doncaster, UK) and two anonymous reviewers improved the quality of the manuscript significantly.

Conflicts of Interest: The authors declare no conflict of interest.

References

1. Blankenburg, H.-J. *Achat*; VEB Deutscher Verlag für Grundstoffindustrie: Leipzig, Germany, 1988; 203p.

2. Zenz, J. *Achate/Agate*; Bode-Verlag: Haltern, Germany, 2005; 656p.

3. Zenz, J. *Achate/Agates II*; Bode-Verlag: Haltern, Germany, 2009; 656p.

4. Zenz, J. *Achate/Agates III*; Bode-Verlag: Haltern, Germany, 2011; 656p.

5. Brückmann, U.F.B. *Abhandlung von Edelsteinen*; Waisenhaus-Buchhandlung: Braunschweig, Germany, 1773.

6. Collini, C. *Tagebuch Einer Reise, Welches Verschiedene Mineralogische Beobachtungen, Besonders Über Die Achate und Den. Basalt Enthält*; C.F. Schwan: Mannheim, Germany, 1776; 582p.

7. Noeggerath, M. On the porosity and colouring of agates, chalcedonies, etc. *Edinb. New Philos. J.* **1850**, *58*, 166–172.

8. Landmesser, M. Das Problem der Achatgenese. *Mitt. Pollichia* **1984**, *72*, 5–137.

9. Daubrée, A. *Synthetische Studien zur Experimental-Geologie*; Vieweg-Verlag: Braunschweig, Germany, 1880; 122p.

10. Nacken, R. Über die Nachbildung von Chalzedon-Mandeln. *Nat. und Volk.* **1948**, *78*, 2–8.

11. Liesegang, R.E. Die Entstehung der Achate. *Zent. für Mineral.* **1910**, *11*, 593–597.

12. Liesegang, R.E. *Die Achate*; Verlag von Theodor Steinkopff: Dresden/Leipzig, Germany, 1915; 122p.

13. Flörke, O.W.; Köhler-Herbertz, B.; Langer, K.; Tönges, I. Water in Microcrystalline Quartz of Volcanic Origin: Agates. *Contrib. Mineral. Petrol.* **1982**, *80*, 324–333. [CrossRef]

14. Fallick, A.E.; Jocelyn, J.; Donnelly, T.; Guy, M.; Behan, C. Origin of agates in volcanic rocks from Scotland. *Nature* **1985**, *313*, 672–674. [CrossRef]

15. Godovikov, A.A.; Ripinen, O.I.; Motorin, S.G. *Agaty*; Nedra: Moskva, Russia, 1987; 368p.

16. Macpherson, H.-G. *Agates*; The Natural History Museum and the National Museums of Scotland: London, UK, 1989; 72p.

17. Moxon, T. On the origin of agate with particular reference to fortification agate found in the Midland Valley, Scotland. *Chem. der Erde* **1991**, *51*, 251–260.

18. Moxon, T. *Agate: Microstructure and Possible Origin*; Terra Publications: Doncaster, UK, 1996; 106p.

19. Moxon, T. Agates: A study of ageing. *Eur. J. Mineral.* **2002**, *14*, 1109–1118. [CrossRef]

20. Moxon, T. *Studies on Agate—Microscopy, Spectroscopy, Growth, High Temperature and Possible Origin*; Terra Publications: Doncaster, UK, 2009; 102p.

21. Moxon, T. A re-examination of water in agate and its bearing on the agate genesis enigma. *Mineral. Mag.* **2017**, *81*, 1223–1244. [CrossRef]

22. Holzhey, G. Vorkommen und Genese der Achate und Paragensemineralien in Rhyolithkugeln aus Rotliegendvulkaniten des Thüringer Waldes. Ph.D. Thesis, TU Bergakademie, Freiberg, Germany, 1993; 132p.

23. Heany, P.J. A proposed mechanism for the growth of chalcedony. *Contrib. Mineral. Petrol.* **1993**, *115*, 66–74. [CrossRef]

24. Pabian, R.K.; Zarins, A. *Banded Agates—Origins and Inclusions*; Educational Circular No. 12; University of Nebraska: Lincoln, RI, USA, 1994; 32p.

25. Ortoleva, P.; Chen, Y.; Chen, W. *Agates, Geodes, Concretions and Orbicules: Self-Organized Zoning and Morphology*; Kruhl, J.H., Ed.; Springer: Heidelberg, Germany, 1994; pp. 283–305.

26. Heaney, P.J.; Davis, A.M. Observation and origin of self-organized textures in agates. *Science* **1995**, *269*, 1562–1565. [CrossRef]

27. Merino, E.; Wang, Y.; Deloule, E. Genesis of agates in flood basalts: Twisting of chalcedony fibers and trace-element geochemistry. *Am. J. Sci.* **1995**, *295*, 1156–1176. [CrossRef]

28. Cross, B.L. *The Agates of Northern Mexico*; Burgess Publishing Division: Edina, MI, USA, 1996; 130p.

29. Götze, J.; Nasdala, L.; Kleeberg, R.; Wenzel, M. Occurrence and distribution of "moganite" in agate/chalcedony: A combined micro-Raman, Rietveld, and cathodoluminescence study. *Contrib. Mineral. Petrol.* **1998**, *133*, 96–105. [CrossRef]

30. Götze, J.; Tichomirowa, M.; Fuchs, H.; Pilot, J.; Sharp, Z. Geochemistry of agates: A trace element and stable isotope study. *Chem. Geol.* **2001**, *175*, 523–541. [CrossRef]

31. Götze, J.; Möckel, R.; Kempe, U.; Kapitonov, I.; Vennemann, T. Origin and characteristics of agates in sedimentary rocks from the Dryhead area, Montana/USA. *Mineral. Mag.* **2009**, *73*, 673–690. [CrossRef]

32. Götze, J.; Martins, M.S.; Czarnobay, J.C. Achate aus Brasilien. *Veröff. Mus. für Nat. Chemnitz* **2010**, *33*, 63–78.

33. Götze, J.; Müller, A.; Polgári, M.; Pál-Molnár, E. Biosignaturen in Achat/Chalcedon–die Rolle von Mikroorganismen bei der Bildung von SiO_2. *Mineralienwelt* **2011**, *22*, 90–96.

34. Götze, J.; Schrön, W.; Möckel, R.; Heide, K. The role of fluids in the formation of agate. *Geochemistry* **2012**, *72*, 283–286. [CrossRef]

35. Götze, J.; Nasdala, L.; Kempe, U.; Libowitzky, E.; Rericha, A.; Vennemann, T. Origin of black colouration in onyx agate from Mali. *Mineral. Mag.* **2012**, *76*, 115–127. [CrossRef]

36. Götze, J.; Gaft, M.; Möckel, R. Uranium and uranyl luminescence in agate/chalcedony. *Mineral. Mag.* **2015**, *79*, 983–993. [CrossRef]

37. Götze, J.; Möckel, R.; Vennemann, T.; Müller, A. Origin and geochemistry of agates from Permian volcanic rocks of the Sub-Erzgebirge basin (Saxony, Germany). *Chem. Geol.* **2016**, *428*, 77–91. [CrossRef]

38. Götze, J.; Möckel, R.; Eulitz, B. "Karbonat-Achat" von Krásný Dvoreček. *Mineralienwelt* **2018**, *4*, 82–87.

39. Götze, J.; Berek, H.; Schäfer, K. Micro-structural phenomena in agate/chalcedony: Spiral growth. *Mineral. Mag.* **2019**, *83*, 281–291. [CrossRef]

40. Moxon, T.; Rios, S. Moganite and water content as a function of age in agate: An XRD and thermogravimetric study. *Eur. J. Mineral.* **2004**, *4*, 693–706. [CrossRef]

41. Petránek, J. Entstehung von gravitations und adhäsionsgebänderten Achaten in Raum und Zeit und in Abhängigkeit vom Klima. *Der Aufschluss* **2006**, *57*, 129–150.

42. Petránek, J. Sedimentäre Achate. *Der Aufschluss* **2009**, *60*, 291–302.

43. Colburn, R.P. *The Formation of Thundereggs*; Geode Kid Productions: Deming, NM, USA, 2008.

44. Dumańska-Słowik, M.; Natkaniec-Nowak, L.; Kotarba, M.J.; Sikorska, M.; Rzymełka, J.A.; Łoboda, A.; Gaweł, A. Mineralogical and geochemical characterization of the "bituminous" agates from Nowy Kościół (Lower Silesia, Poland). *Neues Jahrbuch für Mineralogie Abhandlungen* **2008**, *184*, 255–268.

45. Dumańska-Słowik, M.; Natkaniec-Nowak, L.; Weselucha-Birczyńska, A.; Gaweł, A.; Lankosz, M.; Wróbel, P. Agates from Sidi Rahal, in the Atlas Mountains of Morocco: Gemmological characteristics and proposed origin. *Gems Gemol.* **2013**, *49*, 148–159.

46. Dumańska-Słowik, M.; Powolny, T.; Sikorska-Jaworowska, M.; Gaweł, A.; Kogut, L.; Poloński, K. Characteristics and origin of agates from Płóczki Górne (Lower Silesia, Poland): A combined microscopic, micro-Raman, and cathodoluminescence study. *Spectrochim. Acta Part. A Mol. Biomol. Spectrosc.* **2018**, *192*, 6–15.

47. Clark, R. *South. Dakota's State Gemstone—Fairburn Agate*; Silverwind Agates: Appleton, CT, USA, 2009; 130p.

48. Lyashenko, E.A. Agates of Russia. *Mineral. Alm.* **2010**, *15*, 6–27.

49. Götze, J. Agate–Fascination between Legend and Science. In *Agates III*; Zenz, J., Ed.; Bode-Verlag: Haltern, Germany, 2011; pp. 19–133.

50. Hurst, J.T. *Dryhead Agate*; Agate Treasures-Schatzkammerachate Publishing: Boulder, CO, USA, 2012; 90p.

51. French, M.W.; Worden, R.H.; Lee, D.R. Electron backscatter diffraction investigation of length-fast chalcedony in agate: Implications for agate genesis and growth mechanisms. *Geofluids* **2013**, *13*, 32–44. [CrossRef]

52. Richter, S.; Götze, J.; Niemeyer, H.; Möckel, R. Mineralogical investigation of agates from Cordón de Lila, Chile. *J. Andean Geol.* **2015**, *42*, 386–396.

53. Ottens, B.; Götze, J. *Achatwelt China*; extralapis 51; Christian Weise Verlag: München, Germany, 2016; 110p.

54. Natkaniec-Nowak, L.; Dumańska-Słowik, M.; Pršek, J.; Lankosz, M.; Wróbel, P.; Gaweł, A.; Kowalczyk, J.; Kocemba, J. Agates from Kerrouchen (The Atlas Mountains, Morocco). *Minerals* **2016**, *6*, 77. [CrossRef]

55. Powolny, T.; Dumańska-Słowik, M.; Sikorska-Jaworowska, M.; Wójcik-Bania, M. Agate mineralization in spilitized Permian volcanics from "Borówno" quarry (Lower Silesia, Poland)—Microtextural, mineralogical, and geochemical constraints. *Ore Geol. Rev.* **2019**, *114*, 103–130. [CrossRef]

56. Kigai, I.N. The genesis of agates and amethyst geodes. *Can. Mineral.* **2019**, *57*, 867–883. [CrossRef]

57. Howard, C.B.; Rabinovitch, A. A new model of agate geode formation based on a combination of morphological features and silica sol-gel experiments. *Eur. J. Mineral.* **2017**, *30*, 97–106. [CrossRef]

58. Gliozzo, E.; Cairncross, B.; Vennemann, T. A geochemical and micro-textural comparison of basalt-hosted chalcedony from the Jurassic Drakensberg and Neoarchean Ventersdorp Supergroup (Vaal River alluvial gravels), South Africa. *Int. J. Earth Sci.* **2019**, *108*, 1857–1877. [CrossRef]

59. Zhang, X.; Ji, L.; He, X. Gemological characteristics and origin of the Zhanguohong agate from Beipiao, Liaoning province, China: A combined microscopic, X-ray diffraction, and Raman spectroscopic study. *Minerals* **2020**, *10*, 401. [CrossRef]

60. Pršek, J.; Dumańska-Słowik, M.; Powolny, T.; Natkaniec-Nowak, L.; Toboła, T.; Zych, D.; Skrepnicka, D. Agates from Western Atlas (Morocco)—Constraints from mineralogical and microtextural characteristics. *Minerals* **2020**, *10*, 198. [CrossRef]

61. Moxon, T.; Carpenter, M.A. Crystallite growth kinetics in nanocrystalline quartz (agate and chalcedony). *Mineral. Mag.* **2009**, *73*, 551–568. [CrossRef]

62. Walger, E. Das Vorkommen von Uruguay-Achaten bei Flonheim in Rheinhessen, seine tektonische Auswertung und seine Bedeutung für die Frage nach der Achatbildung. *Jahresber. Mitt. Oberrh. Geol. Ver.* **1954**, *36*, 20–31. [CrossRef]

63. Holzhey, G. Herkunft und Akkumulation des SiO_2 in Rhyolithkugeln aus Rotliegendvulkaniten des Thüringer Waldes. *Geowiss. Mitt. Thüringen* **1995**, *3*, 31–59.

64. Hopkinson, L.; Roberts, S.; Herrington, R.; Wilkinson, J. Self-organization of submarine hydrothermal siliceous deposits: Evidence from the TAG hydrothermal mound, 26°N Mid-Atlantic Ridge. *Geology* **1998**, *26*, 347–350. [CrossRef]

65. Taut, T.; Kleeberg, R.; Bergmann, J. Seifert software: The new Seifert Rietveld program BGMN and its application to quantitative phase analysis. *Mater. Struct.* **1998**, *5*, 57–66.

66. Neuser, R.D.; Bruhn, F.; Götze, J.; Habermann, D.; Richter, D.K. Kathodolumineszenz: Methodik und Anwendung. *Zentralblatt für Geologie und Paläontologie Teil I* **1995**, *1*, 287–306.

67. Götze, J.; Kempe, U. A comparison of optical microscope (OM) and scanning electron microscope (SEM) based cathodoluminescence (CL) imaging and spectroscopy applied to geosciences. *Mineral. Mag.* **2008**, *72*, 909–924. [CrossRef]

68. Götze, J.; Pan, Y.; Stevens-Kalceff, M.; Kempe, U.; Müller, A. Origin and significance of the yellow cathodoluminescence (CL) of quartz. *Am. Mineral.* **2015**, *100*, 1469–1482. [CrossRef]

69. Möckel, R.; Götze, J.; Sergeev, S.A.; Kapitonov, I.N.; Adamskaya, E.V.; Goltsin, N.A.; Vennemann, T. Trace-element analysis by laser ablation inductively coupled plasma mass spectrometry (LA-ICP-MS): A case study for agates from Nowy Kościoł, Poland. *J. Sib. Federal Univ. Eng. Technol.* **2009**, *2*, 123–138.

70. Flem, B.; Müller, A. In situ analysis of trace elements in quartz using Laser ablation inductively coupled plasma mass spectrometry. In *Quartz: Deposits, Mineralogy and Analytics*; Götze, J., Möckel, R., Eds.; Springer Geology: Heidelberg, Germany, 2012; pp. 219–236.

71. Monecke, T.; Bombach, G.; Klemm, W.; Kempe, U.; Götze, J.; Wolf, D. Determination of trace elements in quartz standard UNS-SpS and in natural quartz by ICP-MS. *Geostand. Newsl.* **2000**, *24*, 73–81. [CrossRef]

72. Venneman, T.W.; Morlok, A.; von Engelhardt, W.; Kyser, K. Stable isotope composition of impact glasses from the Nördlinger Ries impact crater, Germany. *Geochim. Cosmochim. Acta* **2001**, *65*, 1325–1336. [CrossRef]

73. Richter-Feig, J.; Möckel, R.; Götze, J.; Heide, G. Investigation of fluids in chalcedony/quartz of agates using Thermogravimetry-Mass-Spectrometry. *Minerals* **2018**, *8*, 72. [CrossRef]

74. Spötl, C.; Vennemann, T.W. Continuous-flow isotope ratio mass spectrometric analysis of carbonate minerals. *Rapid Commun. Mass Spectrom.* **2003**, *17*, 1004–1006. [CrossRef]

75. Moxon, T.; Nelson, D.R.; Zhang, M. Agate recrystallization: Evidence from samples found in Archaean and Proterozoic host rocks, Western Australia. *Aust. J. Earth Sci.* **2008**, *53*, 235–248. [CrossRef]

76. Gilg, H.A.; Morteani, G.; Kostitsyn, Y.; Preinfalk, C.; Gatter, I.; Strieder, A.J. Genesis of amethyst geodes in basaltic rocks of the Serra Geral Formation (Ametista do Sul, Rio Grande do Sul, Brazil): A fluid inclusion, REE, oxygen, carbon, and Sr isotope study on basalt, quartz, and calcite. *Miner. Depos.* **2003**, *38*, 1009–1025. [CrossRef]

77. Gilg, H.A.; Krüger, Y.; Taubald, H.; van den Kerkhof, A.M.; Frenz, M.; Morteani, G. Mineralisation of amethyst-bearing geodes in Ametista do Sul (Brazil) from low-temperature sedimentary brines: Evidence from monophase liquid inclusions and stable isotopes. *Miner. Depos.* **2014**, *49*, 861–877. [CrossRef]

78. Ottens, B.; Götze, J.; Schuster, R.; Krenn, K.; Hauzenberger, C.; Zsolt, B.; Vennemann, T. Exceptional multi-stage mineralization of secondary minerals in cavities of flood basalts from the Deccan Volcanic Province, India. *Minerals* **2019**, *1019*, 351. [CrossRef]

79. Lund, E.H. Chalcedony and quartz crystals in silicified coral. *Am. Mineral.* **1960**, *45*, 1304–1307.

80. Rössler, R. *Der Versteinerte Wald Von Chemnitz*; Museum für Naturkunde Chemnitz: Chemnitz, Germany, 2001; 252p.

81. Rich, P.V.; Rich, T.H.; Fenton, M.A.; Fenton, C.L. *The Fossil Book—A Record of Prehistoric Life*; Dover Publications: Mineola, NY, USA, 2020; 740p.

82. Graetsch, H. Structural characteristics of opaline and microcrystalline silica minerals. *Silica Rev. Mineral.* **1994**, *29*, 209–232.

83. Brewster, D. Über die Ursachen der Farben des irisierenden Achats. *Ann. Phys. und Chem.* **1844**, *137*, 134–138. [CrossRef]

84. Braitsch, O. Über die natürlichen Faser- und Aggregationstypen beim SiO_2, ihre Verwachsungsformen, Richtungsstatistik und Doppelbrechung. *Heidelb. Beiträge Mineral. Petrogr.* **1957**, *5*, 331–372.

85. Lange, P.; Blankenburg, H.-J.; Schrön, W. Rasterelektronenmikroskopische Untersuchungen an Vulkanitachaten. *Z. Geol. Wiss.* **1984**, *12*, 667–681.

86. Miehe, G.; Graetsch, H.; Flörke, O.W. Crystal structure and growth fabric of length-fast chalcedony. *Phys. Chem. Miner.* **1984**, *10*, 197–199. [CrossRef]

87. Folk, R.L.; Pittman, J.S. Length-slow chalcedony; A new testament for vanished evaporates. *J. Sediment. Petrol.* **1971**, *41*, 1045–1058.

88. Flörke, O.W.; Jones, J.B.; Schmincke, H.-U. A new microcrystalline silica from Gran Canaria. *Z. Krist.* **1976**, *143*, 156–165. [CrossRef]

89. Heaney, P.J.; Post, J.E. The widespread distribution of a novel silica polymorph in microcrystalline quartz varieties. *Science* **1992**, *255*, 441–443. [CrossRef]

90. Heaney, P.J.; Veblen, D.R.; Post, J.E. Structural disparities between chalcedony and macrocrystalline quartz. *Am. Mineral.* **1994**, *79*, 452–460.

91. Kingma, K.J.; Hemley, R.J. Raman spectroscopic study of microcrystalline silica. *Am. Mineral.* **1994**, *79*, 269–273.

92. Nasdala, L.; Götze, J.; Gaft, M.; Hanchar, J.; Krbetschek, M. Luminescence techniques in Earth Sciences. In *EMU Notes in Mineralogy*; Beran, A., Libowitzky, E., Eds.; Eötvös University Press: Budapest, Hungary, 2004; Volume 6, pp. 1–49.

93. Correns, C.W.; Nagelschmidt, G. Über Faserbau und optische Eigenschaften von Chalcedon. *Z. für Krist. A* **1933**, *85*, 199–213.

94. Hoffmann, S. *Untersuchungen über den Opalgehalt der Achate*; Schweizerbart: Stuttgart, Germany, 1942; pp. 238–276.

95. Flörke, O.W. Untersuchungen an amorphem und mikrokristallinem SiO_2. *Chem. der Erde* **1962**, *22*, 91–110.

96. Jones, B.; Renaut, R.W. Microstructural changes accompanying the opal-A to opal-CT transition: New evidence from the siliceous sinters of Geysir, Haukadalur, Iceland. *Sedimentology* **2007**, *54*, 921–948. [CrossRef]

97. Moxon, T.; Reed, S.J.B. Agate and chalcedony from igneous and sedimentary hosts aged from 13 to 3480 Ma: A cathodoluminescence study. *Mineral. Mag.* **2006**, *70*, 485–498. [CrossRef]

98. Moxon, T.; Reed, S.J.B.; Zhang, M. Metamorphic effects on agate found near the Shap granite, Cumbria, England: As demonstrated by petrography, X-ray diffraction and spectroscopic methods. *Mineral. Mag.* **2007**, *71*, 461–476. [CrossRef]

99. Iler, R.K. *The Chemistry of Silica: Solubility, Polymerization, Colloid and Surface Properties and Biochemistry*; John Wiley & Sons: New York, NY, USA, 1979; p. 866.

100. Holzhey, G. Mikrokristalline SiO_2-Mineralisationen in rhyolithischen Rotliegendvulkaniten des Thüringer Waldes (Deutschland) und ihre Genese. *Chem. der Erde* **1999**, *59*, 183–205.

101. Tennyson, C. "Struktur" und Farbenspiel des Edelopals. *Lapis* **1979**, *4*, 13–15.

102. Dong, G.; Morrison, G.; Jaireth, S. Quartz textures in epithermal veins, Queensland—Classification, origin, and implication. *Econ. Geol.* **1995**, *90*, 1841–1856. [CrossRef]

103. Weil, J.A. A review of electron spin spectroscopy and its application to the study of paramagnetic defects in crystalline quartz. *Phys. Chem. Miner.* **1984**, *10*, 149–165. [CrossRef]

104. Weil, J.A. A review of the EPR spectroscopy of the point defects in α-quartz: The decade 1982–1992. In *Physics and Chemistry of SiO_2 and the Si-SiO Interface 2*; Helms, C.R., Deal, B.E., Eds.; Plenum Press: New York, NY, USA, 1993; pp. 131–144.

105. Stevens-Kalceff, M.A. Cathodoluminescence microcharacterization of point defects in a-quartz. *Mineral. Mag.* **2009**, *73*, 585–606. [CrossRef]

106. Götze, J. Chemistry, textures and physical properties of quartz—Geological interpretation and technical application. *Mineral. Mag.* **2009**, *73*, 645–671. [CrossRef]

107. Götze, J.; Plötze, M.; Fuchs, H.; Habermann, D. Defect structure and luminescence behaviour of agate—Results of electron paramagnetic resonance (EPR) and cathodoluminescence (CL) studies. *Mineral. Mag.* **1999**, *63*, 149–163. [CrossRef]

108. Mashkovtsev, R.I.; Li, Z.; Mao, M.; Pan, Y. ^{73}Ge, ^{17}O and ^{29}Si hyperfine interactions of the Ge E'_1 center in crystalline SiO_2. *J. Magn. Reson.* **2013**, *233*, 7–16. [CrossRef]

109. SivaRamaiah, G.; Lin, J.R.; Pan, Y. Electron paramagnetic resonance spectroscopy of Fe3+ ions in amethyst: Thermodynamic potentials and magnetic susceptibility. *Phys. Chem. Miner.* **2011**, *38*, 159–167. [CrossRef]

110. Pan, Y.; Hu, B. Radiation-induced defects in quartz. IV. Thermal properties and implications. *Phys. Chem. Miner.* **2009**, *36*, 421–430. [CrossRef]

111. Botis, S.; Nokhrin, S.M.; Pan, Y.; Xu, Y.; Bonli, T. Natural radiation-induced damage in quartz. I. Correlations between cathodoluminescence colors and paramagnetic defects. *Can. Mineral.* **2005**, *43*, 1565–1580. [CrossRef]

112. Nilges, M.J.; Pan, Y.; Mashkovtsev, R.I. Radiation-induced defects in quartz. I. Single-crystal W-band EPR study of an electron irradiated quartz. *Phys. Chem. Miner.* **2008**, *35*, 103–115. [CrossRef]

113. Nilges, M.J.; Pan, Y.; Mashkovtsev, R.I. Radiation-induced defects in quartz. III. EPR, ENDOR and ESEEM characterization of a peroxy radical. *Phys. Chem. Miner.* **2009**, *36*, 63–71. [CrossRef]

114. Pan, Y.; Nilges, M.J.; Mashkovtsev, R.I. Radiation-induced defects in quartz. II. W-band single-crystal EPR study of natural citrine. *Phys. Chem. Miner.* **2008**, *35*, 387–397. [CrossRef]

115. Pan, Y.; Nilges, M.J.; Mashkovtsev, R.I. Multifrequency single-crystal EPR characterization and DFT modeling of new peroxy radicals in quartz. *Mineral. Mag.* **2009**, *73*, 517–535. [CrossRef]

116. Walsby, C.J.; Lees, N.S.; Claridge, R.F.C.; Weil, J.A. The magnetic properties of oxygen-hole aluminum centres in crystalline SiO_2. VI: A stable AlO_4/Li centre. *Can. J. Phys.* **2003**, *81*, 583–598. [CrossRef]

117. Götze, J. Application of cathodoluminescence (CL) microscopy and spectroscopy in geosciences. *Microsc. Microanal.* **2012**, *18*, 1270–1284. [CrossRef]

118. Götze, J.; Plötze, M.; Habermann, D. Cathodoluminescence (CL) of quartz: Origin, spectral characteristics and practical applications. *Mineral. Petrol.* **2001**, *71*, 225–250. [CrossRef]

119. Götze, J.; Hanchar, J. *Atlas of Cathodoluminescence (CL) Microtextures*; GAC Miscellaneous Publication No. 10; Geological Association of Canada: St. John's, NL, Canada, 2018; 248p.

120. Ramseyer, K.; Baumann, J.; Matter, A.; Mullis, J. Cathodoluminescence colours of alpha-quartz. *Mineral. Mag.* **1988**, *52*, 669–677. [CrossRef]

121. Siegel, G.H.; Marrone, M.J. Photoluminescence in as-drawn and irradiated silica optical fibers: An assessment of the role of non-bridging oxygen defect centres. *J. Non-Cryst. Solids* **1981**, *45*, 235–247. [CrossRef]

122. Alonso, P.J.; Halliburton, L.E.; Kohnke, E.E.; Bossoli, R.B. X-ray induced luminescence in crystalline SiO_2. *J. Appl. Phys.* **1983**, *54*, 5369–5375. [CrossRef]

123. Luff, B.J.; Townsend, P.D. Cathodoluminescence of synthetic quartz. *J. Phys. Condens. Matter* **1990**, *2*, 8089–8097. [CrossRef]

124. Perny, B.; Eberhardt, P.; Ramseyer, K.; Mullis, J. Microdistribution of aluminium, lithium and sodium in quartz: Possible causes and correlation with short-lived cathodoluminescence. *Am. Mineral.* **1992**, *77*, 534–544.

125. Götze, J.; Plötze, M.; Graupner, T.; Hallbauer, D.K.; Bray, C. Trace element incorporation into quartz: A combined study by ICP-MS, electron spin resonance, cathodoluminescence, capillary ion analysis and gas chromatography. *Geochim. Cosmochim. Acta* **2004**, *68*, 3741–3759. [CrossRef]

126. Blankenburg, H.-J.; Schrön, W. Zum Spurenelementchemismus der Vulkanitachate. *Chem. Erde* **1982**, *41*, 121–135.

127. Mason, B. Cosmochemistry, Part. I. Meteorites. In *Data of Geochemistry*; Fleischer, M., Ed.; U.S. Geological Survey Professional Paper: Reston, VA, USA, 1979.

128. Walenzak, Z. Geochemistry of minor elements dispersed in quartz (Ge, Al, Ga, Fe, Ti, Li and Be). *Arch. Mineral.* **1969**, *28*, 189–335.

129. Konstatinov, W.M. Uranium bearing lithophysae in acidic extrusive rocks. *Izvestiya Akademii Nauk SSSR Seria Geologika* **1968**, *7*, 43–49. (In Russian)

130. Zielinski, R.A. Uranium mobility during interaction of rhyolitic obsidian, perlite and felsite with alkaline carbonate solution: T = 120 °C, P = 210 kg/cm^2. *Chem. Geol.* **1979**, *27*, 47–63. [CrossRef]

131. Pan, Y.; Li, D.; Feng, R.; Wiens, E.; Chen, N.; Götze, J.; Lin, J. Uranyl binding mechanism in microcrystalline silicas: A potential missing link for uranium mineralization by direct uranyl co-precipitation and environmental implications. *Geochim. Cosmochim. Acta* **2021**, *292*, 518–531. [CrossRef]

132. Peppard, D.F.; Mason, G.W.; Lewey, S. A tetrad effect in the liquid–liquid extraction ordering of lanthanide(III). *J. Inorg. Nucl. Chem.* **1969**, *31*, 2271–2272. [CrossRef]

133. Monecke, T.; Dulski, P.; Kempe, U. Origin of convex tetrads in rare earth element patterns of hydrothermally altered siliceous igneous rocks from the Zinnwald Sn–W deposit, Germany. *Geochem. Cosmochim. Acta* **2007**, *71*, 335–353. [CrossRef]

134. Wood, S.A. The aqueous geochemistry of the rare-earth elements and yttrium. 2. Theoretical predictions of speciation in hydrothermal solutions to 350 °C at saturation water vapor pressure. *Chem. Geol.* **1990**, *88*, 99–125. [CrossRef]

135. Kempe, U.; Götze, J.; Belyatsky, B.V.; Plötze, M. Ce anomalies in monazite, fluorite and agate from Permian volcanics of the Saxothuringian (Germany). *J. Czech. Geol. Soc.* **1998**, *42*, 38.

136. Barsanov, G.P.; Plyusnina, I.I.; Jakovleva, M.E. Specific features of the chemical composition, physical properties and the structure of chalcedony. In *New Data of Minerals*; Nauka: Moscow, Russia, 1981; Volume 28, pp. 3–33. (In Russian)

137. Hoefs, J. *Stable Isotope Geochemistry*, 4th ed.; Springer: Heidelberg, Germany, 1997; 201p.

138. Ingerson, E.; Weshow, R.L. Oxygen isotope fractionation in the system quartz–water. *Geochem. Int.* **1965**, *2*, 691–707.

139. Matsuhisa, Y.; Goldsmith, J.R.; Clayton, R.N. Oxygen isotopic fractionation in the system quarz-albite–anorthite-water. *Geochim. Cosmochim. Acta* **1979**, *43*, 1131–1140. [CrossRef]

140. Harris, C. Oxygen-isotope zonation of agates from Karoo volcanics of the Skeleton Coast, Namibia. *Am. Mineral.* **1989**, *74*, 476–481.

141. Strauch, G.; Nitzsche, H.-M.; Holzhey, G. Isotopenuntersuchungen an Rhyolithen und Achatbildungen. *Neues Jahrb. für Mineral. Abh.* **1994**, *165*, 103–104.

142. Blankenburg, H.J.; Pilot, J.; Werner, C.D. Erste Ergebnisse der Sauerstoffisotopenuntersuchungen an Vulkanitachaten und ihre genetische Interpretation. *Chem. der Erde* **1982**, *41*, 213–217.

143. Harris, C. Oxygen isotope geochemistry of a quartz-agate geode from north-western Namibia. *Commun. Geol. Surv. S.W. Afr. Namib.* **1988**, *4*, 43–44.

144. Holzhey, G. Contribution to petrochemical-mineralogical characterization of alteration processes within the marginal facies of rhyolitic volcanics of lower Permian Age, Thuringian Forest, Germany. *Chem. Erde* **2001**, *61*, 149–186.

145. Fallick, A.E.; Jocely, J.; Hamilton, P.J. Oxygen and hydrogen stable isotope systematics in Brazilian agates. In *Geochemistry and Mineral Formation in the Earth Surface*; Rodriguez-Clemente, R., Tardy, Y., Eds.; Editorial CSIC: Madrid, Spain, 1987; pp. 99–117.

146. Duarte, L.C.; Hartmann, L.A.; Ronchi, L.H.; Berner, Z.; Theye, T.; Massonne, H.J. Stable isotope and mineralogical investigation of the genesis of amethyst geodes in the Los Catalanes gemological district, Uruguay, southernmost Paraná volcanic province. *Miner. Depos.* **2011**, *46*, 239–255. [CrossRef]

147. Rezaei-Kahkhaei, M.; Ansarifar, O.; Ghasemi, H. Geochemistry and oxygen stable isotopes of Reza Abad agates, SE Shahrood, Central Iran: An approach to temperature and formation process. *J. Econ. Geol.* **2019**, *11*, 3.

148. Kita, I.; Taguchi, S. Oxygen isotopic behaviour of precipitating silica from geothermal water. *Geochem. J.* **1986**, *20*, 153–157. [CrossRef]

149. Blankenburg, H.-J.; Thomas, R.; Klemm, W.; Leeder, O. Interpretation der Ergebnisse von Einschlußunersuchungen an den Quarzinkrustaten aus Vulkanitachaten. *Z. Geol. Wiss.* **1990**, *18*, 81–85.

150. Frondel, C. Structural hydroxyl in chalcedony (Type B quartz). *Am. Mineral.* **1982**, *67*, 1248–1257.

151. Graetsch, H.; Flörke, O.W.; Miehe, G. The nature of water in chalcedony and opal-C from Brazilian agate geodes. *Phys. Chem. Miner.* **1985**, *12*, 300–306. [CrossRef]

152. Thomas, R.; Blankenburg, H.-J. Erste Ergebnisse über Einschlussuntersuchungen an Quarzen aus Achatmandeln und –kugeln basischer und saurer Vulkanite. *Z. Geol. Wiss.* **1981**, *9*, 625–633.

153. Commin-Fischer, A.; Berger, G.; Polvè, M.; Dubois, M.; Sardini, P.; Beaufort, D.; Formoso, M. Petrography and chemistry of SiO_2 filling phases in the amethyst geodes from the Serra Geral Formation deposit, Rio Grande do Sul, Brazil. *J. South. Am. Earth Sci.* **2010**, *29*, 751–760. [CrossRef]

154. Ramboz, C.; Pichavant, M.; Weisbrod, A. Fluid immiscibility in natural processes: Use and misuse of fluid inclusion data. *Chem. Geol.* **1982**, *37*, 29–48. [CrossRef]

155. Roedder, E. *Fluid Inclusions*; Reviews in Mineralogy 12; Mineralogical Society of America: Reston, VA, USA, 1984; 645p.

156. Hall, D.L.; Sterner, S.M.; Bodnar, R.J. Freezing point depression of $NaCl–KCl–H_2O$ solutions. *Econ. Geol.* **1988**, *83*, 197–202. [CrossRef]

157. Tripp, R.B. The mineralogy of Warsaw Formation geodes. *Iowa Acad. Sci. Proc.* **1959**, *66*, 350–356.

158. Hayes, J.B. Geodes and concretions from the Mississippian Warsaw Formation. Keokuk region, Iowa, Illinois, Missouri. *Sediment. Petrol.* **1964**, *34*, 123–133.

159. Holzhey, G. Die Paragenese von Mineralen in Rhyolithkugeln (Lithophysen) aus Rotliegendvulkaniten des Thüringer Waldes. *Semana* **2016**, *31*, 25–50.

160. Schmitt-Riegraf, C. Magmenentwicklung und spät- bis post-magmatische Alterationsprozesse permischer Vulkanite im Nordwesten der Nahe-Mulde. *Münstersche Forsch. zur Geol. und Paläontologie* **1996**, *80*, 1–251.

161. Walger, E. Zur Entstehung des Calcitachates. *Fortschr. Mineral.* **1961**, *39*, 360.

162. Landmesser, M. Calcitachat: Zur Deutung eines verblüffenden mineralogischen Phänomens. *Mainz. Nat. Arch.* **1996**, *34*, 9–43.

163. Blankenburg, H.-J.; Unterricker, S.; Eichler, B.; Starke, R.; Stolz, W.; Rösler, H.J. Natürliche Radioaktivität, chemische und Phasenzusammensetzung der Eisenoxide aus Vulkanitachaten. *Chem. Erde* **1986**, *45*, 159–166.

164. Rosemeyer, T. Copper-banded agates from the Kearsarge copper-bearing amygdaloidal lode Houghton county, Michigan. *Rocks Miner.* **2012**, *87*, 352–365. [CrossRef]

165. Moxon, T. Agate in thin section. *Rocks Miner.* **2014**, *89*, 328–339. [CrossRef]

166. Breitkreuz, C. Spherulites and lithophysae—200 years of investigation on hightemperature crystallization domains in silica-rich volcanic rocks. *Bull. Volcanol.* **2013**, *75*, 705–720. [CrossRef]

167. Sakka, S.; MacKenzie, J.D. Relationship between apparent glass transition temperature and liquidus temperature for inorganic glasses. *J. Non-Cryst. Solids* **1971**, *6*, 145–162. [CrossRef]

168. Lofgren, G. Spherulitic textures in glassy and crystalline rocks. *J. Geophys. Res.* **1971**, *76*, 5635–5648. [CrossRef]

169. Breitkreuz, C.; Götze, J.; Weißmantel, A. Mineralogical and geochemical investigation of megaspherulites from Argentina, Germany and USA. *Bull. Volcanol.* **2020**. accepted for publication.

170. Herrington, R.J.; Wilkinson, J.J. Colloidal gold and silica in mesothermal vein systems. *Geology* **1993**, *21*, 539–542. [CrossRef]

171. Jebrag, M. Hydrothermal breccias in vein-type ore deposits: A review of mechanisms, morphology and size distribution. *Ore Geol. Rev.* **1997**, *12*, 111–134. [CrossRef]

172. Haake, R.; Fischer, J.; Reißmann, R. Über das Achat-Amethyst-Vorkommen von Schlottwitz im Osterzgebirge. *Mineralienwelt* **1991**, *2*, 20–24.

173. Chowns, T.M.; Elkins, J.E. The origin of quartz geodes and cauliflower cherts through the silification of anhydrite nodules. *J. Sediment. Petrol.* **1974**, *44*, 885–903.

174. Tucker, M.E. Quartz replaced anhydrite nodules ("Bristol diamonds") from the Triassic of the Bristol District. *Geol. Mag.* **1976**, *113*, 569–574. [CrossRef]

175. Jacka, A.D. Replacement of fossils by length-slow chalcedony and associated dolomitization. *J. Sediment. Petrol.* **1974**, *44*, 421–427.

176. Landmesser, M. Zur Entstehung von Kieselhölzern. *ExtraLapis* **1994**, *7*, 49–80.

177. Milliken, K.L. The silicified evaporate syndrome: Two aspects of silicification history of former evaporate nodules from Southern Kentucky and Northern Tennessee. *J. Sediment. Petrol.* **1979**, *49*, 245–256.

178. Shaub, B.M. *The Origin of Agates, Thundereggs, and Other Nodular Structures*; Agate Publishing Company: Northampton, MA, USA, 1989; 105p.

179. Landmesser, M. Zur Geothermometrie und Theorie der Achate. *Mitt. Pollichia* **1992**, *79*, 159–201.

180. Harder, H. Agates-formation as a multi component colloid chemical precipitation at low temperatures. *Neues Jahrb. Mineral. Mon.* **1993**, *H.1*, 31–48.

181. Williamson, B.J.; Wilkinson, J.J.; Luckham, P.F.; Stanley, C.J. Formation of coagulated colloidal silica in high-temperature mineralizing fluids. *Mineral. Mag.* **2002**, *66*, 547–553. [CrossRef]

182. White, J.F.; Corwin, J.F. Synthesis and origin of chalcedony. *Am. Mineral.* **1961**, *46*, 112–119.

183. Flörke, O.W. Transport and deposition of SiO_2 with H_2O under supercritical conditions. *Krist. Tech.* **1972**, *7*, 159–166. [CrossRef]

184. Oehler, J.H. Hydrothermal crystallization of silica gel. *Geol. Soc. America Bull.* **1976**, *87*, 1143–1152. [CrossRef]

185. Götze, J.; Plötze, M.; Tichomirowa, M.; Fuchs, H.; Pilot, J. Aluminium in quartz as an indicator of the temperature of formation of agate. *Mineral. Mag.* **2001**, *65*, 407–413. [CrossRef]

186. Dennen, W.H.; Blackburn, W.H.; Quesada, A. Aluminum in quartz as a geothermometer. *Contrib. Mineral. Petrol.* **1970**, *27*, 332–342. [CrossRef]

187. Agel, A.; Petrov, I. Im Quarzgitter substituiertes Aluminium als Indikator für dessen Bildungstemperatur. *Eur. J. Miner.* **1990**, *2*, 144.

188. Correns, C.W. The experimental weathering of silicates. *Clay Miner. Bull.* **1961**, *4*, 249–265. [CrossRef]

189. Wirsching, U. Experimente zum Einfluß des Gesteinsglas-Chemismus auf die Zeolithbildung durch hydrothermale Umwandlung. *Contrib. Mineral. Petrol.* **1975**, *49*, 117–124. [CrossRef]

190. Seyfried, W.E.; Bischo, J.L. Low temperature basalt alteration by seawater: An experimental study at 70 °C and 150 °C. *Geochim. Cosmochim. Acta* **1979**, *43*, 1937–1947. [CrossRef]

191. Giggenbach, W.F. Mass transfer in hydrothermal alteration systems—A conceptual approach. *Geochim. Cosmochim. Acta* **1984**, *48*, 2693–2711. [CrossRef]

192. Duplay, J.; Paquet, H.; Kossovskaya, A.; Tard, Y. Estimation de la température de formation des paragenèses saponite-céladonite et glauconite-nontronite dans les altérations sous-marines de basalte, par la méthode des corrélations entre éléments au sein de populations monominérales. *CR Acad. Sci. Paris* **1989**, *309*, 53–58.

193. Klammer, D. Mass change during extreme acid-sulphate hydrothermal alteration of a Tertiary latite, Styria, Austria. *Chem. Geol.* **1997**, *141*, 33–48. [CrossRef]

194. Polgári, M.; Hein, J.R.; Németh, T.; Pál-Molnár, E.; Vigh, T. Celadonite and smectite formation in the Úrkút Mn-carbonate ore deposit (Hungary). *Sediment. Geol.* **2013**, *294*, 157–163. [CrossRef]

195. Götze, J.; Hofmann, B.; Machałowski, T.; Tsurkan, M.V.; Jesionowski, T.; Ehrlich, H.; Kleeberg, R.; Ottens, B. Biosignatures in subsurface filamentous fabrics (SFF) from the Deccan Volcanic Province, India. *Minerals* **2020**, *10*, 540. [CrossRef]

196. Sukheswala, R.N.; Avasia, R.K.; Gangopadhyay, M. Zeolites and associated secondary minerals in the Deccan Traps of western India. *Mineral. Mag.* **1974**, *39*, 658–671. [CrossRef]

197. Mattioli, M.; Cenni, M.; Passaglia, E. Secondary mineral assemblages as indicators of multistage alteration processes in basaltic lava flows: Evidence from the Lessini Mountains, Veneto Volcanic Province, Northern Italy. *Period. Di Mineral.* **2016**, *85*, 1–24.

198. Kryza, R. Bilans chemiczny dla stref mineralizacji agatowej w wulkanitach rejonu Nowego Kościoła (Góry Kaczawskie). *Arch. Mineral.* **1983**, *39*, 125–132.

199. Krauskopf, K.B. Dissolution and precipitation of silica at low temperatures. *Geochim. Cosmochim. Acta* **1956**, *10*, 1–26. [CrossRef]

200. Landmesser, M. Mobility by metastability: Silica transport and accumulation at low temperatures. *Chem. Erde* **1995**, *55*, 149–176.

201. Dietzel, M. Dissolution of silicates and the stability of polysilicic acid. *Geochim. Et Cosmochim. Acta* **2000**, *64*, 3275–3281. [CrossRef]

202. Gránásy, L.; Pusztai, T.; Tegze, G.; Warren, J.A.; Douglas, J.F. Growth and form of spherulites. *Phys. Rev.* **2005**, *72*, 11605–11619. [CrossRef]

203. Krug, H.-J.; Jacob, K.-H. Genese und Fragmentierung rhytmischer Bänderungen durch Selbstorganisation. *Z. Dtsch. Geol. Ges.* **1994**, *144*, 452–460.

204. Bryxina, N.A.; Sheplev, V.S. Auto-oscillation in agate crystallization. *Math. Geol.* **1999**, *31*, 297–309. [CrossRef]

205. Wang, Y.; Merino, E. Self-organizational origin of agates: Banding, fiber twisting, composition, and dynamic crystallizationmodel. *Geochim. Cosmochim. Acta* **1990**, *54*, 1627–1638. [CrossRef]

206. Götze, J.; Stanek, K.; Orozco, G. Auf Achatsuche in Kuba. *Mineralienwelt* **2020**, *3*, 56–63.

207. Goldbaum, J.; Howard, C.; Rabinovitch, A. Spatial Chirp of agate bands. *Minerals* **2019**, *9*, 634. [CrossRef]

208. Mikhailov, A.S.; Showalter, K. Control of waves, patterns and turbulence in chemical systems. *Phys. Rep.* **2006**, *425*, 79–194. [CrossRef]

209. Kumar, D.K.; Steed, J.W. Supramolecular gel phase crystallization: Orthogonal self-assembly under non-equilibrium conditions. *Chem. Soc. Revs.* **2014**, *43*, 2080–2088. [CrossRef]

210. Papineau, D. Chemically oscillation reactions in the formation of botryoidal malachite. *Am. Mineral.* **2020**, *105*, 447–454. [CrossRef]

211. Walger, E.; Mattheß, G.; Von Seckendorff, V.; Liebau, F. The formation of agate structures: Models for silica transport, agate layer accretion, and for flow patterns and flow regimes in infiltration channels. *Neues Jahrb. Mineral. Abh.* **2009**, *186*, 113–152. [CrossRef]

212. Laurs, B.M.; Rossman, G.R. Grape-like "Manakarra" quartz from Sulawesi, Indonesia. *J. Gemmol.* **2018**, *36*, 101–102.

213. Hofmann, B.A. Subsurface filamentous fabrics. In *Encyclopedia of Geobiology*; Reitner, J., Thiel, V., Eds.; Springer: Berlin/Heidelberg, Germany, 2011; pp. 851–853.

214. Thewalt, U.; Dörfner, G. Wie kommt das Moos in den Achat und wie nicht? *Der Aufschluss* **2012**, *63*, 1–16.

215. Reysenbach, A.-L.; Cady, S.L. Microbiology of ancient and modern hydrothermal systems. *Trends Microbiol.* **2001**, *9*, 79–86. [CrossRef]

216. Jones, B.; Renault, R.W. Influence of thermophilic bacteria on calcite and silica precipitation in hot springs with water temperatures above 90 °C: Evidence from Kenya and New Zealand. *Can. J. Earth Sci.* **1996**, *33*, 72–83. [CrossRef]

217. Konhauser, K.O.; Phoenix, V.R.; Bottrell, S.H.; Adams, D.G.; Head, I.M. Microbial-silica interaction in Icelandic hot spring sinter: Possible analogues for some Precambrian siliceous stromatolites. *Sedimentology* **2001**, *48*, 415–433. [CrossRef]

218. Ferris, F.G.; Beveridge, T.J.; Fyfe, W.S. Iron-silica crystallite nucleation by bacteria in a geothermal sediment. *Nature* **1986**, *320*, 609–611. [CrossRef]

219. Parenteau, M.N.; Cady, S.L. Microbial biosignatures in iron-mineralized phototrophic mats at Chocolate Pots Hot Springs, Yellowstone National Park, United States. *PALAIOS* **2010**, *25*, 97–111. [CrossRef]

220. Müller, A.; Polgári, M.; Gucsik, A.; Pál-Molnár, E.; Koós, M.; Veres, M.; Götze, J.; Nagy, S.; Cserháti, C.; Németh, T.; et al. Cathodoluminescent features and Raman spectroscopy of Miocene hydrothermal biomineralization embedded in cryptocrystalline silica varieties, Central Europe, Hungary. In *Micro-Raman Spectroscopy and Luminescence Studies in the Earth and Planetary Sciences*; Gucsik, A., Ed.; American Institute of Physics: Houston, TX, USA, 2009; pp. 207–218.

221. Yang, H.; Coombs, N.; Ozin, G.A. Morphogenesis of shapes and surface patterns in mesoporous silica. *Nature* **1997**, *386*, 692–695. [CrossRef]

222. Glaab, F.; Kellermeier, M.; Kunz, W.; Morallon, E.; Garcia-Ruiz, J.M. Formation and evolution of chemical gradients and potential differences across self-assembling inorganic membranes. *Angew. Chem.* **2012**, *124*, 4393–4397. [CrossRef]

223. Moxon, T.; Petrone, C.M.; Reed, S.J.B. Characterization and genesis of horizontal banding in Brazilian agate: An X-ray diffraction, thermogravimetric and electron microprobe study. *Mineral. Mag.* **2013**, *77*, 227–248. [CrossRef]

224. Haudin, F.; Cartwright, J.H.E.; Braua, F.; De Wit, A. Spiral precipitation patterns in confined chemical gardens. *Proc. Natl. Acad. Sci. USA* **2014**, *111*, 17363–17367. [CrossRef]

225. Noorduin, W.L.; Grinthal, A.; Mahadevan, L.; Aizenberg, J. Rationally designed complex, hierarchical microarchitectures. *Science* **2013**, *340*, 832–837. [CrossRef]

226. Kellermeier, M.; Cölfen, H.; García-Ruiz, J.M. Silica Biomorphs: Complex biomimetic hybrid materials from "sand and chalk". *Eur. J. Inorg. Chem.* **2012**, *32*, 5123–5144. [CrossRef]

227. Dominguez-Bella, S.; Garcia-Ruiz, J.M. Textures in induced morphology crystal aggregates of $CaCO_3$: Sheaf of wheat morphologies. *J. Cryst. Growth* **1986**, *79*, 236–240. [CrossRef]

228. Lenz, G.; Schäfer, K. Spiralen im Achat—Ein biologisches oder ein mineralogisches Phänomen? *Lapis* **2008**, *1*, 21–24.

Publisher's Note: MDPI stays neutral with regard to jurisdictional claims in published maps and institutional affiliations.

 *minerals*

Review

Agate Genesis: A Continuing Enigma

Terry Moxon [1] and Galina Palyanova [2,3,*]

[1]   55 Common lane, Auckley, Doncaster DN9 3HX, UK; moxon.t@tiscali.co.uk

[2]   Sobolev Institute of Geology and Mineralogy, Siberian Branch of Russian Academy of Sciences, 630090 Novosibirsk, Russia

[3]   Department of Geology and Geophysics, Novosibirsk State University, Pirogova str., 2, 630090 Novosibirsk, Russia

*   Correspondence: palyan@igm.nsc.ru

Received: 21 September 2020; Accepted: 21 October 2020; Published: 26 October 2020

Abstract: This review covers the last 250 years of major scientific contributions on the genesis of agates found in basic igneous host rocks. From 1770 to 1955, the genesis question was frequently limited to discussions based on observations on host rock and agate thick sections. Over the next 25 years, experimental investigations examined phase transformations when silica glass and various forms of amorphous silica were heated to high temperatures. This work demonstrated that the change from the amorphous state into chalcedony was likely to be a multi-stage process. The last 40 years has seen modern scientific instrumentation play a key role in identifying the physical and chemical properties of agate. The outcome of this work has allowed limited evidence-based comment on the conditions of agate formation. There is a general consensus that agates in these basic igneous hosts form at <100 °C. However, the silica source and the nature of the initial deposit remain to be proven.

Keywords: agate; chalcedony; XRD; genesis; moganite; crystallite growth; age

1. Introduction

Agate and chalcedony are the compact varieties of silica that are primarily composed of minute crystals of α-quartz. Chalcedony has a fairly uniform colour but is band free while agate is banded chalcedony. Sectioned agates from igneous host rocks generally show the banding as one of two major types with wall lining (sometimes called fortification) banding being the most common type. These agates show initial bands appearing to replicate the supporting cavity wall. Subsequent bands follow with an approximate repetition that frequently continues towards the agate centre. A second type demonstrates horizontal bands that are apparently gravity controlled. Both types can be found in the same agate.

Basalt and andesite are the most widespread agate host rocks but agates can also be found in some fossil wood and sedimentary host rocks, particularly limestone. However, there is no reason why agates in different environments should follow identical genesis routes. Discussion in the present paper is mainly limited to scientific contributions that have been made over the last 250 years on the complex question of agate genesis in basaltic and andesitic host rocks.

The present review is in three parts. From 1770 to 1955, scientific contributions to the genesis question were generally based on host rock observations and agate thick sections. Experimental investigations played a relatively small part and contributions during this period are considered chronologically. During the next 25 years phase transformations in silica glass and amorphous silica were investigated. Many of these studies were not directly concerned with agate but the work has implications for the agate genesis question. Modern scientific instrumentation has played a key role in identifying the physical and chemical properties of agate. In order to avoid information repetition, some relevant pre-1980 work is considered with 1980–2020 in the final discussion. The review includes some contributions by research workers from Russia and Germany, who only published in Russian and German.

2. Scientific Contributions over the Years 1770–1980

2.1. 1770–1900

Agates have been collected and fashioned since the earliest civilisations and many must have wondered about the formation of these beautiful minerals. Collini [1] states, "Not only is the red colour in agate due to iron but it is also the cause of the green, red and brown in moss agate". According to Liesegang [2], Collini [1] was the first to suggest that these colours had their origins from iron compounds that had been transported in circulating waters from the neighbouring host rocks. The iron rich solutions were then able to penetrate the still soft agate. Hardening of the agate prevented further iron solution penetration. Liesegang [2] also cites Lasius as an early worker who suggested silica enters the gas vesicles in solution. In 1931, Liesegang [3] added more experimentation and comment to his earlier observations.

Germany hosts some of the world's finest agates and the agate industry based around Idar Oberstein has produced high quality agate products since the 14th century [4]. This large commercial background accounts for the fact that agate and its genesis was very much the preserve of the German scientists. In 1849, Noeggerath [5] and Haidinger [6] published a series of open letters in which they stated their opposing views on the method of silica solution entry into the gas vesicle. The conference records of the 20th July in the Freunde der Naturwissenschaften, Vienna stated the areas of agreement and their differences. Both agreed that gaseous bubbles are formed in the erupted lavas and the subsequent parallel vesicular flows were due to the movement of the lava. Misshapen amygdales are the result of vapour bubbles meeting within the high viscosity lava; this leads to twin and triplet bubbles. Noeggerath [5] emphasised the agate variations were a feature of the different local environments. He pointed to amygdales in Westphalen where the calcite had been dissolved in the upper reaches but not at the lower levels and claimed the firm rock protected the calcite from dissolution.

Noeggerath [5] asserted that the silicic acid, necessary for the agates, comes from hot springs and appears after the decomposition of the host rock. He noted that agates had been found in compact bedrock while entirely friable rocks can be free from agates. However, it is the method of silica entry into the vesicle that produces major differences of opinion.

Haidinger [6] took a contrary view arguing that the "mountain sweat" carries the separate silica into the gas vesicle. Noeggerath [5] acknowledged that water could pass through the host rock but would not be able to penetrate the impermeable agate layers. As an alternative he pointed to the infiltration canals as means of silica entry and this would allow the continual development of successive layers. These canals are found in some agates and appear to breach the outer layer.

At the end of his open letter of May 1849, Noeggerath [5] writes, "I beg you to compare the permeable and impermeable layers, which allow the art of colouring banded quartz. How is it possible for the silica entry to happen and form successive precipitations from outside when the earliest formed impermeable layers presents a barrier to all further additions?". Haidinger [6] did not answer the question and the infiltration canal was accepted until challenged by Liesegang in 1915 [2]. This interesting infiltration feature is still proposed by some authors as a point of solution input or water exit. The canals feature regularly in agate genesis and are discussed later in the paper.

2.2. 1900–1945

In 1901 Heddle [7] published an important work: "The Mineralogy of Scotland". He had spent many years collecting Scottish agates and the book included a chapter on the question of agate genesis. He suggested that silica could enter the gas vesicle as a result of osmosis with the early deposits acting as a membrane. As a semipermeable membrane and solute concentrations are not mentioned he is really describing diffusion. The desilicified solution then leaves the agate via tubular openings: Noeggerath's [5] infiltration canals.

The first detailed thin section study of agates that we identified was carried out by Timofeev in 1912 [8]. Three of his micrographs are shown in Figure 1. Using a polarizing microscope he was able to reveal the typical chalcedonic pattern with white bands temporarily halting the growth. His work was based upon the agate amygdales found in the Suisari Island basalts in Lake Onega, Karelia, Russia. He proposed a crystallization sequence of chalcedony → quartzine → quartz. Some of the agates had developed a stalactite type growth. Cross sections of the stalactite are described as alternating zones of chalcedony and quartzine separated by a pigment. Some stalactites had the occasional calcite deposit within the agate.

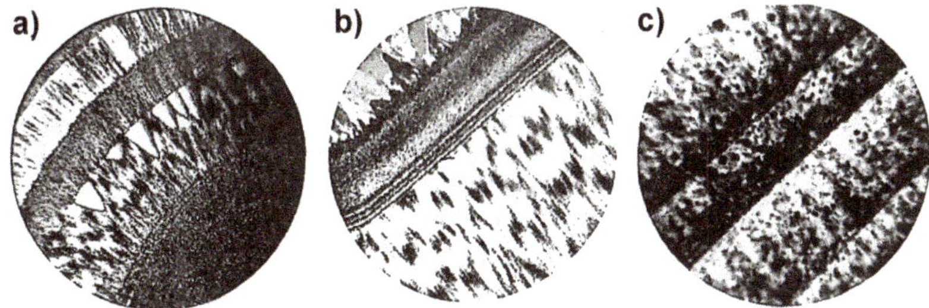

Figure 1. Timofeev micrographs [8]. The original captions describe the (**a**) alternation of zones of quartz (light) and chalcedony (dark); (**b**) the formation of quartzine filaments on the faces of the quartz rhombohedra; and (**c**) zones showing a changing orientation from chalcedony to quartz.

Liesegang [2] had made several publications on rhythmic banding in silica gels and by 1915 he was able to publish his major work Die Achate. Apart from infiltration canals, Liesegang was able to synthesize various agate-like patterns by allowing metal ions to diffuse in silica gels containing suitable anions. The common wall lining agate can be simulated if a single curved line of $Fe^{3+}_{(aq)}$ is drawn across a flat surface of solidified silica gel containing an alkali. A simulated wall lining agate pattern develops due to the ions diffusing across the gel (Figure 2a). Liesegang banding can also be shown when separate partially miscible liquids are poured from a container (Figure 2b). Horizontal bands are readily formed by $OH^-_{(aq)}$ on top of solidified gelatine in a large boiling tube containing $Mg^{2+}_{(aq)}$ or transition metal ions. Liesegang's [2] Die Achate has had a major impact on the discussion of agate genesis and 80 years later it was still being linked with agate genesis.

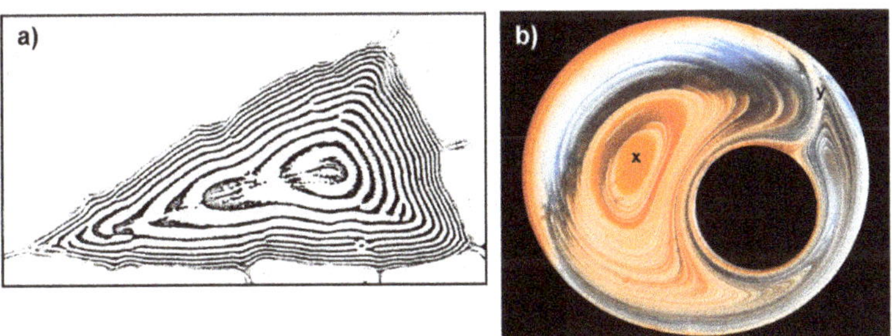

Figure 2. (**a**) Liesegang rings of silver chromate. They are formed when aqueous silver ions slowly diffuse through a gel containing chromate ions producing a precipitate of silver chromate (from Liesegang [3]). (**b**) Liesegang banding that has been produced when samples of blue, orange, and white dyed polyester resin are simultaneously poured round a solid circular disc. The pouring point was at (x) with an apparent exit canal at (y). Disc hole diameter = 2 cm.

Reis [9] published the largest single work on agate and later he produced a further long treatise in which he gave his conclusions on agate genesis: Reis [10]. He was not the first worker to recognise that the wall contact layer was different from the rest of the agate but he was the first to suggest that it was of a different generation. Reis [9,10] opposed the silica entry osmosis mechanism as it depended upon a semipermeable membrane being complete. He believed that the temperature gradient within the lava flow created a sucking effect that drew the silica solutions into the cavity. Iron oxides played a dual role by precipitating the silica and adding colour to the agate. The work of Reiss was largely ignored until Fischer in 1954 [11] added key support for the Reiss hypotheses.

Jessop in 1930 [12] questioned the view that chalcedony was a mixture of quartz and opal. This was based on the fact that the refractive index would not be the approximately constant value found in samples of all geological ages. Jessop [12] proposed that the silica was deposited as a gel containing gas and solid impurities in a solution. As crystallisation advanced from the outside edge, the dissolved substances move with the crystallisation front until a point of saturation is reached. The gaseous or solid impurities are entombed in the crystalline silica and the thickness of the band depends upon the concentration of impurity and the rate of crystallisation. As the solid and gaseous solutes do not necessarily reach saturation at the same point, they allow bands to be partially or totally distinct.

The white bands cannot be stained and this was demonstrated using thin sections. These bands reveal an acicular structure and were parallel to the fibres. The bands that can be dyed are uniform. He proposed that the white bands contain the solid inclusions and the clear bands contain the gaseous bubbles. Such a hypothesis allows dyeing to be a question of relative retentivity and permeability.

In 1931, Liesegang [3] admitted that a pre-existing gel is not always necessary but objected to the length of time taken to form agate via any rhythmic deposition. In this paper, he proposed an alternative mechanism for the formation of the horizontal layering in agate. Layers of varying concentrations of hot sugar solutions can reflect light differently. The salt lakes in Siebenbürgen where the bottom of the lake is 50 °F hotter than the more dilute upper layers show a similar effect. Such variations can also be produced in water glass and he proposed that the horizontal layers are caused by a similar salt-like effect that was eventually "frozen".

Pilipenko in 1934 [13] supported the Noeggerath hypothesis and opposed the Liesegang propositions. Pilipenko [13] was one of a few to base his argument on the results of a detailed study of thin sections, polished sections, and the outer surface of agates. He put forward a hypothesis for the deposition of chalcedony from "aqueous silica solutions" that penetrated the empty chambers in effusive rocks through pores, hair cracks, and supply channels in the walls of voids. He proposed that chalcedony deposition was in a liquid and gaseous medium. He discounted the Liesegang hypothesis on a number of points:

(1)   Some samples of agates are found with a hollow space. This demonstrates that the vesicle is not always filled with a gel;
(2)   The vesicular infill occurs by the layering of the first generation and can be accompanied by metasomatic substitution of crystals, which have been previously precipitated;
(3)   The hypothesis does not explain why, in the same sample, colourless banded chalcedony varies with the pigmented chalcedony;
(4)   The role of pigmentation is not clear with (a) well-formed amethyst crystal centres and (b) brown pigments found in various zones with the disappearance of strong pigment zones;
(5)   In artificial preparations, the deposits are torn and split while in nature this phenomenon is rare.

As an alternative Pilipenko [13] proposed that, "a study of the outer surface of agate reveals a number of pores or openings in the walls of sections which surround the agate". It is presumably these openings that allow the silica to enter the vesicle.

A study of the outer surface of agates in contact with effusive rocks showed the various types of silica deposition in the chambers is grouped into seven types. The first six types of agates are formed in chambers that are not completely filled by silica solutions. In this case, silica deposition occurs horizontally starting at the bottom of the chambers and builds layer upon layer. Silica stalactites and various sagging formations descend from free walls. The seventh type occurs when the chamber is completely filled with a silica solution. Here, deposition of chalcedony goes along the walls of the chamber layer by layer. The number of incoming solutions into the chamber travel through conducting canals and pores of various sizes. Solution inflow equals outflow.

Pilipenko [13] investigated the structure of the infiltration channels and gave great importance to their formation. Infiltration canals develop only within the agate body. In longitudinal sections through the canal, it is clearly shown that traces of agate in the form of a sleeve are pulled out. The sleeves themselves are folded from chalcedony tubes, concentrically embedded into each other.

2.3. 1945–1955

Kuzmin in 1947 [14] included a chapter on agate in a PhD thesis that was based on "Periodic-rhythmical phenomena in Mineralogy and Geology". In 2019, the thesis was published in his memory. He investigated the banded structures of agates and rejected Liesegang's hypothesis of agate formation. As agate banding could form as the wall lining or horizontal type, he proposed that the vesicle silica solutions would be colloidal or molecular leading to the respective wall lining and horizontal banding. He credited molecular or highly diluted solutions for the development of sharply defined horizontal layers in the chamber. However, the so-called infiltration canals in agates were interpreted as a point of "output". He proposed that these exit points occur when the chamber is full of a silica solution and osmotic pressure forces the excess solution to exit via the channels. Once diffusion has started, the channels lower the pressure and allow fresh solutions to diffuse into the cavity. The outward pointing structure shows that they exist as a departure point. The structure of these output channels convincingly shows that the agate structure follows a sequential deposition of rhythmic layers.

In a paper that is more tentative in its suggestions than many of its predecessors, Nacken in 1948 [15] suggests that the silica precursor for agate is the existence of an immiscible liquid state within the magma. Perhaps in an attempt to pre-empt the inevitable criticism, Nacken comments on the problem presented by the melting point of silica at 1600 °C existing within a basaltic magma of 1100 °C. He quotes the work of the Geophysical Laboratory at the Carnegie Institute, Washington, where experiments have been performed on "peculiar acid magma systems" containing excess silica in the molten state. Such systems divide into silica rich and silica poor immiscible mixtures". The silica rich component would lead to drops with a density of 2.3 g/cm^3 within a magma density of 2.6 g/cm^3. This density difference would allow the silica drops to rise and eventually form agates. With silica drops suitably in place Nacken [15] returns to his original synthesis of chalcedony from glass that had been subject to a hydrothermal conversion at 400 °C. The conversion was sensitive to pH, impurities, and the temperature gradient within the glass. When the glass had been converted to chalcedony, there is a volume decrease, which allows further solution attacks along capillary cracks. Any impurities are pushed along with the crystallisation front and precipitate in layers.

Fischer in 1954 [11] reviewed the 20th century German literature on agate genesis and commented on the theories of Liesegang [3], Nacken [15], and Reis [9,10] in some detail. He supported the Reis [9,10] contribution and rejected the Nacken [15] hypothesis on the grounds that a magmatic temperature of 1100 °C would not support a silica melt of 1600 °C. Calcite is often found in agate and at these temperatures calcite would be converted into wollastonite. Liesegang's hypothesis [3] required the existence of a pre-existing gel but his explanation failed to correlate with laboratory findings. Fischer [11] accepted the low solubility of silica in water was a problem but the enormous volumes of water generated and the length of time for the demise of thermal springs are sufficient to support the theories of Reis [9,10].

2.4. 1955–1982

Studies during 1955–1982 allowed modern day instrumentation to play a greater part in the investigation of agate. Potential precursors, such as amorphous silica and silica glass, were examined at high temperature and pressure. Operating in the temperature range of 330–450 °C and pressures of 10–40 kPa, Carr and Fyfe [16] established the conversion of near amorphous silica into quartz. Intermediate silica phases were identified by powder X-ray diffraction (XRD). They established the transition as: near amorphous silica → cristobalite → silica–K → quartz. Silica-K or keatite was first synthesised by Keat [17]. It is very rare in the natural state and has never been identified in agate.

White and Corwin [18] investigated the formation of synthetic chalcedony by heating glass in hydrothermal solutions at 400 °C and 34 MPa. Chalcedony was made by the direct transformation of silica glass or cristobalite. They found that no conversion took place in weakly acid solutions but a complete and rapid conversion occurred under slightly alkaline conditions. The transformation proceeded via cristobalite or keatite.

Experimental investigations on the formation of quartz from amorphous silica were carried out in alkaline hydrothermal conditions with the addition of alkali metal hydroxides or carbonates [19]. Experiments were carried out over a temperature range of 100 and 250 °C with transformations occurring in a matter of days. The results show the change of amorphous silica to quartz went through two intermediate phases of silica-X (a new polymorph of silica) and then cristobalite. The transformation rate was increased by a high temperature and pH.

Ernst and Calvert [20] investigated the changes in porcelanite when heated hydrothermally at temperatures of 300, 400, and 500 °C, 200 MPa pressure. Natural porcelanite from the Monterey Formation, CA, USA, was used in the experiments. The porcelanite was 98% cryptocrystalline with finely crystalline cristobalite together with very minor quartz. Their prime aim was to establish the conversion rates with regard to the eventual formation of chert. The change of cristobalite to quartz is relevant to the question of agate genesis and was found to be zero order.

Colloidal silica has been frequently suggested as the starting material for agate. However, Oehler [21] was the first to carry prolonged heating experiments on colloidal silica solutions. He added HCl to neutralise the alkaline silica colloid bringing the pH from 10 to pH 7 in order to produce a solid gel. The gel was sealed in silver capsules and heated between 100 and 300 °C at 300 MPa pressure for 25–5200 h. The solidified gel crystallised as quartz that was predominantly in the form of chalcedonic spherulites. The spherulites were either length slow or length fast or both. The formation of each type depended upon the conditions. It should be noted that the fibrosity was just in quartz microspheres and was not a synthetic version of an agate amygdale.

Previously, detailed chemical analysis in agate literature was rare but in 1982, Flörke and co-workers [22] investigated several Brazilian agates for their trace metal ion content. Additional work examined the two types of water that are found in agate: free molecular water was low while the silanol water (SiOH) was high. They were able to demonstrate that the wall banding and the horizontal banding in the same agate were clearly of different generations.

3. Scientific Contributions over the Years 1980–2020

3.1. Genesis Contributions in the Early Years

The origins of the silica source, means of transportation, and its physical state on deposition have been discussed over the last 250 years. Unfortunately with the unknowns, every suggestion is met with valid queries or objections. The various problems have been reviewed in detail with Godovikov et al. [23], Goncharov et al. [24], Landmesser [25,26], Moxon [27], and Götze [28].

From 1770 to 1930, the agate genesis hypotheses were largely based on the agate appearance and found in basic igneous host rocks. Our comments are made with hindsight but some suitable techniques were available and rarely applied to the agate genesis question. In particular, there was a general failure to examine agate in the thin section. Sorby [29] was the first person to use a polarizing microscope for the examination of rock thin sections. This initial publication was used to differentiate agate from calcite in the calcareous grit in Yorkshire, UK. The collection held in the Technische Universität Bergakademie, Freiberg, Germany shows that rock thin sections were being studied at various times during the late 19th century [30]. Yet agate thin sections rarely featured in these early years of the agate genesis discussion. The chalcedonic fibrosity and its apparent fibrous growth are shown by a thin section examination (Figure 3).

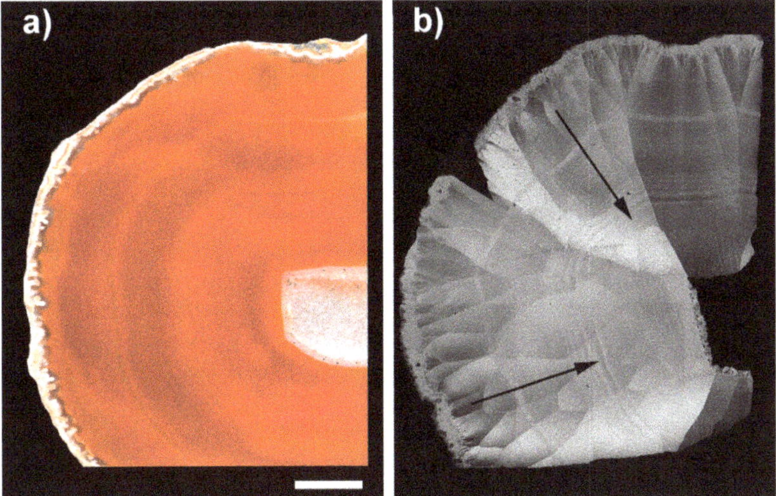

Figure 3. (**a**) A Brazilian agate that has been dyed to a depth of around 2 mm. (**b**) The thin section that was made from a slab just below the dyed surface. Note the wall contact layer shows a different texture to the main microstructure. The arrows indicate the direction of growth leading to a final development of macrocrystalline quartz crystals. Polars crossed. Scale bar = 2 cm.

By 1920, a number of inorganic crystal structures had been investigated using XRD. However as far as we are aware, the first agate investigation using XRD was by Heinz (1930) [31]. Undoubtedly, it was the work of Liesegang [2,3] with his experiments on silica gels that had the greatest influence. His ideas dominated the agate genesis question for much of the 20th century. The simulation of the rhythmic agate patterns in silica gels and in natural settings has ensured that the term "Liesegang bands or rings" will be preserved in perpetuity (Figure 4).

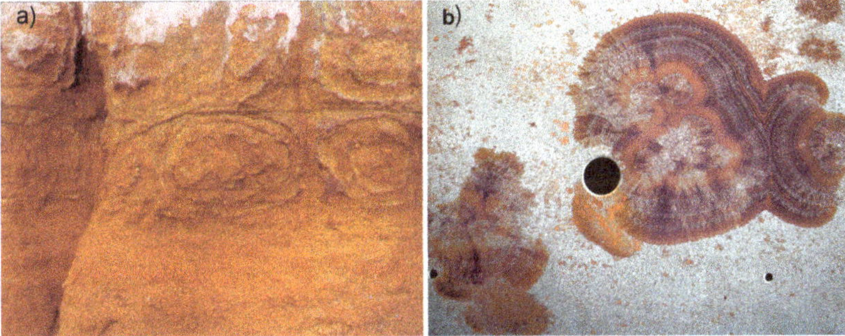

Figure 4. Two natural settings that have developed Liesegang rings. (**a**) An elliptical development in the sandstone cliffs at Hunstanton, Norfolk, England. (**b**) Rust rings that have formed on the support plate of a battery driven clock.

The need to have a pre-existing gel and its reliance of a later diffusion by a suitable cation resulted in some objecting to the Liesegang [2] hypothesis. Ion diffusion across silica gel in a Petri dish and vertical diffusion in a boiling tube produces a clear impression of wall lining and horizontal banding respectively. This initial gelatinous deposit is still favoured by some recent authors. Pabian and Zarins in 1994 [32] invoked genesis from a silica gel deposit with banding caused by the Belousov-Zhabotinsky (B-Z) reaction. The B-Z reaction creates banding oscillations between two chemicals and can produce rhythmic banding in gels [33,34]). The Pabian and Zarins [32] hypothesis is based upon an initial vesicular infill of silica gel and contact is made with alkaline ground water. Based on the B-Z reaction, they propose that this creates a continual electrochemical wave front banding.

Suggestions of an initial gel deposit as a starting material for agate genesis are flawed from the start. A comment was never made about the temporary nature of the generated patterns. After several weeks, the gels become cracked and dehydrated leaving only a coloured silica powder. A further comment is made in the discussion.

3.2. Chalcedony from Silica Glass

During the late 1950s, the electronics industry experienced an increased demand for high quality quartz. There were many attempts to synthesise quartz from various forms of silica glass. As chalcedony was often a byproduct, some data might have potential links to agate genesis. Following on from a previous study, White and Corwin [18] produced synthetic chalcedony from a silica glass that had been subjected to a hydrothermal solution at 400 °C and 34 MPa pressure. Conversion only occurred in weakly alkaline conditions with the transformation proceeding from silica glass → cristobalite → keatite → chalcedony → quartz. During long runs, the glass was completely converted to quartz. Synthetic chalcedony had the same anomalous properties as those of natural quartz: a brown colour in transmitted light and a refractive index lower than quartz. The transformation process is repeated in many changes of amorphous silica to quartz. Unfortunately, the silica glass starting material discounts any direct link with agate genesis.

3.3. Agate Formation Temperature

Götze in 2011 [28] carried out an extensive review of the methods used to study this temperature question. Here, we will deal with just three of the techniques. Oxygen and hydrogen isotope analysis was first used on Scottish Devonian and Tertiary agates by Fallick et al. [35]. Their work demonstrated agate formation was 50 °C. Later, Harris [36] examined Namibian agates and assigned a formation temperature of 120 °C. The Harris data were re-examined by Saunders [37] who placed the agate formation temperature between 39 and 85 °C. A second method is based on the concentration of Al^{3+} ion in macrocrystalline quartz that is sometimes found at the centre of the agate [38]. Here, agate formation temperature was estimated at between 50 and 200 °C. A mineralothermometric study of gas-liquid inclusions in quartz of agates located in andesite-basalts from many deposits in the north-east of Russia showed that the homogenization temperatures were <120–170 °C: Lebedinoe (140–170 °C), Kedon (100–135 °C), Yana (130–135 °C), Yakanvaam (100–120 °C), Takhtayama and Ryrkalaut (110–120 °C) [24].

Most recent investigators into the question of agate genesis accept that agate formation occurs at <100 °C. Any question of agate genesis occurring at temperatures higher than 200 °C has to be totally discounted. Agates when heated at higher temperatures produce irreversible property changes.

A thermogravimetric sketch plot of heated agate powder is drawn in Figure 5a. This typical curve shows three basic changes that have all been investigated using infrared (IR) by Yamagishi et al. [39] and they demonstrated:

(1)   Loosely bound water molecules are lost at <200 °C.
(2)   Tightly bound water and various forms of silanol water are lost up to 800 °C.
(3)   The maximum mass loss is reached at 850 °C.

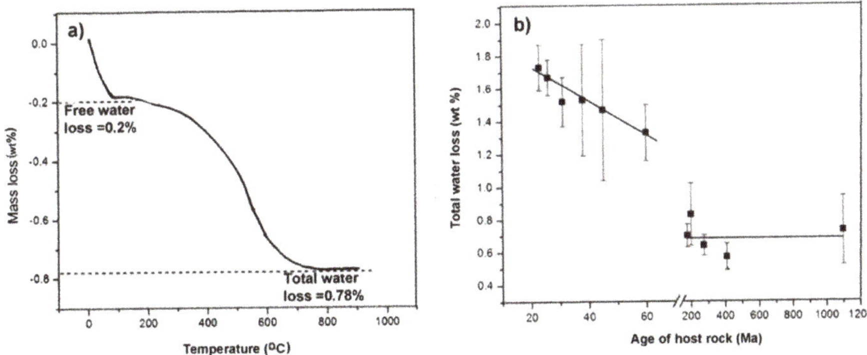

Figure 5. (**a**) A sketch plot of typical TGA (thermo gravimetric analysis) changes when powdered agate is heated to 1000 °C (after Yamagishi et al. [39]). (**b**) Total water loss from agate with respect to age of the host rock. Agate powders (<50 μm) were heated at 1200 °C (after Moxon [40]).

The variations in the total water in agate with respect to host age are shown in Figure 5b. The determined total water took into account the water loss in preparing <50 μm powder, which was then heated to 1200 °C. The plot shows a linear decrease in total water loss in agates over the first 60 Ma years with an approximate constant total water loss in agates from host rocks aged between 180 and 1100 Ma [40].

3.4. The Discovery of Moganite

The application of oxygen and hydrogen isotope analysis of agate formation temperatures by Fallick et al. [35] was a key contribution to the agate genesis enigma. Of equal importance was the discovery of moganite in the cracks and flows of the Mogan formation, Gran Canaria by Flörke et al. [41].

Since 1976, there have been many investigations into the properties of moganite but it was not until 1999 that it was officially recognised as a separate mineral by the International Commission on New Minerals and Mineral Names. Totally pure moganite has never been synthesised nor found pure in the natural state. The samples from Gran Canaria still provide the highest concentrations of moganite. It was the work of Heaney and Post [42] that effectively opened up the moganite story in agate by examining more than 150 samples of agate, chalcedony, chert, and flint. One of the data plots demonstrated chalcedony and agate differences in moganite ranged from 23 to 2 wt%. In a separate study, XRD was used to analyse 10 samples of Mogan moganite; the only mineral phases present were quartz and moganite with the moganite content greater than 80 wt% in half the samples. However, moganite was as low as 52 wt% in the remainder [43].

Raman spectroscopy is also a common technique for determining the moganite content. Götze et al. [44] used the two techniques to examine agate and showed that Raman can give higher values than was determined by XRD. These differences were attributed to the limited contribution to the Bragg reflection due to the small size and distribution of moganite. There have been a number of trace cation and moganite analyses of chalcedony. One was carried out by Petrovic et al. [45] with quartz and moganite (22 wt%) once again the only mineral phases. Al^{3+} at 0.52 wt% produced the largest concentration of the trace metal ions. Moganite is more soluble than quartz and the release of water during the geological time scale results in moganite slowly dissolving and reforming as microcrystalline quartz. The importance of this change is examined in the discussion.

3.5. Cation and Silica Loss from Basalt Host and Their Potential Role in Agate Formation

Suggested mechanisms have frequently been based upon an initial silica solution entering the gas vesicle with the subsequent water loss leaving a deposit of amorphous silica. Process repetition fills the vesicle with silica and eventually agate forms. Noeggerath [5] and Haidinger [6] first introduced aspects of this in 1849. However, the nature of silica in solution and its solubility were unknown until the 1950s. Previously, it had been assumed that aqueous silica existed in the colloidal state together with higher forms of silicic acid. Krauskopf [46] summarised earlier work that had established silicon exists primarily as monomeric silicic acid (H_4SiO_4). He used the yellow colour formation between silica solutions and ammonium molybdate to a great effect. Colorimetric methods could now be used to identify the concentration of $H_4SiO_{4(aq)}$.

Krauskopf [46] was interested in the conditions for silica solubility and the precipitation of amorphous silica. He added to, and summarised, the solubility changes with temperature. Silica solubility was found to be 100–140 ppm at 25 °C; 300–380 ppm at 90 °C. He also demonstrated that the solubility has shown little change between pH values of 1 and 9. Krauskopf [46] was dealing with silica in all sedimentary environments. However, the data clearly has implications for silica transport and deposition into gas vesicles.

The alteration of the host rock has frequently been cited as a possible source of silica. Patterson and Roberson [47] carried out an analysis on the depth of boreholes from the eastern part of Kauai Island, Hawaii. Seven boreholes of varying depth together with fresh rock were analysed for their chemical composition. Water from the host rock was also included in the analysis. Just four metal ions with the highest listed concentrations have been selected for comment (Table 1). The depth of basalt bore holes ranged from 0.6 to 11 m (hole numbers 1 and 7) and the data were compared to fresh basalt (number 8).

Table 1. Trace Analysis of Kauai Basalt Taken from Boreholes at Various Depths.

Analysis of Rock Samples							Water Analysis from the Rock					
Hole	Depth	SiO_2	Fe_2O_3	Al_2O_3	MgO	Na_2O	H_2O	SiO_2	Fe^{3+}	Al^{3+}	Mg^{2+}	Na^+	pH
No	m	(wt%)						(ppm)				
1	0.6	5.7	40.2	25.6	0.65	0.04	18.6	2.9	0.00	0.11	1.9	11	4.6
6	9.5	21.5	30.7	25.1	1.0	0.06	14.6	2.5	0.00	0.12	1.5	8.5	4.7
8	-	45.0	4.3	11.8	12.0	3.1	3.1	42	0.00	0.06	11	21	7.6

Data after Patterson and Roberson [47]. The depths shown in Table 1 are the mid point of the range given in the paper. Hole 8 is fresh basalt.

Kauai Island basalt has been dated at 5.1 Ma [48]. Table 1 shows the changes with borehole depth from a relatively young host. The analysis of water from the rock shows the total loss of Fe^{3+} and demonstrates that this ion has been transported out of the host rock system at an early stage. Such a young island gives an indication of a possible cation and silica mobility from a basalt rock and to a gas vesicle. We are not aware of any agate or chalcedony forming on Kauai but the Patterson and Roberson [47] data has relevance regarding mineral loss in an agate basalt host. Equally important is a chemical analysis of trace metal ions that have been identified in agate and the agate host.

Götze et al. [49] quantified 27 trace metal ions in 18 worldwide agates and host rocks. We selected host and agate data from two Scottish sites and one each from Mexico and Brazil. Agate host rocks for this selection of sites ranged in age from 38 to 412 Ma. Both the Götze et al. [49] and Patterson and Roberson [47] studies identified a larger number of ions, but we limited our speculative comments to those cations that were likely to have the greatest effect on H_4SiO_4. Iron (III) oxide offers colour to many types of agate and the trivalent ions of Fe^{3+} and Al^{3+} would be expected to have the greatest effect on any silica solutions or sols. Magnesium ions are generally the largest concentration of the divalent ions and Na^+ has by far the largest concentration of the monovalent ions.

Montrose and Ardownie Quarry share the same glassy andesite that formed 412 Ma years ago in the Eastern Midland Valley, Scotland. The distance between the two areas is 38 km. Chihuahua is world famous for the high quality agates and, like Brazil, is one of the few countries that mine and export agates. Brazil is by far the world leader in bulk agate export but the agates tend to be very large but dull and unusually, they can be dyed. Trace element analysis in agate and host rock from Mexico, Brazil, and Scotland are shown in Table 2.

Table 2. Cation Concentrations in Agate and Host in Samples from Mexico, Brazil, and Scotland.

Agate Source	Country	Host Rock Type	Age of Host * (Ma)	Fe^{3+}	Al^{3+} (ppm)	Mg^{2+}	Na^+
Chihuahua	Mexico	Andesite	38	6695	nd	5460	25,000
		Agate	-	215 **	925 **	151 **	320 **
Rio Grande do Sol	Brazil	Basalt	135	14,700	nd	nd	19,000
		Agate	-	137	54	10	140
Montrose	Scotland	Andesite	412	33,400	nd	nd	23,000
		Agate	-	70	456	15	230
Ardownie Quarry	Scotland	Andesite	412	37,300	nd	nd	27,000
		Agate	-	36	232	18	210

Data after Götze et al. [49], nd-not determined, ** two colours tested in the agate and the highest concentration is shown. * Citations for host rock age in Moxon and Carpenter [50]. The physical properties of Brazilian agate show its formation was 26–30 Ma.

The two neighbouring Scottish agate sites had similar concentrations of Mg^{2+} and Na^+ and the concentrations of Fe^{3+} in agate are much lower than in the younger host rocks. This is in spite of the older host rocks presently having Fe^{3+} at a greater concentration. The agates from these two areas are attractive but not typically highly coloured with oxides of iron. The concentration of Na^+ in all four host rocks is high and interestingly only relatively small concentrations survive in the agate. Presumably, much remains in solution and is lost with the departing vesicular water.

Two studies have used the silico-molybdate method to investigate the effect of iron ions/compounds on silica solutions and soils. A pressure membrane extractor was used to obtain H_4SiO_4 solutions from various soils [51]. In addition to the effect of iron (III) oxide, they separately added aluminium oxide to the silica solution at pH values 1–12. The addition of iron (III) oxide on the concentration of monosilicic acid resulted in a silica fall from 120 to 80 ppm at pH 4 and 70 ppm at pH 9. Aluminium oxide was more effective in reducing concentrations over the same two pH values producing a respective fall of 35 and 25 ppm.

Moxon [52] examined the separate effects of Fe^{3+} and Mg^{2+} in removing H_4SiO_4 and colloidal SiO_2 from solutions and sols after a short run of 45 min. The various runs were within the normal groundwater pH range of 4–9. A relatively high Fe^{3+}:H_4SiO_4 ratio of 1:3 could only remove 18% while a 1:96 ratio of Fe^{3+}: colloidal SiO_2 removed 90% of the silica. Over a 16 month period, 10, 20, and 40 ppm of $Fe^{3+}_{(aq)}$ and $Mg^{2+}_{(aq)}$ were separate additions to silica sols containing 1600, 3300, and 6800 ppm of silica respectively. These runs included fresh silica controls. In each case the solution pH was adjusted to 8.2. After 16 months, the added cations had been no more effective in precipitating silica than the control.

The solutions were stored in a cupboard at room temperature and reinvestigated after 14 years. All the H_4SiO_4 solutions had reached an equilibrium concentration of 122 ppm. There were a number of gelatinous flocs in suspension, on the side and floor of the containing flask. Apart from one of the 9 solutions tested, the pH fell from 8.2 to 6.1. Samples of the gels were dehydrated by oven heating and examined using XRD that included a new fresh gel control for comparison. In all cases XRD produced one single broad reflection. A full width at half maximum profile intensity (FWHM) was measured. With standard deviation included, all the 14 years old gels showed FWHM values were greater than the new fresh gel control. Unfortunately, FWHM could not differentiate between the Mg^{2+}, Fe^{3+} samples and the original gel control. Nevertheless, over the 14 years some silica reorganisation had taken place in all samples. The new fresh silica gel control showed only very minor unidentifiable peaks in the diffractogram scan from 15 to 36° 2θ. However, a number of the scans did show minor crystalline development peaks 21.5 and 22° 2θ. The XRD signals from the 14-year-old gels (Figure 6b) were compared to those of an agate from New Zealand. The host age was 89 Ma but its agate properties show agate formation was 40 Ma (Figure 6a).

The International Centre for Diffraction Data [53] gives the 100% signal intensity for synthetic tridymite and cristobalite as 21.58 and 21.98° 2θ respectively. There is the possibility of tridymite amongst the signal noise to the left of the cristobalite (Figure 6a) but XRD identification of cristobalite and tridymite together is not straightforward. Additionally, both silica polymorphs show some data variability with samples obtained from natural sources. Agates less than 60 Ma produce a similar collection of peaks and are discussed later. XRD details for Figure 6a: data was obtained from a Bruker D8 diffractometer in the reflection mode; the <10 μm powder was scanned over 17° < 2θ < 25° with a step size of 0.02° 2θ and a scan speed of 20 s/step (after Moxon and Carpenter [50]). Further examples and discussion about this cristobalite signal is described using the samples shown in Figure 13. The data for Figure 6b was obtained from a Philips vertical goniometer with runs over 15° < 2θ < 38°. Full details were given in Moxon [52].

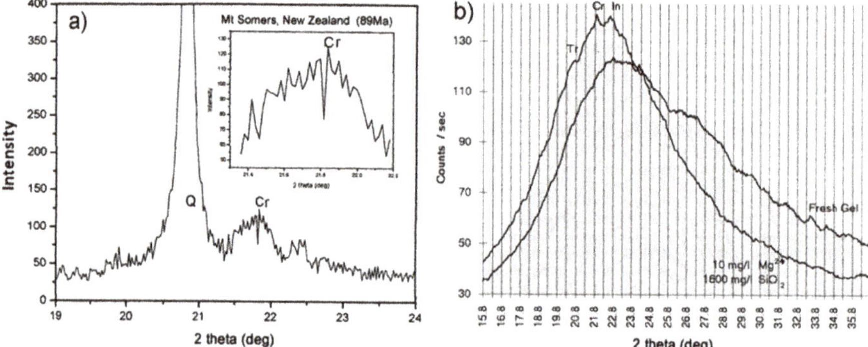

Figure 6. (**a**) An expanded examination of the cristobalite (Cr) and a possible tridymite signal. (**b**) XRD diffractogram of a 14-year-old silica gel and a fresh gel control. The aged gel shows small development peaks that were attributed to tridymite and cristobalite and an unidentified intermediate signal (In) (after Moxon [52]).

3.6. Age of Agate and Host

Unfortunately, there has only been one attempt to date both the age of the agate and volcanic host rock. Yucca Mt., NV, USA was proposed as a nuclear waste storage site and safety concerns have ensured three decades of continual and varied research. Here, chalcedony started development 4 Ma after the formation of the 13 Ma host [54]. Water movement within the mountain was an important part of the investigation. However, the low level of uranium in agates is partially the reason for the lack of any further U-Pb dating.

The availability of a variety of agates from around the world allowed changes of the quartz crystallite size and moganite content to be investigated with respect to host rock age [50]. Agates from 26 host rocks aged between 13 and 3480 Ma were examined using powder XRD to determine the crystallite size and moganite content. A total of 180 and 144 agates were investigated for the respective crystallite size and moganite content determinations. Two of the three agate regions had been involved in later metamorphic events and this had affected their properties [55,56]. The crystallite size (nm) of the remaining 23 regions shows four distinct changes of development (Figure 7a). At (A) the first 60 Ma followed a linear trend; (B) cessation of growth for the next 240 Ma; (C) a small growth spurt over the next 100 Ma; and (D) cessation of growth that continued for the next 700 Ma. The change in moganite content with respect to host rock age is shown in Figure 7b. Here, there was a linear trend over the first 60 Ma with an approximate constancy with agates from older hosts.

Three of the 23 regions were obviously off trend: agates from Brazil, New Zealand and the older agates from Northumbria, England. Brazil and New Zealand agates appeared to have an age of 26 and 30 Ma respectively. Similar comments can be made about the moganite content of the Brazilian and New Zealand agates shown in Figure 7b. The same agates show further evidence of this agate immaturity in the measured various types of water. All show the Brazilian and New Zealand agate regions are not outliers but have clear agate property trends supporting a later agate deposition [40]. Nevertheless for the majority of investigated agates, formation was generally penecontemporaneous with the host rock (Figure 7).

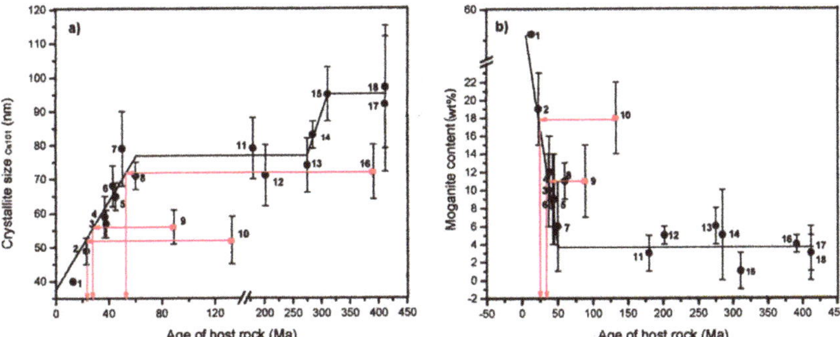

Figure 7. (**a**) Mean crystallite size as a function of age. The plot divides into four age regions: (A) an initial linear growth found in agates from host rocks aged up to 60 Ma: (B) over the next 210 Ma there is minimal growth: (C) growth restarts for 30 Ma: (D) after 300 Ma there is a final cessation of growth up to 1100 Ma. (**b**) Mean moganite content as a function of age. Agates formed from host rocks <60 Ma show a linear decline in the moganite content and little change in the older hosts. The apparent outliers (shown in red) suggest agate formation long after the host age. Agate regions: (1) Yucca Mt., USA; (2) Mt. Warning, Australia; (3) Chihuahua, Mexico; (4) Cottonwood Springs, USA; (5) Washington, USA; (6) Las Choyas, Mexico; (7) Khur, Iran; (8) BTVP, Scotland; (9) Mt Somers, New Zealand; (10) Rio Grande do Sul, Brazil; (11) Semolale, Botswana; (12) Nova Scotia, Canada; (13) Agate Creek, Australia; (14) Thuringia, Germany; (15) Derbyshire, England; (16) East Midland Valley, Scotland; and (17) West Midland Valley, Scotland. A further 8 regions from host rocks aged between 430 and 3750 Ma are not shown as minimal change occurs after 400 Ma. Error bars show one standard deviation. XRD details: Grain size < 10 μm using a Brucker D8 diffractometer in the reflection mode, crystallite size was based on the main (101) quartz reflection recorded on scans $17° < 2\theta < 30°$ with step size of $0.01°\ 2\theta$ and a scan speed of 10 s/step. The changed details for the moganite determination were $17° < 2\theta < 25°$, step size of $0.02°\ 2\theta$ and a scan speed of 20 s/step. The moganite and quartz peak areas were determined by fitting unconstrained Lorentzian functions using the Advanced Fitting Tool in "Origin" Moganite content has been taken as the proportion (in %) of moganite peak area/total area with an estimated resolution of ±2%. Full data is given in Moxon and Carpenter [50].

The changes in moganite content and quartz crystallite size with respect to host age is due to the release of structural silanol water:

$$Si\text{-}OH + HO\text{-}Si \rightarrow Si\text{-}O\text{-}Si + H_2O \tag{1}$$

This age-related change shown in Equation (1) has been established using cathodoluminescence. Released water is then able to dissolve the more soluble moganite and increase the size of recrystallizing quartz crystallites. Collectively, this produces a decrease in the moganite and silanol content with an increase in the age-related quartz crystallite size [57].

3.7. The Role of the Infiltration Canal

This peculiar feature has been noted since the 19th century and has been discussed as a means either of silica solution entry or as a water conduit for eventual water loss. It is not difficult to find these canals in some agates (Figure 8). Equally, it is not difficult to slice an agate into many sections and fail to find any canals. Walger et al. [58] examined the structures of these canals and commented that agates can be found with and without the canal. The Ayr to Girvan railway in Scotland was constructed at the beginning of the 20th century and allowed Smith in 1910 [59] to collect many hundreds of agates. He comments on the rarity of the canals in these Scottish agates. It is clear that the canals are a peculiar and interesting feature but cannot play a universal role in agate genesis.

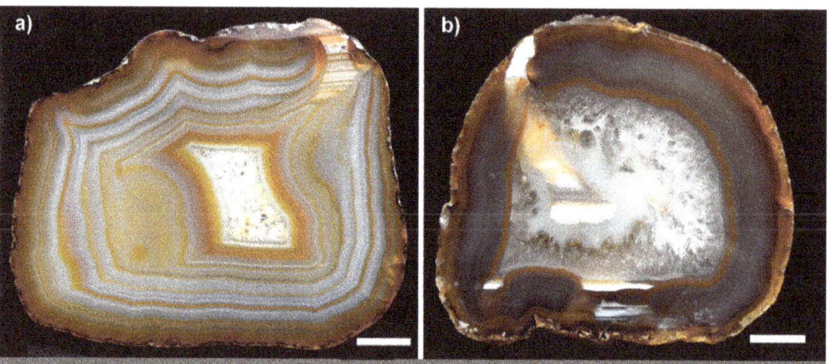

Figure 8. (**a,b**) Two Brazilian agates showing similar horizontal banding in the infiltration canals. Scale bar = 1 cm.

3.8. The Wall-Contact Layer

Agates from basic igneous host rocks always have an initial wall-contact layer that is different from the rest of the agate. This layer varies in thickness, generally 1–3 mm, but may appear to be absent. However, a thin section examination with polars crossed will always demonstrate an initial wall contact layer that is different from the bulk of the agate. These agate contact layers evenly coated the vesicle and did not appear to be influenced by gravity. The Botswana agate in Figure 9a was unusual with an apparent 5 mm contact layer. However, a thin section shows that the wall contact lining was the usual 1–2 mm (x in Figure 9b). The brown layer (y) shows a contrasting micro texture with the central region (z) demonstrating that these two areas were different generations.

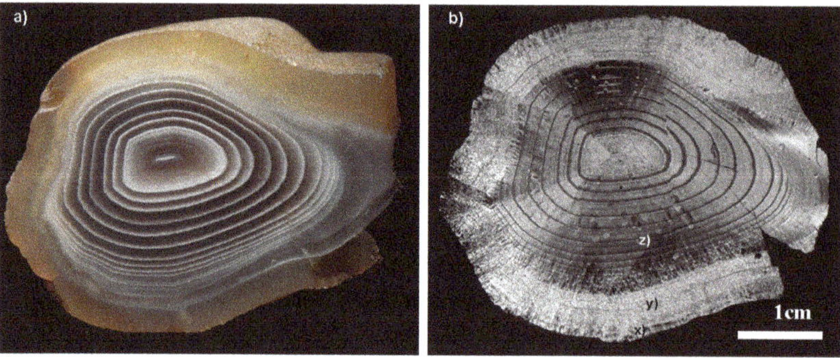

Figure 9. (**a,b**) Botswana agate showing an apparent large wall lining layer. However, a thin section (**b**) shows that there are 3 distinct growth formations with (x) a thin actual wall contact layer that is 0.2 mm; (y) the rest of the brown layer and (z) the black and white agate bulk.

Reis [10,11] proposed this wall-lining was a different generation. Comment has been made about Brazilian agates [60]; bituminous agates [61]; and sedimentary agates [62]. An agate wall lining is also found in the Brazilian amethyst geodes. This led to the proposition that aqueous silica in the forming solutions diffuse through agate fractures [63]. Fractures or even hairline cracks are rare in agate and when observed they are sufficiently significant to be assigned to host rock movement. The genesis implications of this agate contact layer are of prime importance. A first deposit of wall lining would always present a potential barrier for bulk entry of silica into the vesicle.

This wall lining deposit is sometimes apparent in a thick section: as shown by the Brazilian agates in Figure 8b. Notice the near uniformity of the wall lining layers in these agates. It would be remarkable if the first silica solution could fill the gas vesicle and evenly coat the vesicle wall, crystallize, and then allow subsequent silica solutions to enter.

One study characterised nine Brazilian agates that had a lower horizontally banded section and an upper chamber with the normal wall lining pattern [64]. The volume division between the wall lining and horizontal banding varied with different agates. The horizontal banding in these agates is between 30 and 70% of the total. Thin sections of four of these Brazilian agates are shown in Figure 10a–d.

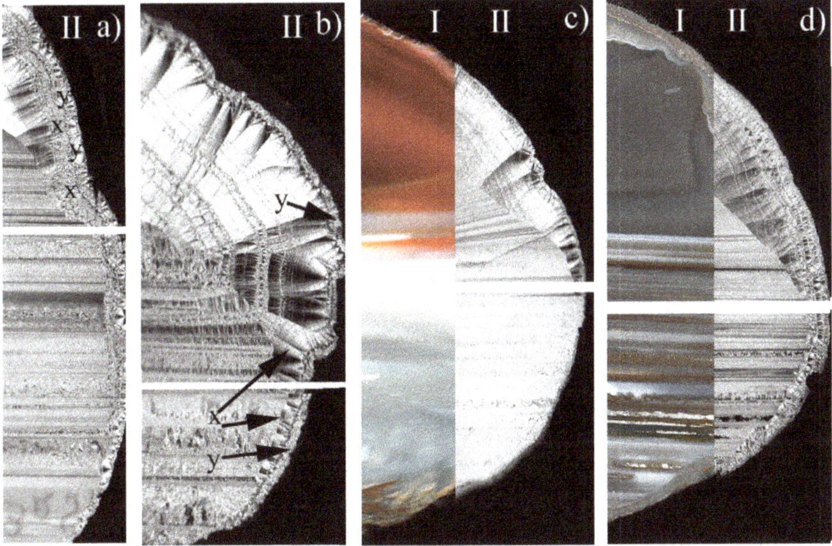

Figure 10. (a–d) The four monochrome micrographs are thin sections of four Brazilian agates with wall lining and horizontal banding, these are labelled as II (**a–d**). There are two thick sections with the original banding (Ic, Id). The discrete layers of horizontal bands are separated from the wall lining sections by the white dividing space. In Brazil slab II (**a**), there are two initial wall lining deposits, labelled x and y: these are reduced in the horizontally banded section. The thickness of the initial wall lining layer in Brazil slab II (**b**) shows tapering in the base of the horizontally banded section. The continuous sheaf-like growth (x) from the wall lining also coats the gap at the edge of the horizontally banded layers. Brazil slab (**c**) shows the initial wall lining deposit is absent in the horizontally banded layers. Brazil slab (**d**) demonstrates four separate wall lining deposits line the cavity wall. Scale of thin sections = 2 cm at the maximum width.

The wall contact layer in Figure 10IIa,b,d was present in the upper and lower chambers but absent in the lower chamber of IIc. There are two wall contact layers shown in agate IIa (labelled x and y). These two layers gradually faded into one at the bottom of the horizontal region. In IIc the wall contact layer was absent in the horizontally banded region. Additionally, these layers got thinner in the horizontal banded region in IId.

The observations on these combined horizontal and wall lining agates suggest the following development. Initially, enough silica is deposited on the empty vesicle floor to form a single horizontal band. At some point, there was a cessation of silica. During a later period fresh silica was deposited and followed by another pause of silica deposition. The process is repeated producing several layers. The evidence for these separate depositions is shown by the differing layer textures that range from chalcedony to prismatic quartz. Clearly this is a stop/start activity. Eventually the upper chamber now had a horizontal base but the silica input continued until sufficient silica had filled the remaining space. These agates were examined using XRD and some areas demonstrated trace cristobalite. The presence of cristobalite in agate supports the development mechanism proposed by Landmesser [25,26] and will be discussed later:

$$\text{amorphous silica} \rightarrow \text{cristobalite/tridymite} \rightarrow \text{cristobalite} \rightarrow \text{chalcedony/moganite} \rightarrow \text{granular quartz} \tag{2}$$

Each of the silica transformations shown in Scheme (2) has a greater density than its predecessor. When the transformation of layers and upper chamber is complete then shrinkage will occur but the larger bulk in the upper chamber will produce the greatest shrinkage. Here, the gap will be wider between the agate and the vesicle wall. Finally, silica deposition fills this wall lining gap. Agate IIb shows a much wider fibrous growth in the upper chamber and with a continuous but thinner fibrosity at the bottom of this chamber where thin horizontal bands have formed. The wall lining falls to around a fifth of the thickness when it fills the contact wall in the lower horizontal bands. Hence, the wall lining layer is the last generation.

3.9. Agate under the Scanning Electron Microscope

Lange et al. [65] and Holzhey [66] used the scanning electron microscope (SEM) to find evidence of the petrographic fibrosity observed in agate thin sections. The fibrosity was not found and instead, a surface composed of globulites was observed. The globulites varied in size between 0.1 and 3 μm [66] or subparticles with sizes ranging from 0.07 to 0.7 μm [66]. Additionally, there have been a number of publications citing the use of the SEM in a support role when investigating agate and chalcedony. Examples include, the effects of heating chalcedony at 500 °C [67]; agates from Permian rocks [68].

Early work had shown age-related trends in the development of quartz crystallite size and the SEM was used to observe differences between agates from young and old host rocks [69]. There are clear trends of increasing globulite size with age (Figure 11). However apart from agates from the youngest and oldest host rocks, there were too many size variables in each micrograph for worthwhile size comparison data. The same study showed host-age related differences between the white bands in agate. Agates from young host rocks will frequently show hints of faint white banding that becomes more intense in agates from host rocks greater 50 Ma. The white bands develop a structural form that is reminiscent of the edges of a stack of plates (Figure 12). Heinz [31] boiled the white and coloured bands in $KOH_{(aq)}$ and found that white bands offered a greater resistance to the $KOH_{(aq)}$: this was credited to an increased opal content. Opal is absent in agate and the SEM shows the observed differences are structural: the "plate" edges offering more resistance to the alkali attack.

3.10. Hydrocarbon Inclusions in Agate

In recent years, various hydrocarbon deposits have been reported in agates. Barsanov and Yakovleva [70] found oil and water in both horizontal layered bands from North Timan in the Russian Urals and in wall lining banded agate from the Kyzal-Tugan deposit, Taldy-Kurgan, Kazakhstan. The identification of hydrocarbons between horizontal agate layers is particularly interesting as the different formations in these horizontal layers is indicative of their long time-related formations.

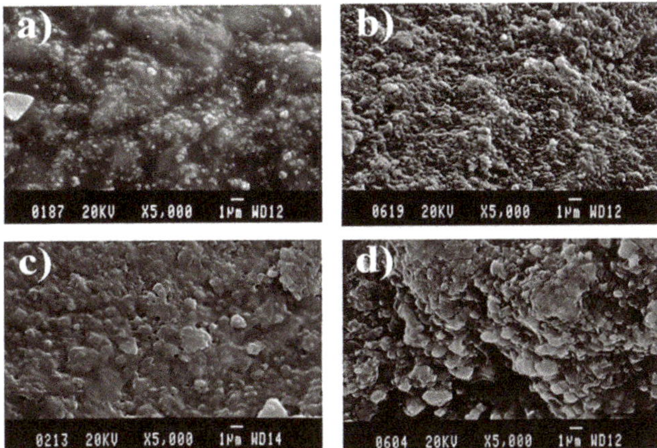

Figure 11. SEM micrographs showing the development in clear bands: (**a**) Rio Grande do Sul, Brazil 135 Ma, (52 nm) 135 Ma; (**b**) Isle of Mull, Scotland, 60 Ma (72 nm); (**c**) Agate Creek, Queensland, Australia 275 Ma (74 nm); and (**d**) Ethiebeaton Quarry, Scotland 412 Ma (96 nm). The stated age is the host age and crystallite size is the mean crystallite size for that particular region. The properties of the Brazilian agate, including crystallite size, indicate a formation age of ~26 Ma. SEM details: small agate slices were coated with gold and examined in a JEOL 820 SEM. The experimental conditions used a 20 kV accelerating voltage with a current of 1 nA.

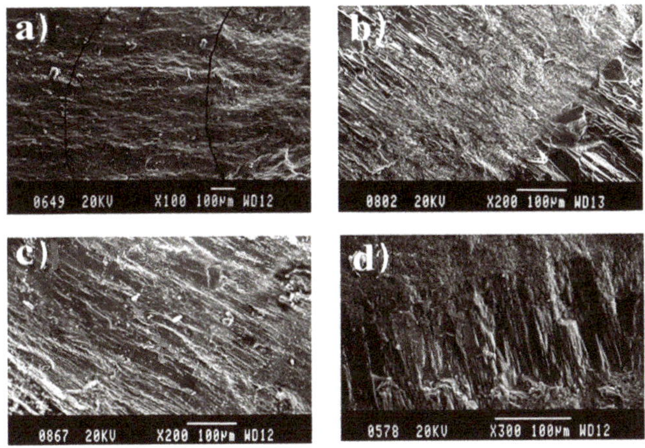

Figure 12. (**a–d**) are the white banded regions from the same agates shown in Figure 11: (**a**) Rio Grande do Sul, Brazil 135 Ma, (52 nm) 135 Ma. The black lines mark a faint white band; (**b**) Isle of Mull, Scotland, 60 Ma (72 nm); (**c**) Agate Creek, Queensland, Australia 275 Ma (74 nm); (**d**) Ethiebeaton Quarry, Scotland 412 Ma (96 nm). The stated age is the host age and crystallite size is the mean crystallite size for that particular region. The properties of the Brazilian agate, including crystallite size, indicate a formation age of ~26 Ma. SEM details as given in Figure 11.

Silaev et al. [71] studied the composition of gases in agates from the epithermal deposits from the Russian Polar Urals. Most of the secretion bodies have a lens shape with intermittent and wedge-shaped protrusions. Using pyrography, they identified the main composition of the inclusions. In order of decreasing concentration these were H_2O, CO_2, CO, CH_4, and N_2. The total content of the gases was in the range of 100–600 µg/g. Hydrocarbons in the fluid inclusions varied from 0.25 to 7 µg/g. The presence of hydrocarbons in agate was confirmed by radioscopic data. As a result of the determinations of the gas–liquid inclusions, they proposed a formation temperature in the range of 100–200 °C.

A third study of hydrocarbons in agates was made using samples from Nowy Kościol Lower Silesia, Poland by Dumeńska-Slowik et al. [61]. They describe the composition of the bitumen as mainly asphaltenes (56%). The remaining fractions in order of composition are saturated hydrocarbons (18%), resins (16%), and aromatic hydrocarbons (10%). Carbon isotope analysis revealed an algal-humic or algal origin.

4. Discussion

Agates are mainly found in basic igneous host rocks and this discussion is limited to the origins of agates from these host rocks. Any solution to the genesis problem requires answers to the following questions.

(a)   What is the formation temperature?
(b)   What is the source of the silica?
(c)   What is the nature of the silica deposit?

(a) Formation temperature: this is the only question that has been answered and the majority of interested workers accept the oxygen and hydrogen isotope analysis of Fallick et al. [35] and Saunders [37]. The collective present evidence shows that the agate formation temperature in these basaltic host rocks was <200 °C and most likely <100 °C. Any proposal of a higher formation temperature has to explain the conflict of evidence found by the various irreversible silanol water losses that are observed between 200 and 850 °C that were described by Yamagishi et al. [37]. All the observed changes would not be observed if agate had formed at a temperature >250 °C. Furthermore, when agate is heated at higher temperatures there are clear visible and further irreversible physical changes, as the agate turns white and porous. Depending upon the temperature, agate is also likely to become fractured.

(b) Silica source: Host rock leaching would be the most popular choice suggested by interested workers. One of us (T.M.) has examined many of the 412 Ma agate-bearing host rocks of the Midland Valley, Scotland. There is ample evidence of weathering/alteration in these old host rocks. Although fresh feldspars were observed, most did show signs of change. The ferromagnesian content was highly altered with thin sections showing indistinct remnants as a common feature. These rocks had the potential to release the silica for agate formation and Fe^{3+} for oxide colouring. However, the most telling feature for some of the Eastern Midland Valley agates is their clear lack of red/brown iron oxide colouring. Conversely, the Burn Anne agates from the Western Midland Valley provide Scotland's most colourful red and yellow agates. Extensive research on the Yucca Mt., NV, USA shows the downward percolation of water is depositing amorphous silica in these 13 Ma Miocene Tuffs. Chalcedony forms in fissures and as a host coating [72]. However, these tuffs are not the typical agate host rocks.

The link between host age and agate crystallinity demonstrates agate formation is mostly a penecontemporaneous deposit of host and agate [50]. This timing would offer support for late hydrothermal solutions or hot springs carrying higher concentrations of H_4SiO_4. Scientists investigating hot spring silica differentiate between two types of silica sinter: a direct deposition from hot solutions and silica released by steam attacking the host rock. Either directly or indirectly these actions are a potential supply of silica for agate formation.

(c) The nature of the silica deposit: There are two potential alternatives for the initial deposit: silica gel or powder. This would be followed by a later transformation into chalcedony. Harder and Flehmig [73] showed that iron, magnesium, and aluminium hydroxides could absorb silica from low concentration of silica solution (0.5 ppm). The silica enriched precipitates formed quartz crystals within days. Elsewhere, quartz crystals have been grown in the laboratory from a solution containing 4.4 ppm of silicic acid at 20 °C [74]. Both these studies require quartz formation to be from solutions that are saturated with respect to quartz but under saturated with respect to amorphous silica. These conditions are limiting and the developing crystals are prismatic quartz and not agate.

Godovikov et al. [23] presented the results of experiments investigating the diffusion processes in silica gel. They used the silicate (office) glue, which is mainly an aqueous solution of Na_2SiO_3 and iron, nickel, or copper sulphate in the experiments. A sparingly soluble salt in solution was poured onto a silica gel containing the same salt. Maximum supersaturation occurs at the interface between solution and gel. At this interface, they observed the appearance of complex branched dendrites. When the degree of supersaturation was minimal, there was a growth of individual full-faceted crystals. They proposed that spherulites arise under conditions of supersaturation. Additionally the authors proposed that the colour of agates has various causes. Essentially, a) uneven distribution of pores and their different shapes in separate layers of agate; b) a different mineral composition in individual zones; and c) finely dispersed inclusions of brightly coloured minerals.

Heaney [75] developed a theoretical model to explain the formation of chalcedony and prismatic quartz in the same sample. This was based on the differences of structure and solubility between chalcedony and prismatic quartz. A final single deposit of macrocrystalline quartz is shown in Figure 3. When this occurs prismatic quartz is usually found around the centre of a sample with a clean break from the surrounding chalcedony. An immediate impression is distinct separate entries and depositions. This is not possible and the Heaney [75] mechanism is based on a weakly polymerised silica solution allowing the formation of chalcedony and leaving a more dilute solution of monomeric silicic acid. Eventually, the very dilute silicic acid solution is now under saturated with respect to amorphous silica but saturated with respect to prismatic quartz: ideal conditions for a final deposit of prismatic quartz.

Over the last century and prompted by the work of Liesegang [2], there has been support for the direct precipitation of silica gel as an agate precursor. Reported direct formations of silica gel are very rare but Naboko and Silnichenko [76] commented on a silica gel (52% silica) that was forming on the solfataras of the Golovnin volcano in Russia. However, gel cannot diffuse through the host rock. Any gel precursor must form within the gas vesicle. To overcome this objection, a hypothesis was proposed by Harder and Flehmig [73] who suggested that a reaction between solutions containing silica sols and $Fe^{3+}{}_{(aq)}$ or $Al^{3+}{}_{(aq)}$ enter the vesicle separately. This mixture produces a gel-like state that could eventual form agate. Present day solutions circulating in host rocks do not contain the high concentrations of these two ions or silica; these would be necessary to generate a silica gel. Dry gels quickly dehydrate and produce a powder of amorphous silica. As a hypothesis, there seems little point in adding one more step to this amorphous silica beginning.

Agate, chalcedony, and chert are all forms of microcrystalline quartz. Although chert commonly shows a granular texture when examined petrographically, patches of chalcedony in chert are not rare. Chalcedony has been identified occurring with amorphous silica in the Yucca Mt, Nevada. Additionally it is amorphous silica (opal-A) that is commonly found around hot springs in New Zealand and Yellowstone Park, USA. Unfortunately, investigators of silica sinter do not tend to differentiate between the various forms of microcrystalline quartz: chalcedony is not often reported. Nevertheless, chalcedony is presently being formed in the Tengchong geothermal area China [77]. Additionally, Marcoux et al. [78] identified black, grey-brown and white chalcedony that formed in the siliceous sinter of the 295 Ma French Massif Central.

Studies on the genesis of chert and sinter have produced a much greater volume of work than agate but similarities and differences between all three provide potential development links for agate investigation. Stages of development for chert [79] and sinter [80] have been shown to follow sequence:

$$\text{amorphous silica} \rightarrow \text{cristobalite/tridymite} \rightarrow \text{cristobalite} \rightarrow$$
$$\text{chalcedony/moganite} \rightarrow \text{granular quartz} \tag{3}$$

Landmesser [25,26] argued the same transformations for agate (2). A comparison of the rate of phase transformations for agate, chert, and sinter demonstrates the major origin differences between these three forms of microcrystalline quartz. Silica sinter discharges from weakly alkaline chloride waters at ≤100 °C. The time scale for completion of the sinter diagenesis is typically 50 ka [81]. However, a faster rate of conversion of sinter into quartz has been achieved when a partial conversion into quartz occurred within weeks [82]. As expected, the completion of each silica sinter transformation step is generally not rapid. For example, a single XRD diffractogram can show evidence of amorphous silica, cristobalite and tridymite.

Moganite is found in sinter, chert, and agate but the survival rates differ greatly. In sinters with ages 20–220 ka the moganite content is at a maximum of 13 wt% whereas Tertiary sinters are generally free of moganite [83]. Chert hosts of the Cretaceous age have been identified with a moganite content of >20 wt% [84]. This contrasts with 56 wt% moganite identified in the Yucca Mt chalcedony and it has been identified in agate from the 1100 Ma Lake Superior host [50]. However, detecting moganite from agate in host rocks >400 Ma using XRD is not always possible.

Chert diagenesis is much more prolonged where surface amorphous silica can survive for 85 Ma. Transformation to cristobalite/tridymite can be 5–10 Ma and is still found in rocks aged 120 Ma but is not observed in rocks >140 Ma. Quartz requires a minimum of 40 Ma at depths of 500 m [85]. The phase transformation changes and moganite differences in sinter, agate, and chert have a link to the particular formation temperatures: silica sinter shows a variable but the most rapid changes with the silica deposits held at 100 °C.

Agate development bears a close link to the age of the host and polymorph precursors have occasionally been identified: cristobalite/tridymite layering in Brazilian agate [22]. Moxon and Carpenter [50] identified cristobalite in agates from Mexico, Iran, Rum, and Scotland: these are all hosts younger than 60 Ma. Cristobalite was also found in the hosts of New Zealand agate (89 Ma) and Brazil (135 Ma): providing further evidence for a formation age much later than their hosts. The basic XRD data identifying the weak cristobalite XRD signals in agates from Brazilian, and Iranian hosts are shown in Figure 13. Cristobalite identification in agate has not been reported in agate from hosts older than 60 Ma elsewhere. As discussed earlier, the properties of Brazilian agate demonstrate a formation at 26–30 Ma.

The prime aim of the Moxon and Carpenter (2009) [50] study was to investigate the moganite to quartz transformation. The quantification of the moganite content was carried out using a Bruker diffractometer with a step size of 0.02° 2θ and a scan rate of 20 s/step. A total of 32 agates were examined from hosts ≤60 Ma (that also included Brazilian and New Zealand agates). Exactly half were free of cristobalite but the remainder gave the cristobalite XRD signal. Three agates from the cristobalite batch were selected for a more intense scan and at least two of the next cristobalite most intense signals were identified at 28.4, 31.5, and 36.1°2θ. The detected cristobalite was noted in the paper but not examined any further.

The same data has been examined in more detail and two of the diffractograms are shown in Figure 13. The Brazilian agate shows the presence of cristobalite and the plot was similar to that produced by the New Zealand agate in Figure 6a. All the agates, apart from those from Brazil and New Zealand, were from hosts younger than 60 Ma.

There was a contrast between the cristobalite peaks shown by the two agates in Figure 13. There were two peaks in the Brazilian agate against the single peak in the Iranian agate. We speculated that the second outer peak could be tridymite and the change from younger to older agate was demonstrating an opal CT → opal C change. The outer peaks of the agates showing this peaked feature were measured and produced a mean value of 21.89(6)° 2θ and a cristobalite mean value of 21.99(4)° 2θ. Data shown in parentheses are ± one standard deviation. The JCPDDS [53] data for synthetic tridymite and synthetic cristobalite were 21.58 and 21.98° 2θ respectively. The identification of tridymite and cristobalite with XRD when found together in the natural state is not straightforward. The regular appearance and consistency of the outer signal in these samples would rule out instrument noise.

Of the 16 agates that showed these two signals, 5 were from Mt Somers New Zealand and 4 in each of the batches from Chihuahua, Mexico, and Khur, Iran. Cristobalite identification in agate has not been found elsewhere in agates older than 60 Ma. Hence, the identification of cristobalite in

Brazilian and New Zealand agates further supports their late agate formation. Dates of the late entry of some secondary minerals into the gas vesicles of the Brazilian Paraná Continental Flood basalts have been summarised by Gilg et al. [86]. Various secondary infills occurred up to 60 Ma after the basalt formation. However, the properties of Brazilian agates (water and moganite content and crystallite size) show a formation age around 100 Ma after host formation.

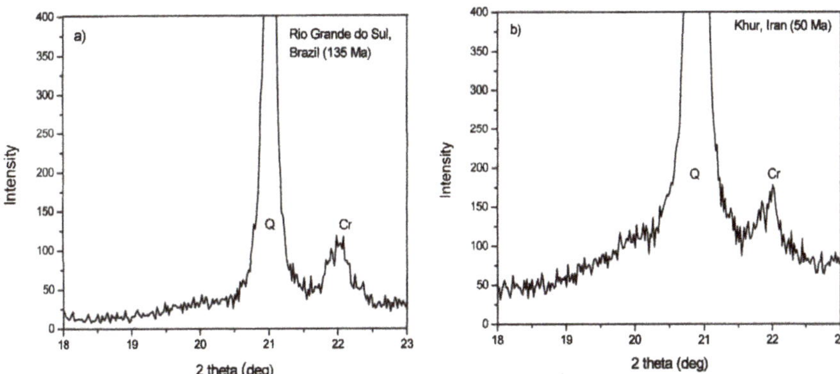

Figure 13. Cristobalite (Cr) identified in agates from Brazil (**a**) and Iran (**b**). The stated ages are the host ages but the properties of Brazilian agate suggest an agate age of 25–30 Ma.

5. Summary

The most prolific source of agate is from basalt and andesite and we concentrated our deliberations on agates from these hosts. Such agates provide by far the greatest scientific interest and involvement. Agates found in acidic igneous and sedimentary rocks are more rare and this is reflected by the smaller number of literature publications.

Until the mid 20th century there were few scientific publications on agate and these were mainly limited to speculative comments based on agate banding and their hosts. Viewed from 2020, clear exceptions would be the early observations and micrographs produced by Timofeev [8]; the detailed thin section examinations of Pilipenko [13] and the gel studies of Liesegang [2,3]. Liesegang's works on gels were still producing links to agate genesis until the end of the last century. However, there is now little support for an explanation of agate formation based upon the concept of metal ion diffusion into a silica gel.

From 1960, industrial requirements for high quality quartz were the driving force for a number of high temperature and pressure investigations on various forms of silica. Unfortunately, none of these experiments had a direct application to agate genesis. Additionally, studies using high temperature/pressure conditions and those under normal earth conditions can proceed by different mechanisms giving conflicting results. Nevertheless, advances in scientific equipment over the last 60 years have allowed a more detailed understanding of the properties and composition of agate.

In these basic igneous host rocks, enough evidence has now been obtained for an agate formation temperature at mostly < 100 °C. The agate silica content was initially assumed to be just fine-grained α-quartz. However, the more recent discoveries of moganite and cristobalite have allowed evidence-based speculation to propose that agate starts as amorphous silica. Over geological time, it progresses to cristobalite before transforming into agate.

It is probable that tridymite forms before cristobalite but the evidence has still to be conclusively found. On initial formation, agate is on a lengthy journey of development. Agates are unusual in ageing over the geological time scale. The changes are due to the release of water and the very slow transformation of moganite into quartz. The result is an increase in quartz crystallite size. SEM examinations demonstrate globular development in the clear bands and further maturity in the white bands. This development in the white bands is often apparent in agate thin sections. White bands in agates older than 180 Ma generally show a more intense white than is found in younger agates (compare the Brazilian and Botswana samples in the respective Figures 8 and 9). A thin section of a 3480 Ma agate from Australia shows that the end of agate development is only reached when fibrous quartz is transformed into microcrystalline granular quartz [55].

Presently, agate genesis research does lack any commercial driving force. The genesis problem has been in the literature since the 18th century and will not be to be solved with a 3-year funding grant. One important incentive for chert and silica sinter research has been the early forms of life that have been trapped in the material. This might change for agate as the Mars Rover units have been investigating silica links to water and microbiological activity and this does include chalcedony.

For the future, agate properties that change w.r.t host age might be a simpler and a cheaper direction of study. In order to make valid comparisons, any such investigation will require sufficient agates from host rocks of various ages. As always, there is the proviso regarding outliers that have formed later than the host. As described in this paper, three of the 26 agate regions demonstrated an earlier but consistent age of formation that was much later than the host.

The most likely area of profitable investigation would involve agate from very young host rocks. We described data from a single sample of Yucca Mt (13 Ma) chalcedony providing a crystallite size and moganite content. While this agate had a crystallite size of 40 nm that is in line with expectations, the 56 wt% moganite did appear to be very high when compared to the 19% moganite in agate from the 23 Ma Mt Warning Australia host. At face value, this would suggest that agates in their very early years of development are exceptionally high in moganite. In which case, extrapolation back to zero time would not be possible. Alternatively, this sample is a freak result and further analyses would produce other data that is more in line with expectations. Where possible a minimum of five samples from the same area should be examined. It would appear that agates from the young host rocks have the most to offer future investigations into agate genesis. In this category, the islands of Hawaii would be a prime source for agate study. These islands have formed as a result of movement across a hot spot in the Earth's mantle. Several islands offer formation dates ranging from 0.43 to 5.1 Ma [48]. It helps that the geology of these islands has been thoroughly investigated by the US Geological Survey and many others.

Author Contributions: Conceptualization, T.M.; methodology, T.M.; writing and editing, T.M. and G.P.; visualization, T.M. and G.P.; supervision, T.M.; translation of works in Russian into English, G.P. All authors have read and agreed to the published version of the manuscript.

Funding: This research was funded by the Ministry of Science and Higher Education of the Russian Federation within the framework of the state assignment of the Sobolev Institute of Geology and Mineralogy of Siberian Branch of Russian Academy of Sciences (topic no. 0330-2016-0001).

Acknowledgments: The agates used in the studies by T.M. have been donated by Glenn Archer (Outback Mining, Australia), Rob Burns, Jeannette Carrillo (Gem Center USA, Inc.), Roger Clark, Nick Crawford, Brad Cross, Robin Field, Gerhard Holzhey, Brian Isfield, Herbert Knuettel (Agate Botswana), Reg Lacon, Brian Leith, Maziar Nazari, Dave Nelson, Leonid Neymark, John Raeburn, John Richmond, Vanessa Tappenden, Bill Wilson, and Johann Zenz. Links between agate properties and host age could not have been investigated without these donations. T.M. is indebted to the Dept. of Earth Sciences, Cambridge University, UK for access to the laboratories. The contributions of co-workers and support staff were essential for the progress made. Thanks are due to Tony Abraham, Michael Carpenter, Ian Marshall, Chris Parish, Chiara Petrone, Stephen Reed, Susana Rios, Paul Taylor, Martin Walker, and Ming Zhang. Gerd Schmid translated the early German papers into English. We thank Evgeniya Svetova for help in finding rare pre-1934 Russian agate articles. We thank three reviewers for their comments after a first submission of the manuscript.

Conflicts of Interest: The authors declare no conflict of interest.

References

1. Collini, C. *Tagebuch Einer Reise, Welches Verschiedene Mineralogische Beobachtungen, Besonders über Die Achate und den Basalt Enthält*; C.F. Schwan: Mannheim, Germany, 1776; p. 582.
2. Liesegang, R.E. *Die Achate*; T. Verlag von Theodor Steinkopff: Dresden/Leipzig, Germany, 1915; p. 122.
3. Liesegang, R.E. Achat Theorien. *Chem Erde* **1931**, *4*, 143–152.
4. Zenz, J. *Fascinating Idar Oberstein in Agates*; Bode: Lauenstein, Germany, 2009; pp. 164–185.
5. Noeggerath, J. Sendschreiben an den K. K. wirklichen Bergrath und Professor Haidinger in Wien, über die Achat-Mandeln in den Melaphyren. *Verh. Nat. Ver. Preuss. Rheinl. Westphalens* **1849**, *6*, 243–260. (In German)
6. Haidinger, W. Versammlungsberichte. In *Berichte über die Mittheilungen von Freunden der Naturwissenschaften in Wien*; Braumüller u. Seidel: Wien, Germany, 1849; pp. 61–69. (In German)
7. Heddle, M.F. *The Mineralogy of Scotland*; David Douglas: Edinburgh, Scotland, 1901; Volume 1, p. 148.
8. Timofeev, V.M. Chalcedony of Sujsar Island. *Proc. Soc. St. Petersburg Nat.* **1912**, *35*, 157–174. (In Russian)
9. Reis, O.M. Einzelheiten über Bau und Entstehung von Enhydros, Kalzitachat und Achat I. *Geognost. Jahresh.* **1917**, *29*, 81–298. (In German)
10. Reis, O.M. Einzelheiten über Bau und Entstehung von Enhydros, Kalzitachat und Achat II. *Geognost. Jahresh.* **1918**, *31*, 1–92. (In German)
11. Fischer, W. Zum Problem der Achatgenese. *N. J. Miner. Abh.* **1954**, *86*, 367–392. (In German)
12. Jessop, R. Agates and cherts of Derbyshire. *Geol. Assoc.* **1930**, *42*, 29–43. [CrossRef]
13. Pilipenko, P.P. Zur Frage der Achat genese. *Bul. Soc. Nat. Mosc.* **1934**, *12*, 279–299. (In German and Russian)
14. Kuzmin, A.M. *Periodic-Rhythmical Phenomena in Mineralogy and Geology*; Scientific & Technical Translations: Tomsk, Russia, 2019; p. 336. (In Russian)
15. Nacken, R. Über die Nachbildung von Chalcedon–Mandeln. *Natur Folk* **1948**, *78*, 2–8.
16. Carr, R.M.; Fyfe, W.S. Some observations on the crystallization of amorphous silica. *Am. Mineral.* **1958**, *43*, 908–916.
17. Keat, P.P. A New Crystalline Silica. *Science* **1954**, *120*, 328–330. [CrossRef] [PubMed]
18. White, J.F.; Corwin, J.F. Synthesis and origin of chalcedony. *Am. Mineral.* **1961**, *46*, 112–119.
19. Heydemann, A. Untersuchungen über die Bildungsbedingungen von Quarz in Temperaturbereich zwischen 100 °C und 250 °C. *Beiträge zur Mineralogie und Petrographie* **1964**, *10*, 242–259. (In German) [CrossRef]
20. Ernst, W.G.; Calvert, S.E. An experimental study of the recrystallization of porcelanite and its bearing on the origin of some bedded cherts. *Am. J. Sci.* **1969**, *267*, 114–133.
21. Oehler, J.H. Hydrothermal crystallization of silica gel. *Geol. Soc. Am. Bul.* **1976**, *87*, 1143–1152. [CrossRef]
22. Flörke, O.W.; Köhler-Herbetz, B.; Langer, K.; Tönges, I. Water in microcrystalline quartz of volcanic origin: Agates. *Contrib. Mineral. Petrol.* **1982**, *80*, 324–333. [CrossRef]
23. Godovikov, A.A.; Ripinen, O.I.; Motorin, S.G. *Agates*; Nedra: Moscow, Russia, 1987; p. 368. (In Russian)
24. Goncharov, V.I.; Gorodinsky, M.E.; Pavlov, G.F.; Savva, N.E.; Fadeev, A.P.; Vartanov, V.V.; Gunchenko, E.V. *Chalcedony of North-East. of the USSR*; Science: Moscow, Russia, 1987; p. 192. (In Russian)
25. Landmesser, M. Mobility by metastability: Silica transport and accumulation at low temperatures. *Chem. Erde* **1995**, *55*, 149–176.
26. Landmesser, M. Mobility by metastability in sedimentary and agate petrology: Applications. *Chem. Erde* **1998**, *58*, 1–22.
27. Moxon, T. *Agate Microstructure and Possible Origin*; Terra Publications: Doncaster, UK, 1996; p. 106.
28. Götze, J. Agate-fascination between legend and science. In *Agates III*; Zenz, J., Ed.; Bode-Verlag: Salzhemmendorf, German, 2011; pp. 19–133.
29. Sorby, H.C. On the microscopical structure of Calcareous Grit of the Yorkshire coast. *Quart. J. Geol. Soc. Lond.* **1851**, *7*, 1–6. [CrossRef]
30. Eberspächer, S.; Lange, J.-M.; Zaun, J.; Kehrer, C.; Heide, G. The Historical Collection of Rock Thin Sections at the Technische Universität Bergakademie Freiberg and Evaluation of Digitization Methods. 2015. Available online: https://www.researchgate.net/publication/273118591e (accessed on 1 September 2020).
31. Heinz, H. Die Entstehung Die Entstehung der Achate, ihre Verwitterung und ihre künstliche Färbung. *Chem. Erde* **1930**, *4*, 501–525. (In German)
32. Pabian, R.K.; Zarins, A. *Banded Agates—Origins and Inclusions. Educational Circular. No. 12*; University of Bebraska: Lincoln, NE, USA, 1994; p. 32.

33. Belousov, B.P. A periodic reaction and its mechanism. In *Sbornik Referatov po Radiatsioanoi Medicine for 1958 Year*; Medgiz: Moscow, Russia, 1959; p. 145. (In Russian)

34. Zhabotinsky, A.M. Periodic liquid phase reactions. *Proc. Acad. Sci. USSR* **1964**, *157*, 392–395. (In Russian)

35. Fallick, A.E.; Jocelyn, J.; Donelly, T.; Guy, M.; Behan, C. Origin of agates in the volcanic rocks of Scotland. *Nature* **1985**, *313*, 672–674. [CrossRef]

36. Harris, C. Oxygen-isotope zonation of agates from Karoo volcanic of the Skeleton Coast, Namibia. *Am. Mineral.* **1989**, *74*, 476–481.

37. Saunders, J.A. Oxygen-isotope zonation of agates from Karoo volcanic of the Skeleton Coast, Namibia: Discussion. *Am. Mineral.* **1990**, *75*, 1205–1206.

38. Götze, J.; Plötze, M.; Tichomirova, M.; Fuchs, H. Aluminium in quartz as an indicator of the temperature of formation of agate. *Miner. Mag.* **2001**, *65*, 407–413. [CrossRef]

39. Yamagishi, H.; Nakashima, S.; Ito, Y. High temperature infrared of hydrous spectra microcrystalline quartz. *Phys. Chem. Miner.* **1997**, *24*, 66–74. [CrossRef]

40. Moxon, T. A re-examination of water in agate and its bearing on the agate genesis enigma. *Miner. Mag.* **2017**, *81*, 1223–1244. [CrossRef]

41. Flörke, O.W.; Jones, J.B.; Schmincke, H.U. A new microcrystalline silica from Gran Canaria. *Zeitschrift für Kristallographie-Crystalline Materials* **1976**, *143*, 156–165. [CrossRef]

42. Heaney, P.J.; Post, J.E. The Widespread Distribution of a Noel Silica Polymorph in Microcrystalline Quartz Varieties. *Science* **1992**, *255*, 441–443. [CrossRef]

43. Zhang, M.; Moxon, T. Infrared absorption spectroscopy of SiO_2-moganite. *Am. Mineral.* **2014**, *99*, 671–680. [CrossRef]

44. Götze, J.; Nasdala, L.; Kleeberg, R.; Wenzel, M. Occurrence and distribution of "moganite" in agate/chalcedony: A combined micro-Raman, Rietveld, and cathodoluminescence study. *Contrib. Mineral. Petrol.* **1998**, *133*, 96–105. [CrossRef]

45. Petrovic, I.; Heaney, P.J.; Navrotsky, A. Thermochemistry of the new silica polymorph moganite. *Phys. Chem. Miner.* **1996**, *23*, 119–126. [CrossRef]

46. Krauskopf, K. Dissolution and precipitation of silica at low temperatures. *Geochim. Cosmochim. Acta* **1956**, *10*, 1–26. [CrossRef]

47. Patterson, S.H.; Roberson, C.E. Weathered basalt in the Eastern part of Kauai, Hawaii. *Prof. Paper 424 C 219 US Geol. Surv.* **1961**, *424*, 195–198.

48. Fleischer, R.C.; McIntosh, C.E.; Tarr, C.L. Evolution on a volcanic conveyor belt: Using phylogeographic reconstruction and K-Ar based ages of the Hawaiian Islands to estimate molecular evolutionary rates. *Mol. Ecol.* **1998**, *7*, 533–545. [CrossRef]

49. Götze, J.; Tichomirow, M.; Fuchs, H.; Pilot, J.; Sharp, Z.D. Chemistry of agates: A trace element and stable isotope study. *Chem. Geol.* **2001**, *175*, 523–541. [CrossRef]

50. Moxon, T.; Carpenter, M.A. Crystallite growth kinetics in nanocrystalline quartz (agate and chalcedony). *Miner. Mag.* **2009**, *73*, 551–568. [CrossRef]

51. Jones, H.P.; Handreck, K.A. Effects of iron and aluminium oxides on silica in solution in soils. *Nature* **1963**, *198*, 852–853. [CrossRef]

52. Moxon, T. The co-precipitation of Fe^{3+} and SiO_2 and its role in agate genesis. *Neues Jahrb. Fur Mineral. Mon.* **1996**, *1*, 21–36.

53. File, P.D. *JCPDS International Centre for Diffraction Data*; ICDD: Newtown Square, PA, USA, 1998.

54. Neymark, L.A.; Amelin, Y.; Paces, J.B.; Peterman, Z.E. U-Pb ages of secondary silica at Yucca Mountain, Nevada: Implications for the paleohydrology of the unsaturated zone. *Appl. Geochem.* **2002**, *17*, 709–734. [CrossRef]

55. Moxon, T.; Nelson, D.R.; Zhang, M. Agate recrystallisation: Evidence found in the Archaen and Proterozoic host rocks, Western Australia. *Aust. J. Earth Sci.* **2006**, *53*, 235–248. [CrossRef]

56. Moxon, T.; Reed, S.J.B.; Zhang, M. Metamorphic effects on agate found near the Shap granite, Cumbria: As demonstrated by petrography, X-ray diffraction spectroscopic methods. *Miner. Mag.* **2007**, *71*, 461–476. [CrossRef]

57. Moxon, T.; Reed, S.J.B. Agate and chalcedony from igneous and sedimentary hosts aged from 13 to 3480 Ma: A cathodoluminescence study. *Miner. Mag.* **2006**, *70*, 485–498. [CrossRef]

58. Walger, E.; Matthess, G.; von Seckendorf, V.; Liebau, F. The formation of agate structure for silica transport, agate layer accretion, and flow atterns and models flow regimes in infiltration canals. *N. Jb. für Geologie und Paläontologie Abh.* **2009**, *186*, 113–152.

59. Smith, J. *Semi-Precious Stones of Carrick*; Kilwining: Spean Bridge, UK, 1910; p. 84.

60. Wang, Y.; Merino, E. Self-organisational origin of agates: Banding fiber twisting, composition and dynamic crystallization model. *Geochim. Cosmochim. Acta* **1990**, *54*, 1627–1638. [CrossRef]

61. Dumeńska-Slowik, M.; Natkaniec-Nowak, L.; Kotarba, M.J.; Sikorska, M.; Rzymełka, J.A.; Łoboda, A.; Gaweł, A. Mineralogical and geochemical characterization of the "bituminous" agates from Nowy Kościol Lower Silesia. *J. Mineral. Geochem.* **2008**, *184*, 255–268.

62. Götze, J.; Möckel, R.; Kempe, U.; Kapitonov, I.; Vennemann, T. Characteristics and origin of agates in sedimentary rocks from the Dryhead area, Montana, USA. *Miner. Mag.* **2009**, *73*, 673–690. [CrossRef]

63. Commin-Fischer, A.; Berger, G.; Polve, M.; Dubois, M.; Sardini, P.; Beaufort, D.; Formoso, M. Petrography and chemistry of SiO_2 filling phases in amethyst geodes from Sierra Geral Formation deposit, Rio Grande do Sul, Brazil. *J. Am. Earth Sci.* **2010**, *29*, 751–760. [CrossRef]

64. Moxon, T.; Petrone, C.M.; Reed, S.J.B. Characterisation and genesis of horizontal banding in Brazilian agate: An X-ray diffraction, thermogravimetric and microprobe study. *Miner. Mag.* **2013**, *77*, 227–248. [CrossRef]

65. Lange, P.; Blankenburg, H.-J.; Schron, W. Rasterelectronmikroscopische Untersuchgen an Vulkanachaten. *Z. Geol. Wiss.* **1984**, *12*, 669–683. (In German)

66. Holzhey, G. Mikrokristalline SiO_2 Minerallisation in rhyolithischen. Rotliegendvulkaniten des Thüringer Waldes (Deutschland) und ihre Genese. *Chem. Erde* **1999**, *59*, 183–205. (In German)

67. Fukuda, J.; Yokoyama, T.; Kirino, Y. Characterization of the states and diffusivity of intergranular water in a chalcedonic quartz by high temperature situ infrared spectroscopy. *Miner. Mag.* **2009**, *73*, 825–835. [CrossRef]

68. Götze, J.; Möckel, R.; Vennemann, T.; Muller, A. Origin and geochemistry of agates in Permian volcanic rocks of the Sub-Erzgebirge basin, Saxony (Germany). *Chem. Geol.* **2016**, *428*, 77–91. [CrossRef]

69. Moxon, T. Agate: A study of ageing. *Eur. J. Mineral.* **2002**, *14*, 1109–1118. [CrossRef]

70. Barsanov, G.P.; Yakovleva, M.E. Mineralogy, macro- and micromorphological features of agates. *New Data Miner.* **1982**, *30*, 3–26. (In Russian)

71. Silaev, V.I.; Shanina, S.N.; Ivanovskii, V.S. Inclusions of oil gases in agate-type secretions: Implications for forecast of the oil- and gas-bearing potential of the Polar Urals. *Dokl. Earth Sci.* **2002**, *383*, 246–252. (In Russian)

72. Whelan, J.B.; Paces, J.B.; Peterman, Z.E. Physical and stable-isotope evidence for the formation of secondary calcite in the unsaturated zone, Yucca Mountain, Nevada. *Appl. Geochem.* **2002**, *17*, 735–750. [CrossRef]

73. Harder, H.; Flehmig, W. Quarzsynthese bei tiefen temperature. *Geochim. Cosmochim. Acta* **1970**, *34*, 295–305. [CrossRef]

74. Mackenzie, F.T.; Gees, R. Quartz synthesis at earth surface conditions. *Science* **1971**, *173*, 533–535. [CrossRef]

75. Heaney, P.J. A proposed mechanism for the growth of chalcedony. *Contrib. Mineral. Petrol.* **1993**, *115*, 66–74. [CrossRef]

76. Naboko, S.L.; Silnichenko. The formation of silica gel on the solfataras of the Golovin volcano on Kunashir Island. *Geochemistry* **1957**, *3*, 253–256. (In Russian)

77. Meixiang, Z.; Wei, T. Surface hydrothermal minerals and their distribution in the Tengchon geothermal area, China. *Geothermics* **1987**, *16*, 181–195. [CrossRef]

78. Marcoux, E.; Le Berre, P.; Cocherie, A. The Meillers Autunian hydrothermal chalcedony: First evidence of a 295 Ma auriferous epithermal sinter in the French Massif Central. *Ore Geol. Rev.* **2004**, *25*, 69–87. [CrossRef]

79. Knauth, L.P. Petrogenesis of chert. *Rev. Mineral.* **1994**, *29*, 233–258.

80. Rodgers, K.A.; Browne, P.R.L.; Buddle, T.F.; Cook, K.L.; Greatrez, R.A.; Hampton, W.A.; Herdianita, N.R.; Holland, G.R.; Lynne, B.Y.; Martin, R.; et al. Silica phases in sinters and residues from geothermal fields of New Zealand. *Earth Sci. Rev.* **2004**, *66*, 1–61. [CrossRef]

81. Herdianita, N.R.; Browne, P.R.L.; Rodgers, K.A.; Campbell, K.A. Mineralogical and textural changes accompanying ageing of silica sinter. *Miner. Depos.* **2000**, *35*, 48–62. [CrossRef]

82. Lynne, B.Y.; Campbell, K.A.; Perry, R.S.; Browne, P.R.L.; Moore, J.N. Acceleration of sinter diagenesis in an active fumarole, Taupo volcanic zone, New Zealand. *Geology* **2006**, *34*, 749–752. [CrossRef]

83. Rodgers, K.A.; Cressey, G. The occurrence, detection and significance of moganite (SiO_2) among some silica sinters. *Miner. Mag.* **2001**, *65*, 157–167. [CrossRef]

such as moganite, opal C, and coloring agents including hematite, goethite, calcite, barite, zeolite, and many others [4,11]. That agates could originate either from the direct precipitation of silica minerals from hydrothermal fluids or the deposition of an amorphous silica gel that subsequently crystallizes or matures by diagenetic processes is still a topic of controversy [11].

The first note about agate occurrences in the vicinity of the Medina oasis (Al Madinah) in Saudi Arabia was mentioned about 3000 BC [12]. Since then, new findings about agate deposits and smaller occurrences were systematically reported [12,13]. Currently, the largest agate deposits in the world are found in Uruguay and Brazil (Rio Grande do Sul province). Less significant occurrences are found in nearly each country of the world [14].

Agates from Morocco, discovered in the 1940s, are considered quite a novelty in the stone market over the few last decades [15–17]. They are found within Triassic basaltic rocks in many places in the High Atlas Mts. (e.g., near Sidi Rahal, Asni, Agouim-Al Hama, and Tizi-n-Tichka) as well as in the Middle Atlas Mts. (e.g., Kerrouchen and Ahouli region). Agate mineralization is pervasively accompanied by other products of hydrothermal activity in this area such as ore deposits of manganese, iron, copper, gold, and lead-zinc (e.g., [18–23]).

Moroccan agates have been systematically investigated by authors of this work for over 12 years. Up to now, reports on agates from Sidi Rahal and Kerrouchen are published [24,25]. In this contribution, the next two occurrences of agates in the Atlas Mts (i.e., Asni and Agouim), are investigated using microscopic methods supported by Raman micro-spectroscopy. The detailed microtextural description, along with the distribution of silica polymorphs and solid inclusions, were provided to put constraints on formation conditions of agates from this region and determine the possible sources of silica-bearing fluids. Moreover, the aim of this report is to highlight both the similarities and differences between agates from these four localities in Morocco to broaden the current knowledge of agate mineralization within the basaltic rocks of the Atlas Mts.

2. Geological Settings and Localizations of Agates Occurrences in Morocco

The Atlas Mountain range is located in the northern part of Africa. It has a length of more than 2000 km, from the Atlantic coast in Morocco up to Kabiska Bay on the coast of the Mediterranean Sea. The range is the continuation of the Alpine orogenic system. The Atlas unit was folded, deformed, dislocated, and upthrusted at the beginning of Paleogene. Later on, it systematically changed during other tectonic activity up to the Pleistocene age [26]. As a result, two main dislocating and fold zones can be distinguished: the inner part with Middle, Upper, Sahara, and Antiatlas [27] and other part with Tellian Atlas and Rif Mountains [28]. These two zones are divided by Morocco Meset (200–1500 m above sea level (m. a.s.l)) and Shott zone (750–1000 m. a.s.l.) and they comprise rocks of various mineral compositions and age.

The High Atlas is built of Paleozoic rocks, mainly quartzites, limestones, and volcanic rocks connected with the Hercynian orogenesis. The upper parts of the High Atlas Mts. are covered by thick series of folded and unfolded Jurassic and Cretaceous limestones as well as lower Paleogene series of limestone-sandstones. Agate occurrences are known from many localities in the High Atlas, as well as from the tectonic forelands of this unit (Figure 1). The richest deposits of agates are in Tizi-n-Tichka-Agouim-Al Hama, Sidi Rahal, and Asni regions, but some new deposits with similar characteristics are found in other parts of the volcanic suite of High Atlas [15].

Sidi Rahal is a small village about 15 km east of Aït Ourir, which became known from a quarry of basalts with agate mineralization. Triassic basaltoides have a thickness up to 3.0 km in that region [15]. This volcanic outcropping is half-moon shaped with a size of 110 × 70 km. Basaltoides frequently contain spherical or almond-shaped geodes filled with SiO_2 group minerals (agates, amethyst, etc.). Similar basaltic bodies are found in other part of the High Atlas Mts. The most common are in the vicinity of Agouim, where they build a mountain range parallel to the main Atlas range or occur along the road to Ouarzazate. Their thickness is up to few hundred meters, and their length is more than 70 km. Agates occur in practically all the basaltic suites in the area and few names are used to describe the localities (Agouim-Al Hama, Tizi-n-Tichka, and Agouim-Sour). Our studied samples were collected in the basaltoids mountain range on the left side of the road connecting Agouim and Lake Ifni close to village Sour. Basaltic rocks outcrop are smaller in the Asni area, where they cover the area up to a few km long and are about 300 m thick. Generally, at least four different basaltic suites seem to be recognized in the High Atlas Mts. One of the most interesting regions for agates is the Agouim–Tizi-n-Tichka area, where agates are found at many places together with carnelian, jasper, and quartz crystals.

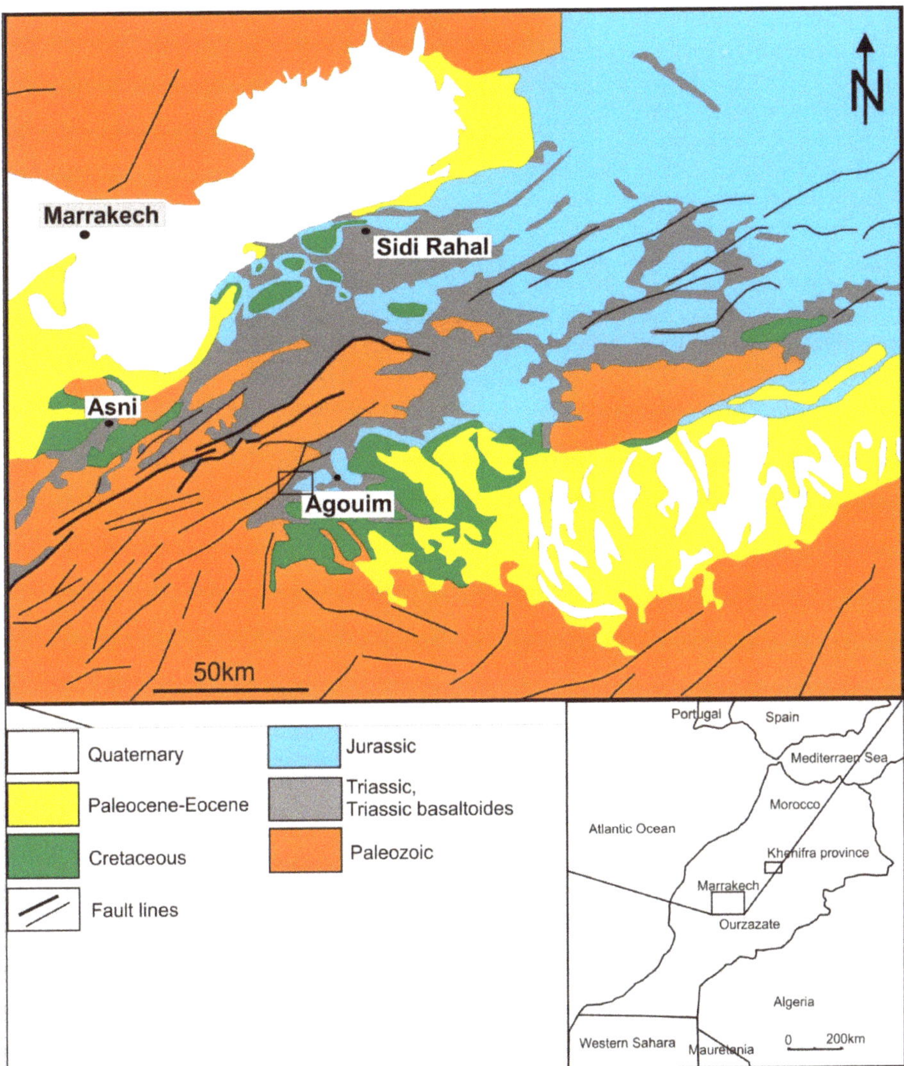

Figure 1. Simplified geological map of the Western Atlas area focusing on the studied agate occurrences (rectangles mark areas of study in Agouim and well as Kerrouchen in Khenifra).

The agate deposit in the Kerrouchen–Ahouli area was found at the beginning of this century [15]. That area is situated in the Khénifra province (Central Morocco), and borders the Middle Atlas on the south, Rif Mountains on the north, and Wadi Muluja on the east. Agate samples are collected there in the Triassic or lower Jurassic basaltoids or their weathered cover. The outcrops of the basaltic rocks cover an area more than 10×20 km in Kerrouchen and a much larger area in Ahouli.

3. Material and Methods

The samples of Moroccan agates from Asni and Agouim were examined using the Motic SNZ-168 stereoscopic microscope with 7.5–50× magnification (Xiamen, China) coupled with a digital camera. The preliminary quality assessment of the material was made, and specific objects were selected for further investigations.

The selected samples were observed in transmitted light using a polarizing microscope (Olympus BX 51 and BA310POL, with a magnification range from 40× to 400×) (Olympus Corporation, Shinjuku Monolith, Tokyo, Japan). The identification of ore minerals under reflected light was made using a polarizing microscope Nicon Eclipse. The microscope was coupled with a DP 12 camera used for photographic documentation.

The backscattered electron observations were performed on polished sections using a FEI Quanta 200 electron microscope equipped with energy-dispersive spectroscopy (FEI Company, Fremont, CA, USA). The system was operated in low-vacuum mode, at 20 kV accelerating voltage and 50 µA current. The working distance was 10 mm. The semi-quantitative analysis of target elements were supported by ZAF correction (Z- atomic number correction, A- absorption correction, F- fluorescence absorption).

The Raman spectra of silica phases and abundant solid inclusions were recorded using a Thermo Scientific DXR Raman microscope (Thermo Scientific, Waltham, MA, USA), equipped with 100–10× magnification objectives, operating in a confocal mode and working in the backscatter geometry. The polished sample pieces were excited with a 532 nm high-power laser. The laser focus diameter was 1–2 µm. The Raman spectra of bituminous matter and Fe compounds were acquired using lower laser power (1–2 mW) in order to avoid heating effects. The spectra were corrected for background by the sextic polynomial method using Omnic software attached to the Thermo Scientific DXR Raman microscope. The identification of mineral phases was supported by CrystalSleuth software (rruff project).

4. Results

4.1. Field and Macroscopic Observations

The agates from Asni and Agouim usually form small specimens, ranging in size from 2–3 cm to a maximum of 15 cm in diameter (Figure 2). During field work, it was possible to find specimens up to 50 cm in size, but they are very rare in these regions. Most of the agates are of lithophile-type rarely occur as vein types. In general, they exhibit monocentric or polycentric banding, with alternate white and grey zones varying in sizes of the individual layers. Other types, rich in red and brown incrustation are quite rare and reach a maximum of 10% of all agates found in these localities. These specimens are mainly found in Asni, where agates were collected in the vicinity of hematite veins crosscutting basaltic rocks.

Agates from Asni are mainly of the vein type (Figure 2A,B). Each has a different shape, size, color, and pattern. As for dominant color, they seem to be similar to Sidi Rahal agates [24]. The agate nodules from Asni are mainly grey-white or red-grey with a characteristic red-orange or pink external rim. In the crustal part of an agate's geode, white-grey chalcedony coexists with colorless quartz. When cut and polished, they reveal various landscape and pseudostalactitic patterns. However, the majority of them are monocentric, with an aperture left in the center or filled with the youngest generation of silica polymorph.

Agates of Agouim, like those from Kerrouchen area [25], are characterized by a wealth of morphological types and structural/textural diversity. They mainly represent monocentric spherical or oval agates, with an aperture filled with idiomorphic rose-orange quartz crystals, or with a central zone stratified horizontally (Figure 2C–F). Many polycentric agates have irregular shapes and the silica matrix alongside coloring oxides, create patterns sometimes similar to stalactite or pseudostalactite, in shades of grey, orange, pink, and red. It is also not uncommon for agates of the vein-type to have reddish-brown colors, which predominate at the outer zones, while lighter, greyish-blue colors dominate in the inner zones, giving an impression of brittle filling.

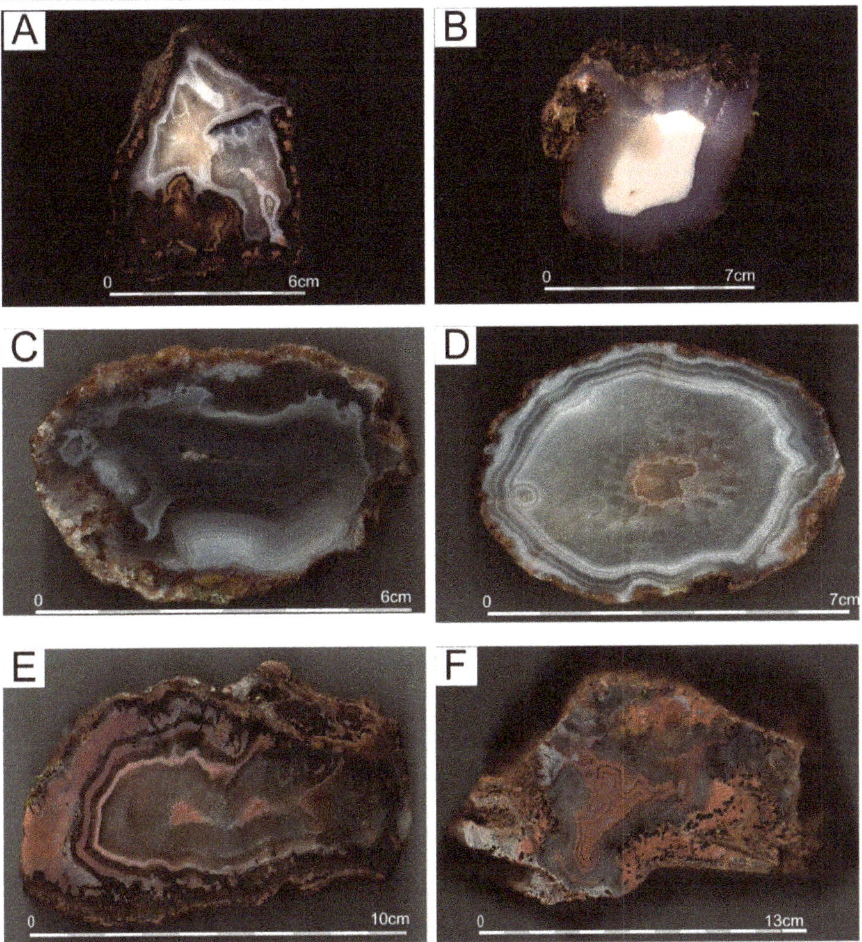

Figure 2. Macrophotographs of examined Moroccan agates: (**A,B**) Agates from Asni showing either polycentric (A) or monocentric fabrics (B). Note the presence of black and red-colored pigments distributed along the external margin of the geode made of polycentric agate. (**C,D**) Monocentric agates from Aguoim marked by wall-lining arrangement of alternating dark and light-colored chalcedonic bands (C). Note the occurrence of a well-developed quartz-rich center (D), as well as abundant hematite pigments distributed along particular chalcedonic layers (**E,F**).

4.2. Microscopic Characteristics in Transmitted and Reflected Light

The silica matrix of agates from Asni and Agouim do not present remarkable differences in terms of microtextural features and an abundance of SiO_2-rich phases. Agates from both localities consist predominantly of various cryptocrystalline silica polymorphs including alternating length-fast (normal chalcedony with c-axis situated perpendicular to the fibers) and length-slow/quartzine (c-axis parallel to the fibers) chalcedony, as well as zebraic chalcedony (length-fast with a helical twisting of fibers along c-axis). These silica phases occur in the form of rounded, bundle-like or polygonal-shaped spherulites, but also appear as fibrous radial aggregates radiating towards the center of the geode (Figure 3A–C). Chalcedonic spherules (especially zebraic chalcedony) are sometimes associated with so-called jigsaw puzzle (mosaic) quartz characterized by sinusoidal, or interpenetrating grain boundaries (Figure 3C) [29,30]. Some of the agates host prismatic quartz crystals, occupying the internal part of the nodule and amounting up to 1 mm in size. Under crossed polarizers, these crystals locally reveal divergent extinction characteristics resulting in a feathery (splitery) or flamboyant appearance, developed both within the single crystals and on the margins of prismatic crystals with a euhedral core (Figure 3D) [30]. Additionally, subparallel wall-lining quartz crystals are locally marked by the presence of μm-sized growth lines, so-called Bambauer quartz [31] (Figure 3E). The growth lines are also frequently impregnated by red, minute hematite pigment. The outer parts of agate nodules comprise tabular and elongated molds filled either with microcrystalline silica associated with intergrowths of calcite and/or barite (Figure 3F,G). Such an alignment of silica phases slightly resembles pseudo-bladed microtexture reported by Dong and colleagues [30] and Moncada and co-workers [32].

Agate nodules are accompanied by secondary hydrothermal phases including celadonite, Fe-oxides/hydroxides, and calcite. Celadonite forms green or bluish-green dull clay masses as well as minute scales (Figure 3H). Fe oxides/hydroxides occur in two generations. The first one is characterized by irregular and ragged crystals randomly distributed within the silica matrix. The second one is found in the outer parts of the nodules and comprises needle-like or capillaceous crystals located perpendicular to chalcedonic-quartz bands (Figure 3H). Conversely, calcite occurs mostly as rhombohedral crystals up 2 mm in size, but some crystals seem to be overprinted by microcrystalline silica phases (Figure 3G).

In reflected light, numerous kidney shaped aggregates of goethite, clearly "rooted" in the host rock, as well as spectacular, large rosette forms of idiomorphic goethite crystals and/or lepidocrocite (higher reflectivity), were found in the silica matrix of agates. The numerous goethite pseudomorphs after quartz/barite crystals were also observed. They were surrounded by lepidocrocite-forming characteristic halo effects around the goethite pseudomorphs (Figure 4A). In agates from Asni and Agouim, the inclusions of pyrite (Figure 4B,C), as well Cu-bearing phases like bornite, chalcopyrite, and covellite were also found. Moreover, the presence of pseudomorphs filled by barite and hematite was noted (Figure 4D). All these inclusions are often accompanied by small grains of titanium oxides (rutile, anatase) and by various types of organic matter, whose presence is confirmed by the effects of fluorescent glow visible under the UV lamp.

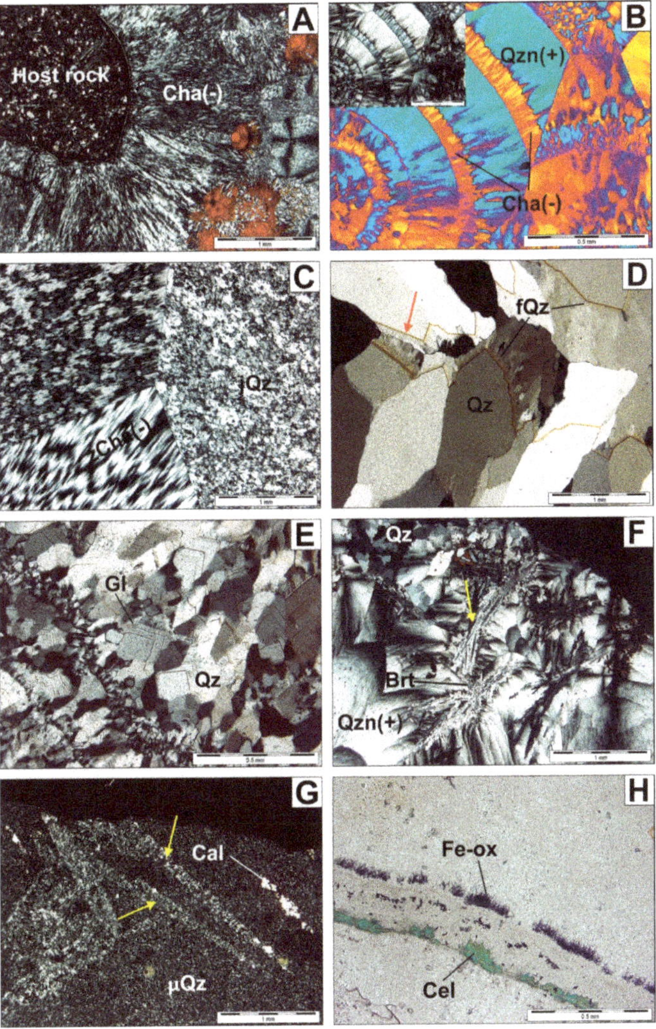

Figure 3. Photomicrographs of agate samples. (**A**) Bundle-like accumulation of chalcedonic spherules and Fe-oxides at the contact with host rocks comprising tabular plagioclases and anhedral pyroxenes—PX. (**B**) Length-fast chalcedony interlayered with length-slow chalcedony (quartzine)—NX with a gypsum plate inserted from the lower left. (**C**) Polygonal-shaped crystals of zebraic chalcedony; note the vanishing chalcedony fibrous texture at the expense of jigsaw puzzle quartz (with interpenetrating grain boundaries)—PX. (**D**) Internal part of the geode occupied by euhedral prismatic quartz crystals with feathery/spilintery appearance; red arrow points to the minute pigment of oxides developed at the edges of quartz crystals—PX. (**E**) Discrete growth lines (so-called Bambauer quartz) developed within quartz crystals. (**F,G**) Plate-like form of silica phases associated with minor barite (almost indistinguishable from quartz due to low birefringence) and calcite. (**H**) Celadonite accompanied by needle-like crystals of Fe-oxides in the external areas of the agate. Abbreviations: Cha(-): length-fast chalcedony; Qzn(+): length-slow chalcedony (quartzine); zCh(-): zebraić chalcedony; fQz: feathery quartz (with undulose extinction); Gl: growth lines; jQz: mosaic quartz (with interpenetrating grain boundaries); Qz: megaquartz (>20 μm in size); μQz: microcrystalline quartz (<20 μm in size); Cel: celadonite; Cal: calcite; Fe-ox: iron oxides/hydroxides.

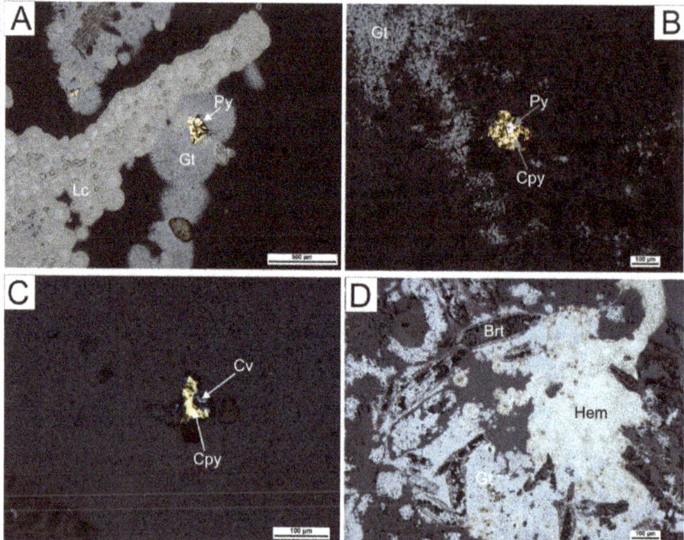

Figure 4. Images in reflected light: (**A**) kidney-shaped clusters of goethite (Gt) and lepidocrocite (Lc) and single pyrite grains (Py); (**B**) pyrite (Py) with chalcopyrite intergrowths in association with goethite; (**C**) chalcopyrite (Cpy) replaced by covellite (Cv); (**D**) large pseudomorphosis after barite filled by hematite (Hem) and goethite (Gt) with remnants of barite (Brt).

4.3. Scanning Electron Microscopy (SEM)

Under SEM observations agates show microporosity of the silica matrix and some fissures were identified. The most micropores and micro-fissures are empty, only locally filled with quartz or carbonates (calcite), forming tiny and well-shaped crystals. Within the silica matrix abundant solid inclusions were identified. Fe compounds occur as thin needles, spherical and irregular forms, which are the most abundant inclusions in Moroccan agates. The vast majority of mineral inclusions are also represented by calcite, which forms individual crystals or compact aggregates. Celadonite and chlorite occur in the outer part of agates, near the host rock. They form fibrous crystals filling the nests within the silica matrix. Barite inclusions were identified only in agates from Agouim. The copper sulfides (i.e., bornite, covellite, and chalcopyrite), as well as sylvite, form small clusters embedded within the silica matrix (Figure 5A,B).

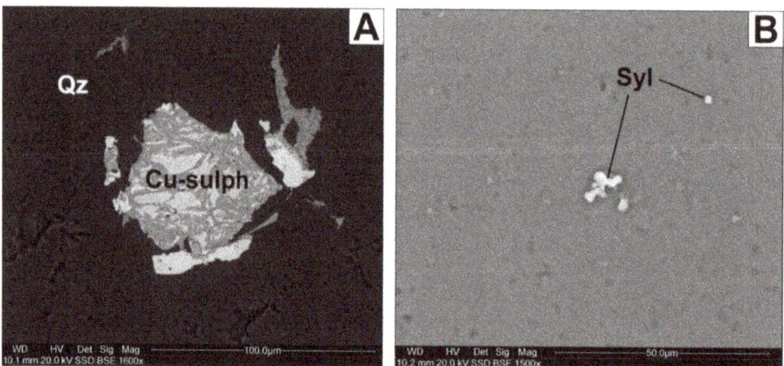

Figure 5. Back Scattered Electrons (BSE) picture of the inclusions in agates from Agouim area (**A**) Cu-compounds (Cu-sulp) in agate matrix. (**B**) Small inclusions of sylvite (Syl) in agate matrix.

4.4. Raman Micro-Spectroscopy (RS)

Raman micro-spectroscopy analyses showed that agates from Asni and Agouim were mainly built of low quartz with diagnostic bands at 463, 357, 210, and 130 cm^{-1}, as well as subordinate moganite with a marker band at 502 cm^{-1} (Table 1) [7]. For a quantitative evaluation of the moganite content in the studied samples, the method proposed by Götze et al. [7] using the relations between intensity of moganite and α-quartz peaks (502 and 463 cm^{-1}) was applied. The estimated local moganite content ranges from 0% to 56%, and 0% to 78% for Asni and Agouim agates, respectively (Table 2).

Table 1. Example of the Raman spectra for all identified minerals.

Raman Bands (cm^{-1})	Solid Phase	References
463 (s), 357 (w), 210 (m), 130 (m)	low quartz	[7]
502 (s)	moganite	[7]
1323 (s), 669 (w), 613 (m), 497 (w), 413 (s), 293 (s), 248 (m), 226 (s)	hematite	[33,34]
1316 (m), 685 (m), 553 (m), 388 (s), 301 (m), 247 (w)	goethite	[33]
610 (s), 443 (s)	rutile	[35]
708–688 (s), 569 (w)	magnesiochromite	[36]
1595 (s), 1362 (s)	bituminous matter	[37,38]
3559 (m), 3534 (m), 1620 (m), 1116 (w), 1075 (w), 959 (m), 700 (s), 547 (s), 456 (m), 399 (w), 382 (w), 277 (m)	celadonite	[39]
1088 (s), 714 (w), 285 (m), 158 (m)	calcite	[40,41]
1142 (w), 989 (s), 649 (w), 619 (w), 465 (m), 456 (w)	barite	[42,43]

Note: s, m, and w refer to strong, medium, and weak intensity bands, respectively.

Table 2. Local moganite content and integral ratio of 502 cm^{-1} and 463 cm^{-1} Raman bands in agate samples from Asni and Agouim.

Analytical Material	Raman Band Integral Ratio I$_{(502)}$/I$_{(463)}$			Local Moganite Content		
	(%)	Mean (%)	Standard Deviation	(wt.%)	Mean (%)	Standard Deviation
Asni (11)	0.0–23.2	13.7	7.2	0.0–56.0	38.3	17.4
Agouim (34)	0.0–65.2	9.1	11.8	0.0–78.0	25.5	18.5

Note: the number of analytical spots are in the parentheses.

Apart from silica phases, agates host abundant mineral and organic inclusions. Goethite and hematite were recognized as major Fe-bearing compounds. Hematite forms characteristic red colored spherical or irregularly-shaped aggregates. Its presence is proved by its Raman bands at 1323, 669, 613, 497, 413, 293, 248, and 226 cm^{-1}. The bands at 613, 413, 293, and 248 cm^{-1} are due to Fe–O symmetric bending vibrations, while the bands at 669, 497, and 226 cm^{-1} originate from Fe–O symmetric stretching vibrations (Table 1) [33,34]. The most intensive band at ca. 1323 cm^{-1}, attributed to the second-order 2LO mode with 2Eu symmetry due to defects in its lattice, is diagnostic for disordered hematite [44]. Goethite forms orange-colored aggregates irregularly scattered within the agate matrix. It is evidenced by marker bands at 1316, 685, 553, 388, 301, and 247 cm^{-1}. The most intensive band at 388 cm^{-1} is assigned to Fe–O–Fe/–OH symmetric stretching vibrations [33]. The other bands at 685 and 553 cm^{-1} area assigned to Fe–O symmetric stretching and Fe–OH asymmetric stretching vibrations, respectively. The intensive line at 301 cm^{-1} corresponds to Fe–OH symmetric bending vibrations [33]. The broad 2LO band at 1316 cm^{-1} is characteristic of various Fe-oxide and Fe-hydroxide with disordered structures [44].

Raman spectra recorded for dark brown and black inclusions hosted in agates revealed the presence of two broad bands found in the regions of 708–688 cm^{-1} and ca. 569 cm^{-1}, which seemed to be diagnostic of spinel group mineral (i.e., magnesiochromite [36]) (Table 1). The most intensive band at ca. 708–688 cm^{-1} was attributed to $\nu1(A_{1g})$ symmetric stretching vibrations of the BO_6 groups (where B = Cr, Al, Fe), whereas the weak band at 569 cm^{-1} was due to $\nu3(F_{2g})$ mode [36].

Rutile manifested its presence by two major bands found at 610 and 443 cm^{-1}, which were assigned to A1g and Eg modes, respectively [35].

Local mineral inclusions are accompanied by bituminous matter, whose presence is proved by its two (first-order) marker bands at 1595 and 1362 cm^{-1} (Figure 6, Table 1) [37,38]. Based on the shape of the Raman spectrum recorded for bituminous matter, it was found, that the most intensive and broad band at 1595 cm^{-1} in the case of low temperature samples is assigned to the disordered D2 band, whereas the band at 1362 cm^{-1} is called a disordered carbon band (D1) [45]. The relative intensities of these two bands, their width and the height of the saddle suggest that bituminous matter represents low-grade carbonaceous material with features typical of amorphous carbon [45,46].

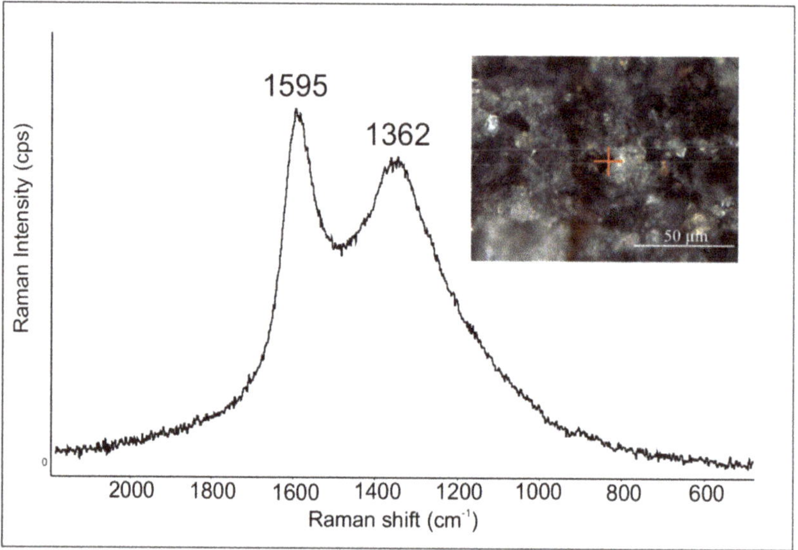

Figure 6. Microphotograph and Raman spectrum of low-grade carbonaceous material in the range of 2200–500 cm^{-1}. Analytical spot was marked by red cross.

At the outermost parts of agate nodules, green colored and irregularly-shaped aggregates are frequently found. The Raman spectrum recorded for these inclusions revealed the presence of numerous bands found at 3559, 3534, 1620, 1116, 1075, 959, 700, 547, 456, 399, 382, and 277 cm^{-1}, diagnostic of celadonite (Figure 7, Table 1) [39]. In the spectral region of 300–800 cm^{-1} the band is mainly due to Si–O–R and R–O-bending vibrations (where R is an octahedral cation). The bands at 1116, 1075, and 959 cm^{-1} are assigned to in-plane Si–O stretching vibrations [39]. The two narrow bands at 2559 and 3534 cm^{-1} and one broad band ca. 1620 cm^{-1} are due to hydroxyl stretching and bending vibrations, respectively [39].

Moreover, agates from Agouim contain inclusions of carbonates and sulfides. The presence of calcite is evidenced by its marker bands at 1088, 714, 285, and 158 cm^{-1}. The most intense $\nu1$ band at 1088 cm^{-1} corresponds to the symmetric stretching vibrations of CO_3 group (Table 1). The Raman $\nu4$ line at 714 cm^{-1} is due to asymmetric bending vibrations [cf. 40]. The lower wavenumber region of calcite at 285 and 158 cm^{-1} reflects lattice modes [41].

Barite forms tabular crystals randomly distributed in agate matrix. It is clearly marked by bands at 1142, 989, 649, 619, 465, and 456 cm^{-1} (Table 1). The dominant $\nu1$ band of barite was found at 989 cm^{-1} and arose from the symmetric stretching of SO_4. The $\nu2$ bands at ca. 465–456 cm^{-1} and $\nu4$ band at ca. 649–619 cm^{-1} were assigned to in-plane and out-of-plane bending vibrations, respectively [42,43].

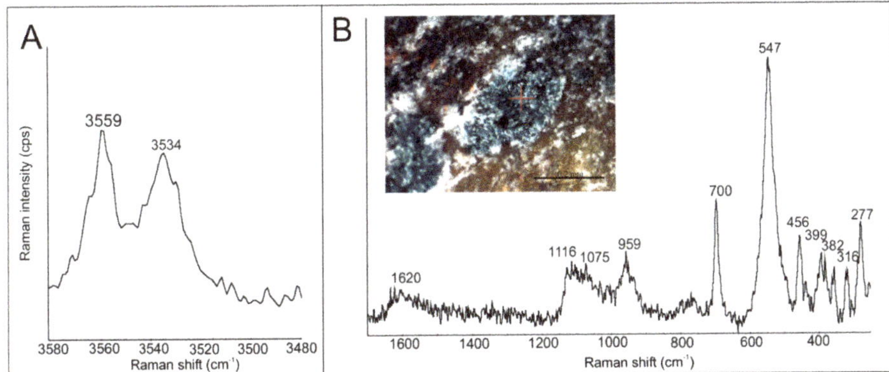

Figure 7. Microphotograph and Raman spectra of celadonite in the range of 3580–3480 cm^{-1} (**A**) and 1700–250 cm^{-1} (**B**). Analytical spot was marked by red cross.

5. Discussion

Agates from Agouim and Asni in Western Atlas Mountains are mainly built of length-fast and length-slow chalcedony with local admixture of moganite (from 0% up to 78%). The moganite contents in agates, estimated with Raman micro-spectroscopy, seem to be too high for agates hosted in Triassic rocks [8,47]. The significantly overestimated mean values of moganite result from the Raman measurements in individual points, where moganite locally predominates over low quartz. According to Götze and colleagues [7], determination of moganite content using Raman spectroscopy may be overestimated as compared with the results provided by X-ray powder diffraction. These variations may be due to the presence of moganite nanocrystals or moganite nano-range lamellae that are not large enough to contribute to the Bragg reflection, but may simultaneously induce the effect of Raman scattering [7,48]. The occurrence of alternating crystals of length-fast and length-slow chalcedony gives a clue for remarkable fluctuations of physicochemical conditions during agates formation. The former silica polymorph is indicative of neutral to acidic and/or sulphate-poor environments, whereas the latter is believed to be associated with alkaline/sulfur-rich solutions [49]. However, it should be noted that the above discrimination was later undermined by Keene [50]. The abundance of growth lines (so-called Bambauer quartz) also invokes rapid changes in pH and/or silica content in low-temperature parental fluids [10,51], that were necessary for agate formation. Additionally, the abundance of so-called, pseudobladed silica could be explained by metasomatic replacement of pre-existing plate-like barite and/or calcite by microcrystalline quartz [30,32]. The development of this fabric could be induced by boiling-related conditions [32], referring to the rapid drop of temperature and pressure that result in the formation of coexisting fluid-rich and vapor-rich phases. The presence of boiling could be further responsible for the formation of ore-related phases, such as Fe and Cu sulfides found within the silica matrix of agates. The feathery and jigsaw puzzle quartz microtextures found within agate nodules provide, in turn, strong evidence for the recrystallization of pre-existing massive chalcedony or amorphous silica [52–54].

The abundance of hematite points to the high activity of iron in silica bearing medium. Furthermore, needle-like arrangement of Fe-oxides (Figure 3F) along chalcedonic bands originated from self-purification processes, whereas irregular or oval-shaped crystals of these phases possibly represent primary impurities [2]. The presence of celadonite, found in the outer parts of agate nodules, reflects oxidizing near-surface conditions and argues for the influence of slowly-circulating K- and Mg-rich solutions during the initial stage of agate formation [55,56]. The origin of poorly ordered and low-grade solid carbonaceous matter, forming irregular accumulations within the silica matrix of agates, is probably connected with the low-hydrothermal activity of SiO$_2$-bearing fluids (i.e., meteoric waters). Kříbek et al. [57] noted that precipitation of hydrothermal carbonaceous matter occurred due to the supersaturation of C–O–H fluids (containing CH_4 or CO_2) with carbon at the deeper or middle crustal levels.

6. Conclusions

The agates from Agouim and Asni display exceptionally high diversity in microtextural features as well as accessory compounds, forming solid inclusions. They exhibit a wealth of colors, morphological features, and patterns. They are similar to agates occurring in other Moroccan localities, such as Sidi Rahal and Kerrouchen-Ahouli. All gems contain assemblages of similar solid inclusions, which are mainly represented by hematite, goethite, rutile, as well as barite, copper sulfides, and carbonates, which were probably formed during the post-magmatic processes of the hydrothermal stage and/or the tectonic processes which affected the area of High Atlas during the Alpine orogen system. Asni and Agouim are located close to the main Atlas mountain ridge as well as to the main tectonic lines, so the hydrothermal solution had a greater opportunity to alter agates in that area, compared to Kerrouchen or Sidi Rahal regions, which are located outside the main tectonic zones [18,22,23].

Agates from the Atlas Mts. are hosted by the same type of volcanic rocks (basalt), which were assembled during the Triassic period [24,25]. The main feature, which distinguishes agates from Sidi Rahal compared to agates from the other localities, is the presence of opal CT (disordered α- cristobalite, α-tridymite) which was not confirmed in agates from Agouim-Asni-Kerrouchen areas. The existence of opal CT in agates from Sidi Rahal could be attributed to the transformation of amorphous silica deposited within basaltoic rocks during the initial stage of agate formation following the reaction: opal A → opal CT → opal C → chalcedony [58]. Hence, it is assumed that agates from Sidi Rahal exhibiting the presence of remnants of primary silica, that are not fully recrystallized into chalcedony, could be the younger than gems from other regions of Morocco. However, the presence of quite elevated moganite content (av. 25–38 wt.%, Table 2) in the agates from the study area infers that their formation was closely related to the syn- and post-volcanic alterations of the host rocks. According to Moxon and Ríos [8], moganite tends to recrystallize into more stable and water-poor α-quartz after silica deposition within particular cavities. Hence, this silica polymorph is nearly absent in agates found within relatively old volcanos (i.e., of Early-Permian age; see [59]), but could still be present in relatively young host basaltoids, such as those found in the vicinity of Asni and Aguim, but also Sidi Rahal and Kerrouchen.

Author Contributions: Field work, as well as manuscript writing by J.P.; microscopic study by D.S. and D.Z.; EDS study and writing by L.N.-N.; figure editing, review and manuscript editing by T.P.; Raman spectroscopy by M.D.-S. and T.T. All authors have read and agreed to the published version of the manuscript.

Funding: This work was supported by research grant no 16.16.140.315 from the AGH University of Science and Technology.

Acknowledgments: The mineral collectors, Jacek Szczerba, Mariusz Jarzyński, and Andrzej Kuźma are greatly acknowledged for providing Moroccan agate samples for the study. We would like to thank the anonymous reviewers as well as handling editor Luke Wu for their useful and friendly comments which helped to improve the manuscript.

Conflicts of Interest: The authors declare no conflicts of interest.

References

1. Götze, J.; Plötze, M.; Tichomirowa, M.; Fuchs, H.; Pilot, J. Aluminum in quartz as an indicator of the temperature of formation of agate. *Mineral. Mag.* **2001**, *65*, 407–413. [CrossRef]
2. Götze, J.; Möckel, R.; Kempe, U.; Kapitonov, I.; Vennemann, T. Characteristics and origin of agates in sedimentary rocks from the Dryhead area, Montana, USA. *Mineral. Mag.* **2009**, *73*, 673–690. [CrossRef]
3. Götze, J. *Agate-Fascination between Legend and Science*; Agates Zenz, J., III, Ed.; Bode-Verlag: Salzhemmendorf, Germany, 2011; pp. 19–133.
4. Moxon, T.; Nelson, D.R.; Zhang, M. Agate recrystallization: Evidence from samples found in Archaean and Proterozoic host rocks, Western Australia. *Aust. J. Earth Sci.* **2006**, *53*, 235–248. [CrossRef]

5. Heaney, P.J. A proposed mechanism for the growth of chalcedony. *Contrib. Mineral. Petrol.* **1993**, *115*, 66–74. [CrossRef]

6. Graetsch, H. Structural characteristics of opaline and microcrystalline silica minerals. *Rev. Mineral. Geochem.* **1994**, *29*, 209–232.

7. Götze, J.; Nasdala, L.; Kleeberg, R.; Wenzel, M. Occurrence and distribution of "moganite" in agate/chalcedony: A combined micro-Raman, Rietveld, and cathodoluminescence study. *Contrib. Mineral. Petrol.* **1998**, *133*, 96–105. [CrossRef]

8. Moxon, T.; Ríos, S. Moganite and water content as a function of age in agate: An XRD and thermogravimetric study. *Eur. J. Mineral.* **2004**, *16*, 269–278. [CrossRef]

9. Götze, J.; Nasdala, L.; Kempe, U.; Libowitzky, E.; Rericha, A.; Vennemann, T. The origin of black colouration in onyx agate from Mali. *Miner. Mag.* **2012**, *76*, 115–127. [CrossRef]

10. Richter, S.; Götze, J.; Niemeyer, H.; Möckel, R. Mineralogical investigations of agates from Cordón de Lila, Chile. *Andean Geol.* **2015**, *42*, 386–396.

11. French, M.W.; Worden, R.H.; Lee, D.R. Electron backscatter diffraction investigation of length-fast chalcedony in agate: Implications for agate genesis and growth mechanisms. *Geofluids* **2013**, *13*, 32–44. [CrossRef]

12. Sobczak, N.; Sobczak, T. *Big Encyclopedy of Gemstones*; Polskie Wydawnictwa Naukowe Publishing House: Warsaw, Poland, 1998; p. 422. (In Polish)

13. Manecki, A. *Agates and Silicites: Genesis of Beauty-a Beauty of Genesis*; AGH—The University of Science and Technology Publishing House: Krakow, Poland, 2015; p. 90. (In Polish)

14. Żaba, J. *Ilustrated Dictionary of Rocks and Minerals*; Videograf, II: Katowice, Poland, 2003; p. 503. (In Polish)

15. Jahn, S.; Bode, R.; Lyckberg, P.; Medenbach, O.; Lierl, H.-J. *Marokko. Land der Schönen Mineralien und Fossilien*; Bode-Verlag GmbH: Salzhemmendorf, Germany, 2003; p. 535. (In Germany)

16. Zenz, J. *Achate*; Bode-Verlag GmbH: Salzhemmendorf, Germany, 2005; p. 656. (In Germany)

17. Zenz, J. *Achate-Schätze*; Bode-Verlag GmbH: Salzhemmendorf, Germany, 2009; p. 656. (In Germany)

18. Bakarat, A. Gold mineralization in the Ourika Gneiss (High Atlas of Marrakech, Morocco): Mineral paragenesis and fluid P-T-X evolution. *Arabian J. Geosci.* **2015**, *8*, 3207–3222. [CrossRef]

19. Ilmen, S.; Alansari, A.; Bajddi, A.; Ennaciri, A.; Maacha, L. Contribution à l'étude géologique du gisement à Cu, Zn, Pb et Ag-Au d'Amensif (District minier d'Amizmiz, Haut Atlas occidental, Maroc). *Int. J. Innov. Appl. Stud.* **2014**, *6*, 768–783. (In French)

20. Lafforgue, L.; Barbarand, J.; Missenard, Y.; Brigaud, B.; Saint-Bézar, B.; Yans, J.; Dekoninck, A.; Saddiqi, O. Origin of the Bouarfa Manganese Ore Deposit (Eastern High Atlas, Morocco): Insights from Petrography and Geochemistry of the Mineralization. Mineral resources in a sustainable world. In Proceedings of the 13th SGA Biennial Meeting 2015, Nancy, France, 24–27 August 2015; pp. 1949–1952.

21. Rddad, L.; Bouhlel, S. The Bou Dahar Jurassic carbonate-hosted Pb–Zn–Ba deposits (Oriental High Atlas, Morocco): Fluid-inclusion and C–O–S–Pb isotope studies. *Ore Geol. Rev.* **2016**, *72*, 1072–1087. [CrossRef]

22. Essalhi, M.; Mrani, D.; Essalhi, A.; Toummite, A.; Ali-Ammar, H. Evidence of a high quality barite in Drâa-Tafilalet region, Morocco: A nonupgraded potential. *J. Mater. Environ. Sci.* **2018**, *9*, 1366–1378.

23. Verhaert, M.; Bernard, A.; Saddiqi, O.; Dekoninck, A.; Essalhi, M.; Yans, J. Mineralogy and genesis of the polymetallic and polyphased low grade Fe–Mn–Cu ore of Jbel Rhals deposit (Eastern High Atlas, Morocco). *Minerals* **2018**, *8*, 39. [CrossRef]

24. Dumańska-Słowik, M.; Natkaniec-Nowak, L.; Wesełucha-Birczyńska, A.; Gaweł, A.; Lankosz, M.; Wróbel, P. Agates from Sidi Rahal, in the Atlas Mountains of Morocco: Gemological characteristics and proposed origin. *Gems Gemol.* **2013**, *49*, 148–159. [CrossRef]

25. Natkaniec-Nowak, L.; Dumańska-Słowik, M.; Pršek, J.; Lankosz, M.; Wróbel, P.; Gaweł, A.; Kowalczyk, J.; Kocemba, J. Agates from Kerrouchen (the Atlas Mountains, Morocco): Textural Types and Their Gemmological Characteristics. *Minerals* **2016**, *6*, 77. [CrossRef]

26. Ellero, A.; Ottria, G.; Malusà, M.G.; Ouanaimi, H. *Structural Geological Analysis of the High Atlas (Morocco): Evidences of a Transpressional Fold-Thrust Belt, Chapter 9 in Sharkov E.*; Tectonics-Recent Advances; InTech: Rijeka, Croatia, 2012; p. 323.

27. Choubert, G.; Marais, J. Geologie du Marocco. In Proceedings of the 19th International Geological Congress, Algiers, Algeria, 8–15 September 1952.

28. Laville, E. Role of the Atlas Mountains (northwest Africa) within the African-Eurasian plate-boundary zone: Comment. *Geology* **2002**, *30*, 1–95. [CrossRef]

29. Lovering, T.G. *Jasperoid in the United States*; its Characteristics, Origin, and Economic Significance (No. 710); USGS: Reston, VA, USA, 1972; p. 176.

30. Dong, G.; Morrison, G.; Jaireth, S. Quartz textures in epithermal veins, Queensland; classification, origin and implication. *Econ. Geol.* **1995**, *90*, 1841–1856. [CrossRef]

31. Bambauer, H.U.; Brunner, G.O.; Laves, F. Beobachtungen über Lamellen bau an Bergkristallen. *Z. Krist.* **1961**, *116*, 173–181. (In German) [CrossRef]

32. Moncada, D.; Mutchler, S.; Nieto, A.; Reynolds, T.J.; Rimstidt, J.D.; Bodnar, R.J. Mineral textures and fluid inclusion petrography of the epithermal Ag–Au deposits at Guanajuato, Mexico: Application to exploration. *J. Geochem. Explor.* **2012**, *114*, 20–35. [CrossRef]

33. Legodi, M.; de Waal, D. The preparation of magnetite, goethite, hematite and maghemite of pigment quality from mill scale iron waste. *Dyes Pigment.* **2007**, *74*, 161–168. [CrossRef]

34. Hanesch, M. Raman spectroscopy of iron oxides and (oxy)hydroxides at low laser power and possible applications in environmental magnetic studies. *Geophys. J. Int.* **2009**, *177*, 941–948. [CrossRef]

35. Balachandran, U.; Eror, N.G. Raman spectrum of titanium dioxide. *J. Solid State Chem.* **1982**, *42*, 276–282. [CrossRef]

36. Kharbish, S. Raman spectroscopic features of Al- Fe^{3+}- poor magnesiochromite and Fe^{2+}- Fe^{3+}- rich ferrian chromite solid solutions. *Mineral. Petrol.* **2017**, *112*, 245–256. [CrossRef]

37. Beyssac, O.; Goffe, B.; Petitet, J.-P.; Froigneux, E.; Moreau, M.; Rouzaud, J.-N. On the characterization of disordered and heterogeneous carbonaceous materials by Raman spectroscopy. *Spectrochim. Acta Part A* **2003**, *59*, 2267–2276. [CrossRef]

38. Starkey, N.A.; Franchi, I.A.; Alexander, C.M.O. A Raman spectroscopic study of organic matter in interplanetary dust particles and meteorites using multiple wavelength laser excitation. *Meteorit. Planet. Sci.* **2013**, *48*, 1800–1822. [CrossRef]

39. Ospitali, F.; Bersani, D.; Di Lonardo, G.; Lottici, P. 'Green earths': Vibrational and elemental characterization of glauconites, celadonites and historical pigments. *J. Raman Spectrosc.* **2008**, *39*, 1066–1073. [CrossRef]

40. Buzgar, N.; Apopei, A.I. The Raman study of certain carbonates. *Geol. Tomul L* **2009**, *2*, 97–112.

41. Gunasekaran, S.; Anbalagan, G.; Pandi, S. Raman and infrared spectra of carbonates of calcite structure. *J. Raman Spectrosc.* **2006**, *37*, 892–899. [CrossRef]

42. Dimova, M.; Panczer, G.; Gaft, M. Spectroscopic study of barite from the Kremikovtsi deposit (Bulgaria) with implication for its origin. *Ann. Geol. Penins. Balk.* **2006**, *67*, 101–108. [CrossRef]

43. White, S.N. Laser Raman spectroscopy as a technique for identification of seafloor hydrothermal and cold seep minerals. *Chem. Geol.* **2009**, *259*, 240–252. [CrossRef]

44. Marshall, C.P.; Marshall, A.O. Raman Hyperspectral Imaging of Microfossils: Potential Pitfalls. *Astrobiology* **2013**, *13*, 920–931. [CrossRef] [PubMed]

45. Kouketsu, Y.; Mizukami, T.; Mori, H.; Endo, S.; Aoya, M.; Hara, H.; Nakamura, D.; Wallis, S. A new approach to develop the Raman carbonaceous material geothermometer for low-grade metamorphism using peak width. *Isl. Arc* **2014**, *23*, 33–50. [CrossRef]

46. Aoya, M.; Kouketsu, Y.; Endo, S.; Shimizu, H.; Mizukami, T.; Nakamura, D.; Wallis, S. Extending the applicability of the Raman carbonaceous-material geothermometer using data from contact metamorphic rocks. *J. Metamorph. Geol.* **2010**, *28*, 895–914. [CrossRef]

47. Constantina, C.; Moxon, T. Agates from Gurasada, Southern Apuseni Mountains, Romania: An XRD and Thermogravimetric study. *Carpathian J. Earth Environ. Sci.* **2010**, *5*, 89–99.

48. Zhang, M.; Moxon, T. Infrared absorption spectroscopy of SiO_2-moganite. *Am. Mineral.* **2014**, *99*, 671–680. [CrossRef]

49. Folk, R.L.; Pittman, J.S. Length-slow chalcedony; a new testament for vanished evaporites. *J. Sediment. Res.* **1971**, *41*, 1045–1058.

50. Keene, J.B. Chalcedonic quartz and occurrence of quartzine (length-slow chalcedony) in pelagic sediments. *Sedimentology* **1983**, *30*, 449–454. [CrossRef]

51. Rykart, R. *Quarz-Monographie*; Ott Verlag: Thun, Switzerland, 1989; p. 462. (In German)

52. Fournier, R.O. The behavior of silica in hydrothermal solutions. *Rev. Econ. Geol.* **1985**, *2*, 45–60.

53. Saunders, J.A. Silica and gold textures in bonanza ores of the Sleeper Deposit, Humboldt County, Nevada; evidence for colloids and implications for epithermal ore-forming processes. *Econ. Geol.* **1994**, *89*, 628–638. [CrossRef]

54. Yilmaz, T.I.; Duschl, F.; Di Genova, D. Feathery and network-like filamentous textures as indicators for the re-crystallization of quartz from a metastable silica precursor at the Rusey Fault Zone, Cornwall, UK. *Solid Earth* **2016**, *6*, 1509–1536. [CrossRef]

55. Odin, G.S.; Desprairies, A.; Fullgar, P.D.; Bellon, H.; Decarreau, A.; Fröhlich, F.; Zelvelder, M. Nature and geological significance of celadonite. In *Green Marine Clays: Oolitic Ironstone Facies, Verdine Facies, Glaukony Facies and Celadonite-Bearing Rock Facies-A Comparative Study*; Odin, G.S., Ed.; Elsevier: Amsterdam, The Netherlands, 1988; pp. 337–392.

56. Baker, L.L.; Rember, W.C.; Sprenke, K.F.; Strawn, D.G. Celadonite in continental flood basalts of the Columbia River Basalt Group. *Am. Mineral.* **2012**, *97*, 1284–1290. [CrossRef]

57. Kříbek, B.; Sýkorová, I.; Machovič, V.; Knésl, I.; Laufek, F.; Zachariáš, J. The origin and hydrothermal mobilization of carbonaceous matter associated with Paleoproterozoic orogenic-type gold deposits of West Africa. *Precambrian Res.* **2015**, *270*, 300–317. [CrossRef]

58. Moxon, T.; Carpenter, M.A. Crystallite growth kinetics in nanocrystalline quartz (agate and chalcedony). *Mineral. Mag.* **2009**, *73*, 551–568. [CrossRef]

59. Powolny, T.; Dumańska-Słowik, M.; Sikorska-Jaworowska, M.; Wójcik-Bania, M. Agate mineralization in spilitized Permian volcanics from "Borówno" quarry (Lower Silesia, Poland) –microtextural, mineralogical, and geochemical constraints. *Ore Geol. Rev.* **2019**, *114*, 103–130. [CrossRef]

Article

Gemological Characteristics and Origin of the Zhanguohong Agate from Beipiao, Liaoning Province, China: A Combined Microscopic, X-ray Diffraction, and Raman Spectroscopic Study

Xuemei Zhang [1], Lei Ji [2] and Xuemei He [1,*]

[1] School of Gemology, China University of Geosciences, Beijing 100083, China; xmzhang@cugb.edu.cn
[2] State Key Laboratory of Geological Processes and Mineral Resources, China University of Geosciences, Beijing 100083, China; lji@cugb.edu.cn
* Correspondence: hexuemei@cugb.edu.cn; Tel.: +86-10-82321502

Received: 20 February 2020; Accepted: 28 April 2020; Published: 30 April 2020

Abstract: The Zhanguohong agate from Beipiao (Liaoning province, China), which occurs in the intermediate–felsic volcanic breccias of the Early Cretaceous Yixian Formation, generally shows massive and banded structures, with red, yellow, and/or white layers or zones. Little research has been done on its mineralogical and gemological characteristics or its genesis. In this study, we present petrographic and spectroscopic constraints on the mineral composition and micro-texture of the silica matrix, as well as the ferruginous inclusions within the agates, in order to deduce the origin of the Zhanguohong agate. According to the microscopic observations, sandwich-like interlayered micro-granular quartz, fibrous chalcedony, and jigsaw quartz bands are common in the banded agates. X-ray diffraction (XRD) and Raman spectroscopic analyses revealed that all of the samples were mainly composed of α-quartz and moganite, with minor hematite and goethite. The moganite content (17–54 wt%) of the silica matrix decreases by varying degrees from the outermost to the innermost part of the banded agates. The crystal defects and ferric iron in the microcrystalline silica grains probably contributed to the moganite crystallization. The red, yellow, and orange zones are rich in hematite, goethite, and their mixtures, respectively. The ore-forming fluids fluctuated between acidic and alkaline within a temperature range of 100–200 °C and at a sustained positive Eh. Combined with the field observations, these results suggest that the multiperiod precipitation of the agates probably resulted from the episodic volcanic activity during the Early Cretaceous lithospheric extension in eastern China.

Keywords: Zhanguohong agate; moganite; Fe compounds; micro-textures; Raman spectroscopy; Beipiao

---

1. Introduction

Agate, generally defined as banded chalcedony, is mainly composed of silica phases (i.e., α-quartz, moganite, opal-CT, opal-C, and opal-A) [1–4], with minor impurities, such as Fe compounds, sulfates, and carbonates [4–6]. The different colors and patterns of agates make each specimen unique from the others. Although agates have been found in igneous, metamorphic, and sedimentary rocks all over the world [7–10], their genesis is still not completely understood. Several studies have reported that the formation of agates is probably associated with syn- and/or post-volcanic alteration or weathering of volcanic wall rocks [6,9,11].

Agates mainly exhibit microcrystalline fibrous, granular, and jigsaw textures with different contents of moganite [1,3,12] which belongs to the monoclinic group. The structure of moganite has been described as the alternate stacking of layers of left- and right-handed α-quartz corresponding to a periodic Brazil-law twinning on the unit-cell scale [13,14]. As a metastable phase, moganite can be easily transformed into α-quartz, given sufficient time or by changing ambient conditions [2,15,16]. Rodgers and coworkers have suggested that moganite, as a part of the silica sinter maturation sequence, can be derived from the crystallization of noncrystalline and pseudocrystalline opaline silica [17,18]. Moganite has been found in different environments [19–23], including lunar meteorites [24].

Agate deposits are extensively distributed throughout China, such as in Taiwan [25], Hebei [26,27], Liaoning [28,29], Inner Mongolia [30], Heilongjiang, Sichuan, and Yunnan [31]. As one of the most famous kinds of agates, the Zhanguohong agate, which is mainly red and yellow in color and is found locally in Beipiao and Xuanhua, draws a great deal of attention from the public. Detailed studies have been done on the Zhanguohong agate in Xuanhua, Hebei province, including investigations of its microstructural characteristics, mineral composition, coloration mechanism, and ore genesis [26,27,32]. In general, the agates occurred as amygdales in the intermediate–mafic volcanic rocks of the Middle Jurassic Tiaojishan Formation, which is equivalent to the Lanqi Formation in west Liaoning [27,32]. Elaborate fieldwork was conducted to clarify the geological characteristics of the different types of agates and their volcanic wall rocks in the Fuxin district, Liaoning province [29,33]. Similar conclusions were obtained, which emphasized the key roles of the vesicles and fractures in volcanic rocks in the formation of agate deposits. Occurring in the intermediate–felsic volcanic breccias of the Early Cretaceous Yixian Formation in the Beipiao district, Liaoning province, the Zhanguohong agate has been highly neglected. Little attention has been paid to its mineralogical and gemological characteristics, let alone its genesis [28,34]. Through microscopic observations, X-ray diffraction (XRD) analysis, and Raman spectroscopic analysis, this study presents detailed information on the mineralogical and gemological characteristics of the Zhanguohong agate in the Beipiao district. Combined with field investigations, we try to provide an insight into the origin of the Zhanguohong agate.

2. Geological Setting

The agate deposit is located near Cunzhuyingzi village, Beipiao, Liaoning province, China, and it is adjacent to the northeastern part of the Jurassic Beipiao basin in the east Yanshan belt, northern North China Craton (NCC). The NEE striking Lingyuan–Beipiao–Shahe fault zones and the NNE striking Chaoyang–Yaowangmiao fault zones intersect in this area (Figure 1).

Two rock units are present in the Beipiao district: a Cretaceous sedimentary–volcanic unit and a pre-Cretaceous unit. The latter is comprised of Archean crystalline basement rocks and Jurassic sedimentary–volcanic rocks [35,36].

The Archean crystalline basement rocks are mainly composed of granulite, gneiss, and leptynite, with some supracrustal enclaves. It has been suggested that these rocks underwent amphibolite–granulite facies metamorphism at ~2.8 Ga, which was overprinted by upper greenschist facies metamorphism at ~2.5 Ga [35]. Anshan-type iron deposits occur in these rocks.

The Jurassic volcanic and terrestrial coal-bearing clastic rocks in the Beipiao basin, >1.6 km in thickness, have traditionally been divided, from the base upward, into the Xinglonggou, Beipiao, Haifanggou, Lanqi, and Tuchengzi formations [35–37]. The Yanshan movement was defined based on detailed work near Beipiao and the Western Hills of Beijing [38,39]. It is characterized by stratigraphic unconformities and intense intracontinental compression deformation, e.g., folds and thrusts, of the Middle–Late Jurassic and even the earliest Cretaceous volcanic and sedimentary strata (e.g., the Haifanggou, Lanqi, and Tuchengzi formations and their coeval strata elsewhere in the Yanshan belt) [40,41]. However, whether the Early Jurassic and earliest Middle Jurassic strata (e.g., the Xinglonggou and Beipiao formations and the other equivalent formations) in the northern NCC were formed in rift basins or in flexural basins is still controversial [42–44].

The Cretaceous sedimentary–volcanic unit includes, from bottom to top, the Early Cretaceous Yixian, Jiufotang, and Fuxin formations and the Late Cretaceous Sunjiawan Formation [35–37]. The Jehol Biota, which is defined by its characteristic *Eosestheria–Ephemeropsis–Lycoptera* assemblage, is widely distributed in this unit [45]. Unconformably overlying the Tuchengzi Formation, the Yixian Formation consists primarily of basalt in the lower part, and andesite, andesitic rhyolite, and their corresponding volcaniclastic rocks, with minor conglomerate, sandstone, and shale interlayers, in the upper part. The Zhanguohong agate deposits locally occur in these beds. The accurate age of the Yixian Formation has long been a matter of debate in early paleontological and biostratigraphic studies [45–47]. Recently, radio-isotropic age dating of these extrusive rocks, which were generally thought to be formed by the intermittent crevasse eruptions of volcanoes in rift basins, reveals a conclusive Early Cretaceous age (135–120 Ma) [48–50]. The Jiufotang and Fuxin formations are predominantly comprised of shale, siltstone, and sandstone, with subordinate conglomerate and coal-bearing interlayers, which indicates a lacustrine depositional environment. In the uppermost part of this unit, the Sunjiawan Formation is predominantly comprised of conglomerate, with minor sandstone, siltstone, and shale interlayers. The Lower Cretaceous strata were formed in rift basins due to the intense lithospheric extension and thinning in eastern China, and throughout east Asia, in the Late Mesozoic [51–57].

With prolonged lithospheric extension, the Cretaceous intermediate, felsic, and alkaline intrusive rocks were extensively emplaced to the east of our study area ([58–60] and references therein). Additionally, Permian granite and granitic diorite and Jurassic andesitic porphyrite locally occur to the west of our study area.

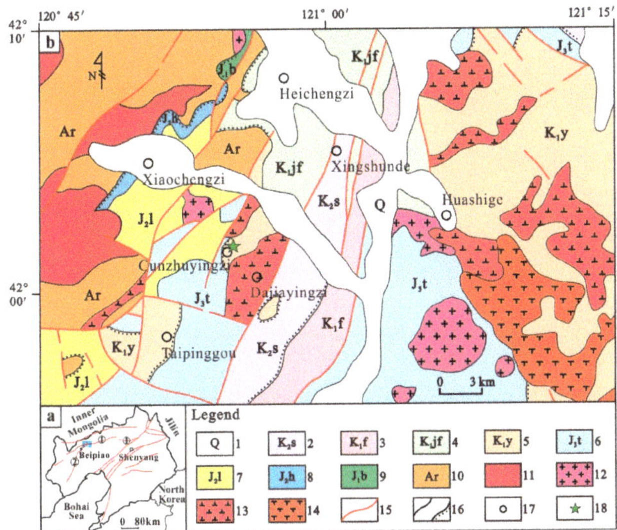

Figure 1. (**a**) Schematic tectonic map showing the distribution of the major faults in Liaoning, China (modified after [35,36]); (**b**) geological map of the Cunzhuyingzi agate deposit and its adjacent areas (modified after [35,36]). 1 Quaternary sediments; 2–5 Cretaceous sedimentary–volcanic rocks (2 Sunjiawan Fm; 3 Fuxin Fm; 4 Jiufotang Fm; 5 Yixian Fm); 6–9 Jurassic sedimentary–volcanic rocks (6 Tuchengzi Fm; 7 Lanqi Fm; 8 Haifanggou Fm; 9 Beipiao Fm); 10 Archean basement rocks; 11 Permian–Jurassic intrusive rocks; 12–14 Cretaceous intrusive rocks (12 granitoid rocks; 13 diorite porphyrite; 14 syenite porphyry); 15 fault; 16 conformity and unconformity; 17 toponym; 18 sample location. ① Lingyuan–Beipiao–Shahe fault; ② Chaoyang–Yaowangmiao fault. Fm: Formation.

An ore-bearing zone has been found in the Yixian Formation near Cunzhuyingzi village, which generally strikes E–W and dips gently (30°) to the N (Figures 1 and 2a). It extends for up to 4 km with various widths (<150 m) [34]. Interbedded by andesite to the bottom and rhyolitic andesite to the top, the intermediate–felsic volcanic breccias with tuffaceous matrix constitute the wall rock of the agates. In general, the agates occur as veins (Figure 2b,c) and as irregular shapes (Figure 2d–f) in the fractures and pores of these pyroclastic rocks. They have various colors and weights ranging from the gram to kilogram scales (Figure 2).

Figure 2. Outcrop photographs of the Zhanguohong agate. (**a**) Mining pit near Cunzhuyingzi village; (**b**,**c**) vein type agates; (**d**–**f**) irregularly-shaped agates with banded (**d**,**e**) and massive (**f**) structures.

3. Materials and Methods

Fifteen agate samples (Z–1 to Z–15) with different shapes and colors were collected and studied, especially irregularly-shaped examples (Figure 3). Featuring red, yellow, and/or white colors, the majority of the samples show banded structures (Figures 2c–e and 3), while the others exhibit massive structures (Figures 2f and 3). Macrocrystalline quartz grains are common at the center of the banded agates (Figures 2d and 3) and sometimes act as middle layers (Figures 2e and 3).

The XRD analysis was carried out by a Rigaku D/max-rA type 12 kW rotating target diffractometer with a Cu-Kα radiation (40 kV, 100 mA) and a graphite monochromator at the Beida Micro-Structure Analytical Lab of Beijing, China. Measurements were conducted in the range between 10–40° 2θ with a scan rate of 2°/min.

Raman spectra of the silica matrix and inclusions were acquired between 100 and 1500 cm^{-1} using the Horiba LabRAM HR Evolution equipped with a Peltier-cooled charged-coupled device (CCD) detector, edge filters, and a Nd-YAG laser (100 mW, 532 nm) at the School of Gemology, China University of Geosciences, Beijing, China. The confocal hole was fixed at 100 µm, and the diffraction grating was 600 grooves/mm. The laser focusing and sample viewing were operated through 100×, 50×, and 10× objective lens with a polaroid. With this configuration, the spatial resolution was <1 µm, and the spectral resolution was determined to be ~1 cm^{-1}. The spectrometer calibration was set using the 520.5 cm^{-1} band of a silicon wafer. The spectra were recorded at a laser power of 50 mW with 3 s acquisition time and twice accumulation.

Figure 3. Photograph showing the characteristics of the Zhanguohong agate samples collected from the Cunzhuyingzi deposit.

The spectra of the silica phases were background corrected by the polynomial method using the LabSpec software (Version LabSpec 6, Horiba Ltd., Kyoto, Japan) and were normalized by the strongest band (465 cm^{-1}) intensity normalization method [61]. The curve-fitting algorithm in the OriginPro sofware (Version OriginPro 2018 SR1, OriginLab, Northampton, MA, USA) was used to fit the pre-processed spectra in the 400–600 cm^{-1} range. The spectra were deconvoluted with Fourier Self-Deconvolution (Pro), and only two peaks were found at 502 cm^{-1} (moganite) and 465 cm^{-1} (quartz). The peak parameters (e.g., position, width, and area) were obtained by fitting the spectra with Lorentzian functions constrained by a constant minimum baseline.

4. Results

4.1. Microscopic Observations

The Zhanguohong agates in the Beipiao district are mainly composed of microcrystalline silica phases with different micro-textures (i.e., micro-granular quartz, fibrous chalcedony, and jigsaw quartz) and macro-quartz, with minor Fe compounds (Figures 4 and 5). Two distinct types of agates with banded and massive structures were observed.

Concentric and rhythmic bands resulting from changes in composition and/or texture are common in the banded agates, which are emphasized by the Fe compounds (Figure 4a–g). Here, the terms micro-granular quartz, fibrous chalcedony, jigsaw quartz, and macro-quartz bands are used to describe the microscopic characteristics of the banded agates. The macro-quartz (~0.05–1 mm in size) bands are preferentially located in the center of the banded agates (Figure 4a,b,f,g) and occasionally occur as interlayers among the microcrystalline silica bands (Figure 4e). Sandwich-like interlayered micro-granular quartz (~5–20 μm in size) and fibrous chalcedony bands comprise the majority of the concentric and rhythmic bands (Figure 4c–g), with a few jigsaw quartz band interlayers (Figure 4c,d,f,g). However, some banded agates are predominant with micro-granular quartz bands or fibrous chalcedony bands (Figure 4a,b). Near the boundary between the volcanic wall rock and the agate, banded and/or blocky macro-quartz and micro-granular quartz aggregates were observed (Figure 4b). The complex changes in texture (e.g., size and shape) (Figure 4a–g) suggest that the formation of the banded agates may have been a multi-stage process. Spherical and radiated structures are locally present in some of the banded agates (Figure 4f,g). The former, which is thought to be the result of nucleation and growth starting in the innermost part of a free space [6], is composed of micro-granular quartz grains (Figure 4f), and the latter of microcrystalline silica aggregates (i.e., fibrous chalcedony and jigsaw-like quartz) (Figure 4g). At least two periods of silica precipitation occurred, based on the occurrence of microcrystalline silica veins which cross-cut the concentric and rhythmic bands (Figure 4a,c). However, without concentric and rhythmic bands, massive agates are predominantly comprised of micro-granular quartz grains and/or fibrous chalcedony aggregates (Figure 4h).

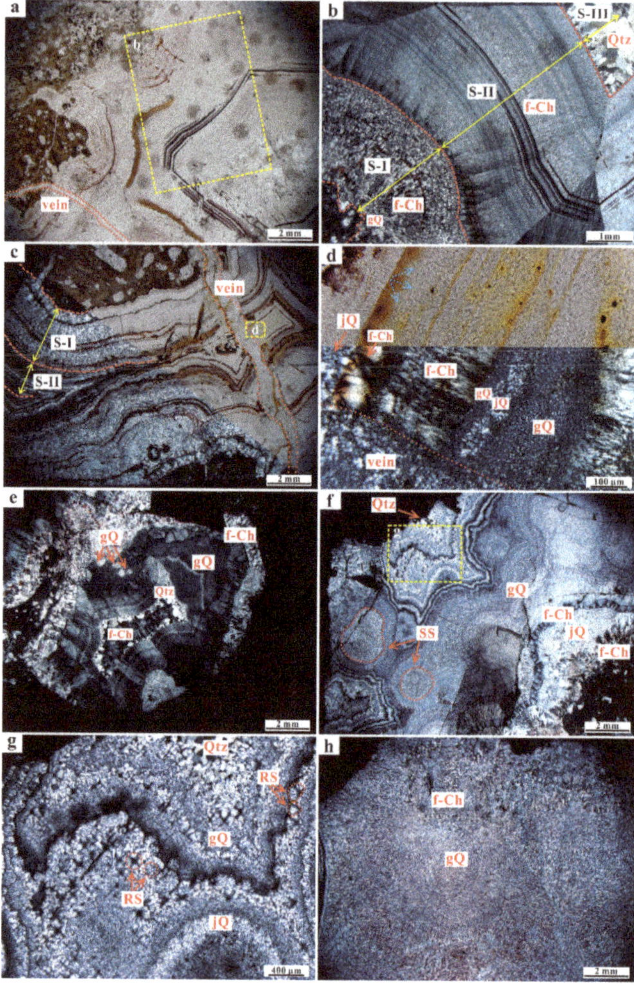

Figure 4. Microscopic characteristics of banded (**a–g**) and massive (**h**) Zhanguohong agates. (**a,b**) Multi-stage (S-I to S-III) precipitation of silica polymorphs; (**c,d**) complex changes in grain sizes and shapes of the multiple silica phases precipitated in two stages (S-I and S-II) and diffusion of the Fe compounds. Note that a typical sandwich-like banded agate comprising gQ, jQ, and f-Ch interlayers is shown in photograph d. The blue arrows indicate the diffusive direction of the Fe compounds; (**e**) a Qtz interlayer in the sandwich-like banded agate; (**f,g**) radiated and spherical structures outlined by dotted red lines; (**h**) massive agate virtually impregnated with Fe compounds. Photograph a, the upper right part of photograph c, and the upper part of photograph d are under plane-polarized light (PPL), while the other parts and photographs are under cross-polarized light (CPL). gQ: micro-granular quartz; jQ: jigsaw quartz; f-Ch: fibrous chalcedony; Qtz: macro-quartz; RS: radiated structure; SS: spherical structure; Vein: microcrystalline silica vein.

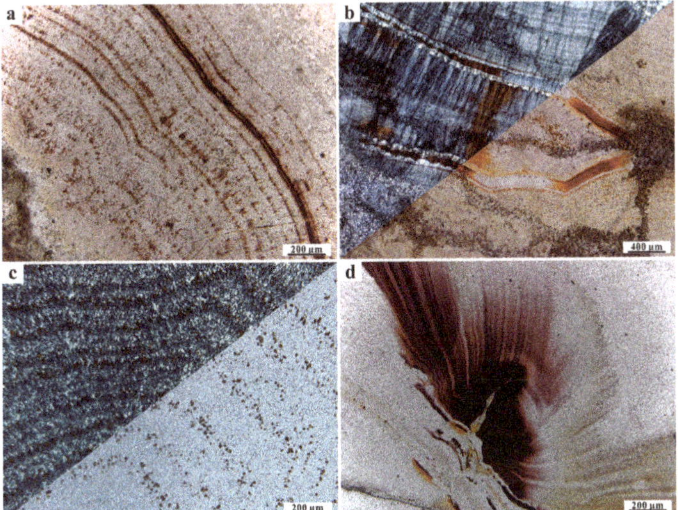

Figure 5. (**a–c**) Characteristics of the Fe-bearing zones consisting mainly of short prismatic (**a**), spherical (**b**), and granular (**c**) ferruginous inclusions, and (**d**) migration of Fe compounds. Photographs a and d, and the lower right parts of photographs b and c are under PPL, while the other parts and photographs are under CPL.

Red and yellow zones and bands in the agates are highly impregnated with Fe-oxides and Fe-hydroxides, which were identified using Raman spectroscopic analysis (as discussed later in Section 4.3.2). The Fe compounds are generally present as short prismatic (Figure 5a), spherical (Figure 5b), and granular (Figure 5c) inclusions distributed perpendicular to the growth direction of the agate. The majority of these inclusions form Fe-rich layers (Figure 5a,c), while the others are scattered randomly in the silica matrix (Figure 5b). However, in some of the red and yellow zones, the ferruginous particles are too small to identify under the optical microscope (Figure 5b). In addition, the transmission of Fe compounds is common in agates (Figures 4d and 5d).

4.2. XRD Analysis

Two samples (Z–4 and Z–9) were chosen for XRD analysis (Figure 6). Their moganite contents were significantly different. It was low in sample Z–9, which was demonstrated by the negligible moganite peaks; while it was relatively high in sample Z–4, but was still low compared with the α-quartz content. The moganite contents of samples Z–4 (11%) and Z–9 (7%) were estimated by the peak area ratio method [16]. Hematite and albite were also present as minor phases in sample Z–4. The existence of the latter phase was probably due to the entrainment of pieces of wall rocks (Figures 3 and 4a). Pseudocrystalline and amorphous silica phases were not detected. None was goethite.

4.3. Raman Spectroscopic Analysis

Compared with XRD analysis, in-situ Raman spectroscopy is a more effective method of distinguishing between different silica phases at the micron-scale [62,63]. Both the silica matrix and the ferruginous inclusions in the Zhanguohong agates were analyzed using Raman spectroscopy. The relative Raman calibration model proposed by Götze et al. (1998) [3] was used to evaluate the moganite content of the silica matrix.

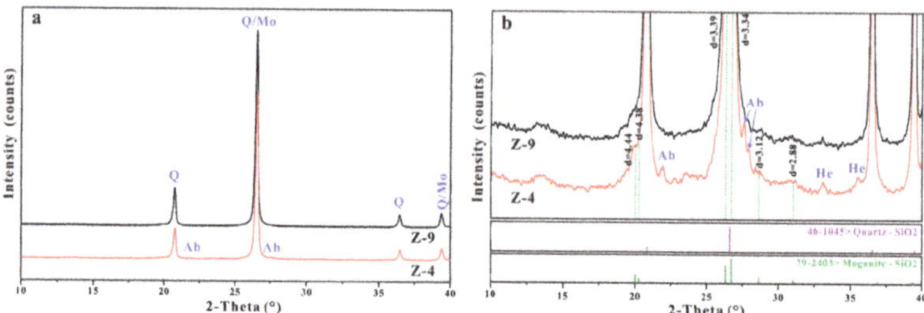

Figure 6. (**a**) XRD patterns of samples Z–4 and Z–9; (**b**) enlarged detail of part of the XRD patterns, with pdf cards of Q and Mo at the bottom [14]. Q: α-quartz; Mo: moganite; Ab: albite; He: hematite.

4.3.1. Silica Matrix

The silica matrix consists of α-quartz and moganite. However, hematite and goethite also appear in the Raman spectra, which is attributed to the presence of fine-grained Fe-bearing particles in the matrix (Figure 7). Pseudocrystalline and amorphous silica phases were not detected in any of the samples. Marker bands were assigned for the silica phases (i.e., α-quartz and moganite) and Fe compounds (i.e., hematite and goethite), and are listed in Table 1.

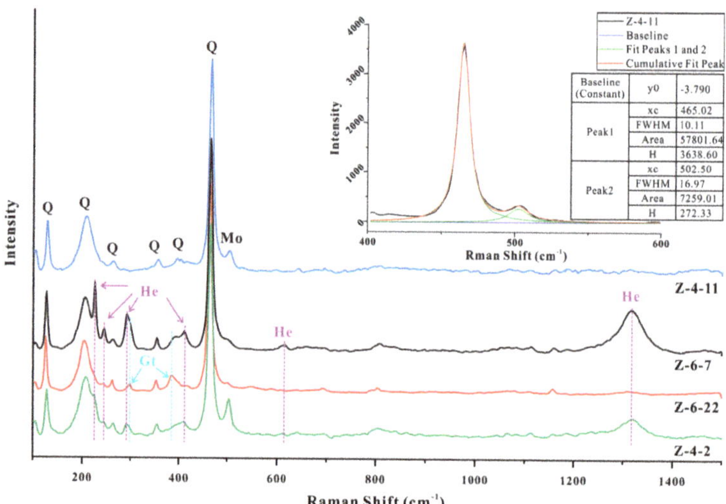

Figure 7. Representative Raman spectra of the silica matrix of the agates. The peaks of He and Gt are indicated by purple and blue dashed vertical lines, respectively. Note that the Lorentzian fit for the 465 cm^{-1} (Q) and 502 cm^{-1} (Mo) bands of the curve Z–4–11 is shown in the upper right corner. Gt: goethite. The other abbreviations are the same as in Figure 6.

Table 1. Raman frequencies of hematite, goethite, moganite, and α-quartz.

Hematite [64–66]	Goethite [65,66]	Moganite [62]	α-Quartz [62]	
ν_1 (cm^{-1})	ν_1 (cm^{-1})	ν_1 (cm^{-1})	ν_1 (cm^{-1})	Mode Symmetry
		129	128	$E_{(LO+TO)}$
		141		
227		220	206	A_1
246	243			
		265	265	$E_{(LO+TO)}$
293	299			
		317		
			355	A_1
		370		
		377		
	394	398	394	$E_{(TO)}$
411	418		401	$E_{(LO)}$
		432		
		449	450	$E_{(TO)}$
	483	463	464	A_1
498		501		
			511	$E_{(LO)}$
	550			
610				
	685	693	696	$E_{(LO+TO)}$
		792		
			796	$E_{(TO)}$
			808	$E_{(LO)}$
		833		
		950		
	993	978		
		1058		
			1069	$E_{(TO)}$
		1084	1085	A_1
		1171	1162	$E_{(LO+TO)}$
		1177		
			1230	$E_{(LO)}$
1320				

Figures 8–10 illustrate the spatial distribution of the moganite content, which ranges from 17 to 54 wt%, in the Zhanguohong agates (samples Z–2, –4, –6, –8, and –12). Detailed microscopic observations were conducted before the Raman spectroscopic analysis to ensure that the Fe-bearing zones were excluded as much as possible. The micro-granular quartz contains up to 54 wt% moganite near the wall rock (Figure 8, point 1), whereas the moganite content is much lower among the micro-granular quartz aggregates, which have a spherical structure (Figure 10b,d). In general, the majority of the moganite contents are between 20 and 45 wt% (Figures 8–10). With the consistent precipitation of silica phases, the moganite content fluctuates and decreases with different degrees in the banded agates (Figures 8 and 9). Moreover, this variation in moganite content is found in microcrystalline silica grains formed in different periods as well (Figure 10c,d), which may reflect the physicochemical differences in their parent fluids [2].

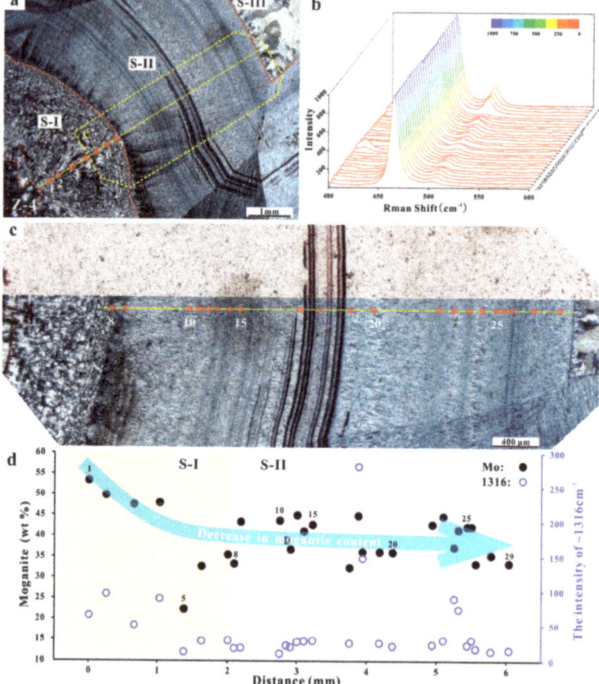

Figure 8. Spatial distribution of moganite in banded agate (sample Z–4) with multi-stage (S-I to S-III) silica precipitation revealed by Raman spectroscopic analysis. (**a,c**) Microstructures and distribution of the analysis points (red dots with white numbers); (**b**) 3D spectra between 400 and 600 cm^{-1} for each analysis point; (**d**) moganite content and intensity of the peak at ~1316 cm^{-1}. The upper part of photograph c is under PPL, while Photograph a and the lower and middle parts of photograph c are under CPL.

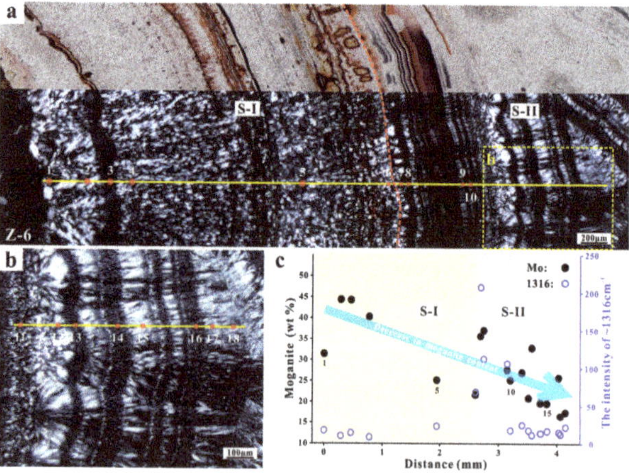

Figure 9. Spatial distribution of moganite in banded agate (sample Z–6) with two-stage (S-I and S-II) silica precipitation revealed by Raman spectroscopic analysis. (**a,b**) Microstructures and distribution of the analysis points (red dots with white numbers); (**c**) moganite content and intensity of the peak at ~1316 cm^{-1}. The upper part of photograph a is under PPL, while photograph b and the lower and middle parts of photograph a are under CPL.

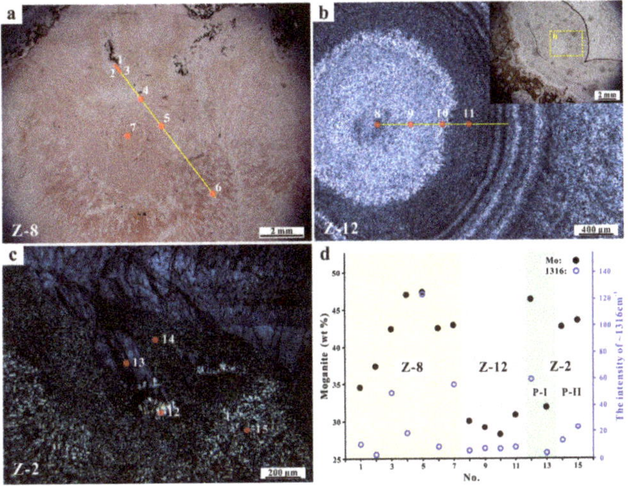

Figure 10. Spatial distribution of moganite in different samples (Z–2, –8, and –12). (**a–c**) Microstructures and distribution of the analysis points (red dots with white numbers); (**a**) Massive agate (PPL); (**b**) banded agate with spherical structure (CPL). The inset photograph shows the general microstructural characteristics under PPL; (**c**) Agate with two periods of silica precipitation (CPL); (**d**) moganite content and intensity of the peak at ~1316 cm^{-1}.

In the matrix, hematite is evidenced by marker bands at 225, 246, 411, 499, 610, and 1316 cm^{-1}, while goethite is evidenced by maker bands at 298 and 394 cm^{-1} (Figure 7). Several studies have suggested that ferric iron acts as an agent in the formation of moganite [2–4]. In order to determine whether there is a correlation between ferric iron and moganite, this study qualitatively analyzed the intensity of the band at ~1316 cm^{-1} (hematite) in the normalized spectra (Figure 8d, Figure 9c, and Figure 10d). This band was chosen due to its relatively high intensity and lack of overlap with other bands. Because of the superimposition of the ferruginous inclusions (faintly red, orange, and/or yellow fine grains under the microscope) underlying the sample surface, some of the Raman spectroscopic analysis points (i.e., the open circles in Figure 11) were omitted when we determined the correlation. After this omission, a positive correlation was obtained between the intensity of Raman peak ~1316 cm^{-1} and the moganite content of the silica matrix (Figure 11).

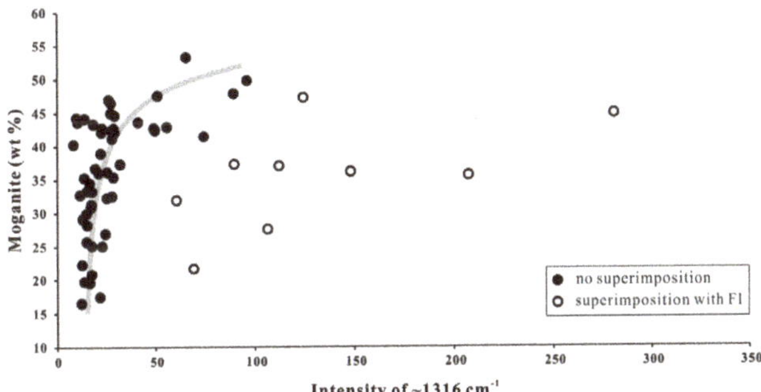

Figure 11. Moganite content (M) versus intensity of the Raman band at ~1316 cm^{-1} (I) for the silica matrix. Note that the grey line shows the positive correction between M and I. The analysis points with and without the superimposition of ferruginous inclusions (FI) are marked with open and solid circles, respectively.

4.3.2. Ferruginous Inclusions

The ferruginous inclusions are found to be mixtures of hematite, goethite, α-quartz, and moganite (Figure 12). The spectrum of hematite displays bands at 225, 246, 292, 411, 499, 610, and 1316 cm^{-1}. The bands at 225 and 499 cm^{-1} result from the A_{1g} vibrational mode (Fe–O symmetric stretching vibrations), and the band at 1316 cm^{-1} corresponds to the second-order *2LO* mode with *$2E_u$* symmetry. The other bands were assigned to the E_g vibrational mode (Fe–O symmetric bending vibrations) [65,66]. Goethite is evidenced by bands at 242, 297, 394, 416, 546, 682, and 995 cm^{-1} [65,66]. The spectrum of goethite consists of two strong bands at 394 cm^{-1} (Fe–O–Fe/–OH symmetric stretching vibrations) and 297 cm^{-1} (Fe–OH symmetric bending vibrations) [6,67]. The bands at 546 and 682 cm^{-1} originate from Fe–OH asymmetric stretching vibrations and Fe–O symmetric stretching vibrations, respectively [67]. The extra bands at 126, 205, 263, 356, 464, and 502 cm^{-1} come from the silica phases.

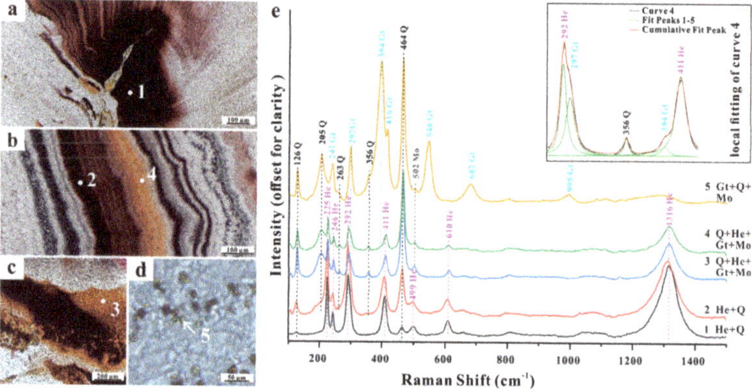

Figure 12. (**a–d**) Microphotographs (PPL) and (**e**) Raman spectra of the ferruginous particles. The peaks of He, Gt, Q and Mo are indicated by purple, blue, black, and grey dashed vertical lines, respectively. Note that the Lorentzian fit for curve 4 in the 270–430 cm^{-1} range is shown in the upper right corner. The abbreviations are the same as in Figures 6 and 7.

The dark red inclusions consist mainly of hematite with minor α-quartz (Figure 12, points 1 and 2), and the presence of goethite makes their color more orangish (Figure 12, points 3 and 4). Furthermore, the high goethite content of these inclusions leads to a yellow appearance (Figure 12, point 5).

5. Discussion

5.1. Moganite in the Zhanguohong Agate

The Raman spectroscopic analysis revealed that the moganite contents of the Zhanguohong agates in the Beipiao district vary significantly (17–54 wt%, Figures 8–10), which is in agreement with the findings that the moganite contents are significantly different for the different samples, and even for different zones of the same sample collected from other places [20,62,68,69]. In general, the moganite content decreases and fluctuates with different degrees in the banded agates during the progressive precipitation of silica (Figures 8 and 9). Moganite crystallization appears to be favorable in the initial stage (Figures 8 and 9), which indicates that moganite formation may be attributed to the high structural defects generated by rapid silica growth from a strongly supersaturated solution [70–72]. With progressive precipitation, the degree of silica supersaturation and the crystallization rate of silica gradually decrease [70]. As a consequence, the density of the defects decreases, which results in the decreasing moganite contents observed in the Zhanguohong agates (Figures 8 and 9). However, the moganite contents also exhibit significant fluctuations (Figures 8 and 9), which are likely caused by local oscillations in the degree of silica supersaturation [73,74].

The relatively low moganite content (Figure 9, point 1) at the very beginning of stage I can be ascribed to the recrystallization of the pre-existing silica phases [6] next to the wall rock where alteration is easier. Alternatively, it can be attributed to the mixing of the micro-granular quartz with the underlying macro-quartz grains. In addition, the moganite content deviates slightly from the trendline at the end of stage I (e.g., points 5–7 in Figure 8, and points 5 and 6 in Figure 9), and sometimes it is slightly lower than that in stage II (Figures 8 and 9). This low moganite content may be due to alteration or recrystallization after the initial precipitation of the microcrystalline silica grains, which results in their jigsaw-like micro-textures with irregularly shaped boundaries, particularly along the direction of silica growth (Figures 8a and 9a).

Compared with the Raman spectroscopic results, the XRD analysis indicated much lower moganite contents (~10 wt%) for the Zhanguohong agates (Figure 6, Figure 8, Figure 9, and Figure 10). This difference can be explained by the occurrence of nano-sized moganite, which makes a limited contribution to the Bragg reflection [3]. In addition, the mixture of macro-quartz grains and wall rock in XRD analysis further reduces the moganite content (Figures 2 and 6).

It has been suggested that moganite crystallization is favored by the alkaline fluid combined with high activities of ferric iron [2,4,5]. The positive correlation between the moganite content and the intensity of ~1316 cm^{-1} peak (Figure 11) provides evidence that ferric iron probably plays an important role in moganite crystallization.

The statistics for the moganite contents of microcrystalline silica samples (i.e., agate, chalcedony, chert, and flint) determined using XRD and Raman spectroscopy in the literature [2,10,16,23,30,31, 68,69,75,76] and in this study (Figure 13) reveal that the proportion of moganite increases as the volcanic wall rocks become younger, that is, from Devonian to Neogene, based on the change in the maximum moganite content for each period. This is consistent with the findings of Moxon and his coworkers [10,16,75]. It should be noted that the ages of the Phanerozoic rocks are approximate ages. For example, if the volcanic wall rocks formed in the Cretaceous (145.5–65.5 Ma), the median value (105.5 Ma) was taken as the approximate age. It has been suggested that moganite is thermodynamically less stable [15] and recrystallizes readily to α-quartz over millions of years [2] or in the presence of water (e.g., hydrothermal activity and weathering) [16,17,20]. The decay rate (~0.218 wt%/Ma) of moganite could be evaluated. We suppose that the wide variation in the moganite contents of the Zhanguohong agates is more likely due to the initial difference rather than to heterogeneous recrystallization or alteration after its precipitation.

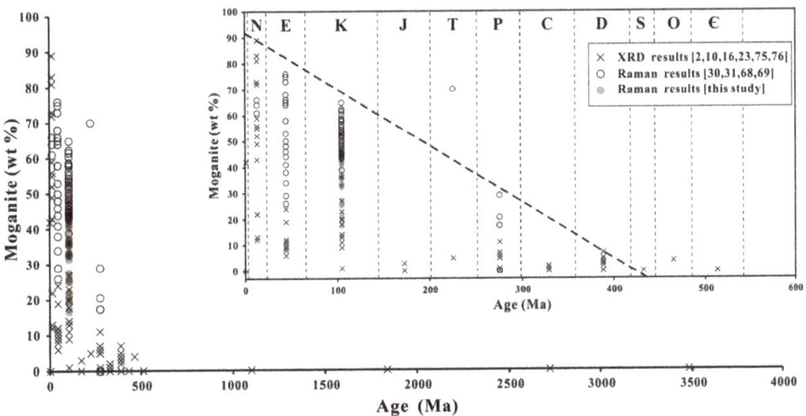

Figure 13. Summary of the moganite contents of the microcrystalline silica samples and the ages of their volcanic wall rocks. The inset picture provides detailed information on the moganite content and the ages of the Phanerozoic rocks. Note that the ages of the Phanerozoic rocks are approximate ages.

5.2. Coloration Mechanism

Yellow, red, and orange bands or zones are common in the Zhanguohong agates. These colored zones are rich in red and/or yellow, relatively coarse (μm-scale) inclusions with short prismatic (Figure 5a), spherical (Figure 5b), and granular (Figure 5c) shapes, or contain extremely fine colored grains. The Raman spectroscopic analysis reveals that these inclusions are mixtures of Fe compounds (i.e., hematite and goethite) and silica phases (Figure 12), which suggests that these Fe-bearing minerals and silica phases simultaneously precipitated from the solution. In general, the red and yellow zones consist largely of hematite- and goethite-bearing particles (i.e., coarse inclusions and fine grains), respectively. To some extent, when they mix together, the color will turn orange (Figure 12). In other words, as the hematite content increases and the goethite content decreases, the color of the agate changes from yellow, to orange, to red.

Fe compounds were also detected by the XRD analysis (Figure 6). The absence of goethite in the XRD patterns may be ascribed to its low content, which was below the detection limit. Low as their contents are, hematite and goethite have high tinting strengths [77], which results in the abundance of colored zones in the Zhanguohong agates.

Some of the goethite- and hematite-bearing inclusions comprise several colored (yellow, orange, and red) bands parallel to the sandwich-like interlayers composed of microcrystalline silica with different micro-textures (Figure 5a,c), which could result from the self-purification process during silica crystallization [4,6,7]. The presence of spherical inclusions (Figure 5b) indicates that the fluids were locally gelatinous. It has been suggested that the Fe compounds in agates may originate from the weathering products of primary Fe-bearing phases [6,78]. Some of the Fe-bearing particles in the Zhanguohong agates are large and irregular (Figures 4c and 5d), which imply an allogenic origin. Overlain by the Yixian Formation, the Archean crystalline basement rocks are rich of iron deposits. It is reasonable to infer that these rocks act as the provenance of the Fe compounds in the agates. However, to ascertain the origin of these Fe compounds, further more detailed research (e.g., trace element and isotope geochemistry studies) needs to be conducted on the Fe-bearing minerals hosted in both the Archean crystalline basement rocks and the agates.

5.3. Origin of the Zhanguohong Agates

The formation temperature of agates hosted in volcanic rocks is still controversial [9,78]. A temperature range of 50–200 °C (probably >150 °C) was reported based on fluid inclusion and oxygen isotope studies [7,78,79]. With a high moganite content, agates are thought to be formed at a relatively high temperature (100–200 °C) ([72] and references therein). Considering their high moganite contents (17–54 wt%) (Figure 11), Zhanguohong agates in the Beipiao district are more likely to have been precipitated from ore-forming fluids at temperatures of 100–200 °C.

It was suggested that ferruginous layered silicates (i.e., nontronite and lembergite) prefer to crystallize from Si- and Fe-bearing solutions with a negative Eh [80–82]. The absence of ferruginous layered silicates in the Zhanguohong agates indicates that the ore-forming fluids were oxidizing. Furthermore, the extensive occurrence of ferric minerals (hematite and goethite) in the agates confirms this oxidizing condition.

Experimental studies on synthetic Fe-bearing minerals reveal that the formation of hematite and goethite is strongly pH dependent in the temperature range of 100–200 °C at low pressures [83]. Hematite, goethite, and their mixtures can be precipitated in acidic, alkaline, and alkaline-neutral conditions, respectively [83,84]. With hematite, goethite and their mixtures respectively, the red, yellow, and orange layers in banded agates imply pH fluctuations between acidic and alkaline during the formation of the agates.

The agates in Beipiao generally exhibit concentric and rhythmic bands, and sometimes these bands are cut by microcrystalline silica veins (Figure 4a,c), which indicates multiple periods of fluid activity during the formation of the agates. This is also confirmed by the existence of interlayered macro-quartz bands at hand specimen and microscopic scales (Figures 2e and 4e), which are thought to have been formed during the final stage of silica precipitation in each period. Field investigations reveal that intense episodic volcanic activity occurred in Beipiao during the Early Cretaceous lithospheric extension in eastern China. The corresponding magmatic fluids probably greatly contributed to the multiperiod precipitation of the agates. During the intermittent crevasse eruptions of these volcanoes in the Cretaceous rift basins near Beipiao, a series of intermediate–felsic volcanic breccias were formed, which contained plenty of irregular cavities and syn-kinematic fractures. These cavities and fractures provided sufficient space for the migration of the fluids and the precipitation of the agates.

The Si-rich fluids derived from the syn- or post- magmatic activity migrated upward from a deep level and may have extracted Fe-bearing minerals from the Archean crystalline basement wall rocks. When the fluids flowed into the intermediate–felsic volcanic breccias in the Yixian Formation, which were at or near the Earth's surface, the confining pressure dropped significantly and the temperature decreased to lower than 200 °C. Consequently, the ore-forming fluids became highly supersaturated with respect to silica [1], and microcrystalline silica precipitated rapidly on the walls of the cavities and fractures. With progressive precipitation, the degree of silica supersaturation, the crystallization rate of silica, and the density of the defects in the silica matrix decreased, which resulted in the decreasing moganite content and the increasing size of the quartz grains, from microcrystalline (micro-granular, fibrous, and/or jigsaw quartz) to macrocrystalline in the banded agates (Figure 14). However, local oscillations in silica supersaturation [73,74] induced fluctuations in the moganite content (Figure 14a). In addition, Fe-bearing inclusions were precipitated along some bands parallel to the sandwich-like interlayers composed of microcrystalline silica grains with different textures under sustained oxidizing conditions, which resulted from the self-purification process during silica crystallization [4,6,7] (Figure 14b). As the pH conditions of the fluids fluctuated between acidic and alkaline, the distinct red, yellow, and white bands in the Zhanguohong agates were formed (Figure 14).

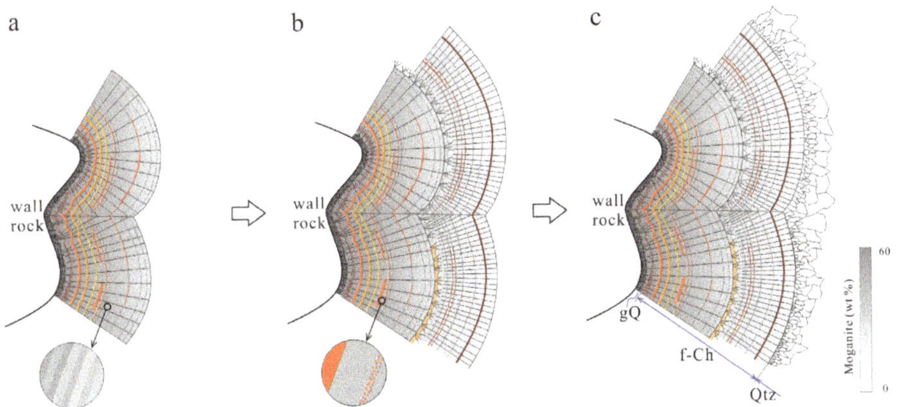

Figure 14. Model of the formation process of the banded Zhanguohong agates. (**a,b**) Rapid precipitation of gQ and f-Ch. The enlarged pictures in the lower parts of photographs a and b show the fluctuations in the moganite content and the zonations of the Fe-bearing particles, respectively; (**c**) growth of defect-free quartz crystals (Qtz) without moganite. The abbreviations are the same as in Figure 4.

6. Conclusions

We conclude that the Zhanguohong agates occurring in the intermediate–felsic volcanic breccias of the Early Cretaceous Yixian Formation in the Beipiao district exhibit banded and massive structures. Featuring red, yellow and/or white colors, they are predominantly composed of α-quartz and moganite, with minor hematite and goethite.

The moganite content (17–54 wt%) of the silica matrix fluctuated and decreased with varying degrees during the progressive precipitation of silica. The crystal defects and ferric iron in the microcrystalline silica grains probably contributed to the formation of moganite.

The ferruginous inclusions are mixtures of hematite, goethite, and silica phases. The red, yellow, and orange zones are rich in hematite, goethite, and their mixtures, respectively.

Also considering the results of previous studies, we infer that the Zhanguohong agates probably precipitated under sustained oxidizing conditions at temperatures of 100–200 °C and that the pH condition of the ore-forming fluids fluctuated between acidic and alkaline. The multiperiod precipitation of the agates probably resulted from the episodic volcanic activity during the intense Early Cretaceous lithospheric extension in eastern China.

Author Contributions: Investigation, X.Z.; Writing—original draft preparation, X.Z. and L.J.; Writing—review and editing, X.Z., L.J. and X.H.; Supervision, X.H.; Project administration, X.H. All authors have read and agreed to the published version of the manuscript.

Funding: This research was funded by the project of "Geology of mineral resources in China" from China Geological Survey (Grant No. DD20160346, DD20190379) and the project of "Beihong agate" from China National standardizing committee (Grant No. 20161367-T-334).

Acknowledgments: The authors are indebted to T. Moxon (visiting researcher from Cambridge University) and Wang Shiqi (Peking University) for their helpful discussions and suggestions. We also thank Xie Tianqi (Western University) for her assistance with the XRD analysis, Yuan Ye (China University of Geosciences, Beijing) for his help with the Raman spectroscopic analysis, and Hou Ce (Fourth Bureau of Geology and Mineral Resources of Liaoning province) for his help with the fieldwork. We are grateful to the anonymous Reviewers for their constructive and helpful comments, which significantly improved the manuscript.

Conflicts of Interest: The authors declare no conflict of interest.

References

1. Flörke, O.W.; Köhler-Herbertz, B.; Langer, K.; Tönges, I. Water in microcrystalline quartz of volcanic origin: Agates. *Contrib. Mineral. Petrol.* **1982**, *80*, 324–333. [CrossRef]

2. Heaney, P.J. Moganite as an indicator for vanished evaporites: A testament reborn? *J. Sediment. Res.* **1995**, *A65*, 633–638.

3. Götze, J.; Nasdala, L.; Kleeberg, R.; Wenzel, M. Occurrence and distribution of "moganite" in agate/chalcedony: A combined micro-Raman, Rietveld, and cathodoluminescence study. *Contrib. Mineral. Petrol.* **1998**, *133*, 96–105. [CrossRef]

4. Dumańska-Słowik, M.; Natkaniec-Nowak, L.; Wesełucha-Birczyńska, A.; Gaweł, A.; Lankosz, M.; Wróbel, P. Agates from Sidi Rahal, in the Atlas Mountains of Morocco: Gemological characteristics and proposed origin. *Gems Gemol.* **2013**, *49*, 148–159. [CrossRef]

5. Dumańska-Słowik, M.; Powolny, T.; Sikorska-Jaworowska, M.; Gaweł, A.; Kogut, L.; Poloński, K. Characteristics and origin of agates from Płóczki Górne (Lower Silesia, Poland): A combined microscopic, micro-Raman, and cathodoluminescence study. *Spectrochim. Acta, Part A* **2018**, *192*, 6–15. [CrossRef] [PubMed]

6. Powolny, T.; Dumańska-Słowik, M.; Sikorska-Jaworowska, M.; Wójcik-Bania, M. Agate mineralization in spilitized Permian volcanics from "Borówno" quarry (Lower Silesia, Poland)—microtextural, mineralogical, and geochemical constraints. *Ore Geol. Rev.* **2019**, *114*, 103130. [CrossRef]

7. Götze, J.; Möckel, R.; Kempe, U.; Kapitonov, I.; Vennemann, T. Characteristics and origin of agates in sedimentary rocks from the Dryhead area, Montana, USA. *Mineral. Mag.* **2009**, *73*, 673–690. [CrossRef]

8. Heaney, P.J. A proposed mechanism for the growth of chalcedony. *Contrib. Mineral. Petrol.* **1993**, *115*, 66–74. [CrossRef]

9. Moxon, T. Agate: A study of ageing. *Eur. J. Mineral.* **2002**, *14*, 1109–1118. [CrossRef]

10. Moxon, T.; Nelson, D.R.; Zhang, M. Agate recrystallisation: Evidence from samples found in Archaean and Proterozoic host rocks, western Australia. *Aust. J. Earth Sci.* **2006**, *53*, 235–248. [CrossRef]

11. Götze, J.; Schrön, W.; Möckel, R.; Heide, K. The role of fluids in the formation of agates. *Geochemistry* **2012**, *72*, 283–286. [CrossRef]

12. Saunders, J.A. Silica and gold textures in bonanza ores of the Sleeper deposit, Humboldt county, Nevada: Evidence for colloids and implications for epithermal ore-forming processes. *Econ. Geol.* **1994**, *89*, 628–638. [CrossRef]

13. Miehe, G.; Graetsch, H.; Flörke, O.W. Crystal structure and growth fabric of length-fast chalcedony. *Phys. Chem. Miner.* **1984**, *10*, 197–199. [CrossRef]

14. Miehe, G.; Graetsch, H. Crystal structure of moganite: A new structure type for silica. *Eur. J. Mineral.* **1992**, *4*, 693–706. [CrossRef]

15. Gíslason, S.R.; Heaney, P.J.; Oelkers, E.H.; Schott, J. Kinetic and thermodynamic properties of moganite, a novel silica polymorph. *Geochim. Cosmochim. Acta* **1997**, *61*, 1193–1204. [CrossRef]

16. Moxon, T.; Ríos, S. Moganite and water content as a function of age in agate: An XRD and thermogravimetric study. *Eur. J. Mineral.* **2004**, *16*, 269–278. [CrossRef]

17. Rodgers, K.A.; Cressey, G. The occurrence, detection and significance of moganite (SiO_2) among some silica sinters. *Mineral. Mag.* **2001**, *65*, 157–167. [CrossRef]

18. Rodgers, K.A.; Browne, P.R.L.; Buddle, T.F.; Cook, K.L.; Greatrex, R.A.; Hampton, W.A.; Herdianita, N.R.; Holland, G.R.; Lynne, B.Y.; Martin, R.; et al. Silica phases in sinters and residues from geothermal fields of New Zealand. *Earth-Sci. Rev.* **2004**, *66*, 1–61. [CrossRef]

19. Bourli, N.; Kokkaliari, M.; Iliopoulos, I.; Pe-Piper, G.; Piper, D.J.W.; Maravelis, A.G.; Zelilidis, A. Mineralogy of siliceous concretions, Cretaceous of Ionian zone, western Greece: Implication for diagenesis and porosity. *Mar. Petrol. Geol.* **2019**, *105*, 45–63. [CrossRef]

20. Heaney, P.J.; Post, J.E. The widespread distribution of a novel silica polymorph in microcrystalline quartz varieties. *Science* **1992**, *255*, 441–443. [CrossRef]

21. Heaney, P.J.; McKeown, D.A.; Post, J.E. Anomalous behavior at the *I2/a* to Imab phase transition in SiO_2-moganite: An analysis using hard-mode Raman spectroscopy. *Am. Mineral.* **2007**, *92*, 631–639. [CrossRef]

22. Saminpanya, S.; Sutherland, F.L. Silica phase-transformations during diagenesis within petrified woods found in fluvial deposits from Thailand–Myanmar. *Sediment. Geol.* **2013**, *290*, 15–26. [CrossRef]

23. Schmidt, P.; Bellot-Gurlet, L.; Leá, V.; Sciau, P. Moganite detection in silica rocks using Raman and infrared spectroscopy. *Eur. J. Mineral.* **2014**, *25*, 797–805. [CrossRef]

24. Kayama, M.; Tomioka, N.; Ohtani, E.; Seto, Y.; Nagaoka, H.; Götze, J.; Miyake, A.; Ozawa, S.; Sekine, T.; Miyahara, M.; et al. Discovery of moganite in a lunar meteorite as a trace of H_2O ice in the Moon's regolith. *Sci. Adv.* **2018**, *4*, eaar4378. [CrossRef]

25. Chen, Q.L.; Yuan, X.Q.; Jia, L. Study on the vibrational spectra characters of Taiwan blue chalcedony. *Spectrosc. Spect. Anal.* **2011**, *31*, 1549–1551, (In Chinese with English abstract).

26. Meng, G.Q.; Chen, M.H.; Jiang, J.L.; Chen, S. Structural characteristic and cause of colour of "Zhanguohong" agate from Xuanhua, Hebei province. *J. Gems Gemmol.* **2016**, *18*, 28–34, (In Chinese with English abstract).

27. Xu, W.H.; Xu, X.C.; Yang, L.L.; Wu, H.Y. Ore-geology and metallogenesis of "Zhanguohong" agate from Xuanhua, Hebei province. *J. Gems Gemmol.* **2017**, *19*, 1–11, (In Chinese with English abstract).

28. Gao, Y.Q. Geological features and metallogenetic conditions of agate deposits in Cunzhuyingzi village. *Jilin Agric.* **2011**, *12*, 265. (In Chinese)

29. Wang, Y.L.; Zhong, M.S.; Jia, C.; Yan, N.; Wang, Z.L.; Wang, J.; Zhou, C.S. Metallogeneic regularity and prospecting direction of agate in volcanic rocks of Fuxin area, western Liaoning. *Geol. China* **2011**, *38*, 1179–1187, (In Chinese with English abstract).

30. Zhou, D.Y.; Chen, H.; Lu, T.J.; Ke, J.; He, M.Y. Study on the relationship between the relative content of moganite and the crystallinity of quartzite jade by Raman scattering spectroscopy, infrared absorption spectroscopy and X-ray diffraction techniques. *Rock Miner. Anal.* **2015**, *34*, 652–658, (In Chinese with English abstract).

31. Lu, Z.Y.; He, X.M.; Lin, C.L.; Jin, X.Y.; Pan, Y.M. Identification of Beihong agate and Nanhong agate from China based on chromaticity and Raman spectra. *Spectrosc. Spect. Anal.* **2019**, *39*, 2153–2159, (In Chinese with English abstract).

32. Xu, C.; Liang, R.; Yang, H.X.; Zhao, J. Prospecting of agate deposit in Zhangjiakou, Hebei province. *J. Gems Gemmol.* **2017**, *19*, 17–24, (In Chinese with English abstract).

33. Qi, J.Y. Metallogenic characteristics and prospecting orientation of the agate deposits in western Liaoning, China. *Geol. Resour.* **2014**, *23*, 135–137, (In Chinese with English abstract).

34. Fourth Bureau of Geology and Mineral Resources of the Liaoning Province (FBGMRLP). *Prospecting Report of Agate Deposits Near Cunzhuyingzi Village, Beipiao, Liaoning Province*, 2011; unpublished report in Chinese.

35. Bureau of Geology and Mineral Resources of the Liaoning Province (BGMRLP). *Regional Geology of Liaoning Province*; Geological Publishing House: Beijing, China, 1989; pp. 1–856, (In Chinese with English summary).

36. Yang, X.D.; Li, X.Y. *Stratigraphy of Liaoning Province*; China University of Geosciences Press: Wuhan, China, 1997; pp. 1–247. (In Chinese)

37. Mi, J.R.; Xu, K.Z.; Zhang, C.B.; Chang, J.P.; Yao, P.Y. Mosozic strata in Beipiao, Liaoning province. *J. Jilin Univ. Earth Sci. Ed.* **1980**, *4*, 18–37. (In Chinese)

38. Wong, W.H. Geological structures in Beipiao, Jehol. *Bull. Geol. Surv. China* **1928**, *11*, 1–23. (In Chinese)

39. Wong, W.H. Crustal movement in eastern China. Proccedings of the 3th Pan-Pacific Scientific Congress, Tokyo, Japan, 30 October–11 November 1926.

40. Zhao, Y.; Zhang, S.H.; Xu, G.; Yang, Z.Y.; Hu, J.M. The Jurassic major tectonic events of the Yanshanian intraplate deformation belt. *Geol. Bull. China* **2004**, *23*, 854–863, (In Chinese with English abstract).

41. Zhang, C.H.; Li, C.M.; Deng, H.L.; Liu, Y.; Liu, L.; Wei, B.; Li, H.B.; Liu, Z. Mesozoic contraction deformation in the Yanshan and northern Taihang mountains and its implications to the destruction of the North China Craton. *Sci. China Earth Sci.* **2011**, *54*, 798–822. [CrossRef]

42. Davis, G.A.; Zheng, Y.D.; Wang, C.; Darby, B.J.; Zhang, C.H.; Gehrels, G. Mesozoic tectonic evolution of the Yanshan fold and thrust belt, with emphasis on Hebei and Liaoning provinces, northern China. In *Paleozoic and Mesozoic Tectonic Evolution of Central Asia: From Continental Assembly to Intracontinental Deformation*; Hendrix, M.S., Davis, G.A., Eds.; Geological Society of America Memoir: Boulder, CO, USA, 2001; Volume 194, pp. 171–197.

43. Liu, S.F.; Su, S.; Zhang, G.W. Early Mesozoic basin development in north China: Indications of cratonic deformation. *J. Asian Earth Sci.* **2013**, *62*, 221–236. [CrossRef]

44. Meng, Q.R.; Wei, H.H.; Wu, G.L.; Duan, L. Early Mesozoic tectonic settings of the northern North China Craton. *Tectonophysics* **2014**, *611*, 155–166. [CrossRef]

45. Grabau, A.W. *Stratigraphy of China, Part II: Mesozoic*; Geological Survey of China: Beijing, China, 1928; pp. 1–774.

46. Chen, P.J.; Jin, F. *The Jehol Biota*; Press of University of Science and Technology of China: Hefei, China, 1999; pp. 1–342, (In Chinese with English summary).

47. Zhou, Z.H.; Barrett, P.M.; Hilton, J. An exceptionally preserved Lower Cretaceous ecosystem. *Nature* **2003**, *421*, 807–814. [CrossRef]

48. Chen, J.S.; Li, B.; Yao, Y.L.; Liu, M.; Yang, F.; Xing, D.H.; Li, W.; Wang, Y. Comparison between Mesozoic volcanic rock strata in northeast of Liaoning–South of Jilin and Yixian Formation in west of Liaoning. *Acta Geol. Sin.* **2016**, *90*, 2733–2746, (In Chinese with English abstract).

49. Xu, D.B.; Li, B.F.; Chang, Z.L.; Zhang, J.B.; Cai, H. A study of the U-Pb isotope age and the sequence of the Cretaceous volcanics and coal-searching in Fuxin-Zhangwu-Heishan area, west Liaoning province. *Earth Sci. Front.* **2012**, *19*, 155–166, (In Chinese with English abstract).

50. Zhang, H.; Liu, X.M.; Zhang, Y.Q.; Yuan, H.L.; Hu, Z.C. Zircon U-Pb ages of bottom and top parts of the Zhangjiakou Formation in Lingyuan area (west Liaoning province) and Luanping area (north Hebei province) and their significance. *J. China Univ. Geosci.* **2005**, *16*, 115–129.

51. Fan, W.M.; Zhang, H.F.; Baker, J.; Jarvis, K.E.; Mason, P.R.D.; Menzies, M.A. On and off the North China Craton: Where is the Archaean keel? *J. Petrol.* **2000**, *41*, 933–950. [CrossRef]

52. Gao, S.; Rudnick, R.L.; Yuan, H.L.; Liu, X.M.; Liu, Y.S.; Xu, W.L.; Ling, W.L.; Ayers, J.; Wang, X.C.; Wang, Q.H. Recycling lower continental crust in the North China Craton. *Nature* **2004**, *432*, 892–897. [CrossRef] [PubMed]

53. Gao, S.; Zhang, J.F.; Xu, W.L.; Liu, Y.S. Delamination and destruction of the North China Craton. *Chin. Sci. Bull.* **2009**, *54*, 3367–3378. [CrossRef]

54. Griffin, W.L.; Andi, Z.; O'Reilly, S.Y.; Ryan, C.G. Phanerozoic evolution of the lithosphere beneath the Sino-Korean Craton. In *Mantle Dynamics and Plate Interactions in East Asia*; Flower, M.F.J., Chung, S.L., Lo, C.H., Lee, T.Y., Eds.; American Geophysical Union: Washington, DC, USA, 1998; Volume 27, pp. 107–126.

55. Liu, J.L.; Davis, G.A.; Ji, M.; Guan, H.M.; Bai, X.D. Crustal detachment and destruction of the keel of North China Craton: Constraints from Late Mesozoic extensional structures. *Earth Sci. Front.* **2008**, *15*, 72–81. [CrossRef]

56. Menzies, M.A.; Fan, W.M.; Zhang, M. Palaeozoic and Cenozoic lithoprobes and the loss of > 120 km of Archaean lithosphere, Sino-Korean Craton, China. *Geol. Soc. Lond. Spec. Publ.* **1993**, *76*, 71–81. [CrossRef]

57. Menzies, M.A.; Xu, Y.G. Geodynamics of the North China Craton. In *Mantle Dynamics and Plate Interactions in East Asia*; Flower, M.F.J., Chung, S.L., Lo, C.H., Lee, T.Y., Eds.; American Geophysical Union: Washington, DC, USA, 1998; Volume 27, pp. 155–165.

58. Lin, J.Q.; Tan, D.J.; Chi, X.G.; Bi, L.J.; Xie, C.F.; Xu, W.L. *Mesozoic Granites in Jiao–Liao Peninsula*; Science Press: Beijing, China, 1992; pp. 1–208. (In Chinese)

59. Sun, J.F.; Yang, J.H. Early Cretaceous A-type granites in the eastern North China Block with relation to destruction of the craton. *J. China Univ. Geosci. Chin. Ed.* **2009**, *34*, 137–147, (In Chinese with English abstract).

60. Wu, F.Y.; Lin, J.Q.; Wilde, S.A.; Zhang, X.O.; Yang, J.H. Nature and significance of the Early Cretaceous giant igneous event in eastern China. *Earth Planet. Sci. Lett.* **2005**, *233*, 103–119. [CrossRef]

61. Cooper, J.B. Chemometric analysis of Raman spectroscopic data for process control applications. *Chemometr. Intell. Lab.* **1999**, *46*, 231–247. [CrossRef]

62. Kingma, K.J.; Hemley, R.J. Raman spectroscopic study of microcrystalline silica. *Am. Mineral.* **1994**, *79*, 269–273.

63. Ilieva, A.; Mihailova, B.; Tsintsov, Z.; Petrov, O. Structural state of microcrystalline opals: A Raman spectroscopic study. *Am. Mineral.* **2007**, *92*, 1325–1333. [CrossRef]

64. Ciobotă, V.; Salama, W.; Jentzsch, P.V.; Tarcea, N.; Rösch, P.; El Kammar, A.; Morsy, R.S.; Popp, J. Raman investigations of Upper Cretaceous phosphorite and black shale from Safaga district, Red Sea, Egypt. *Spectrochim. Acta Part A* **2014**, *118*, 42–47. [CrossRef]

65. De Faria, D.L.A.; Venâncio Silva, S.; De Oliveira, M.T. Raman microspectroscopy of some iron oxides and oxyhydroxides. *J. Raman Spectrosc.* **1997**, *28*, 873–878. [CrossRef]

66. Oh, S.J.; Cook, D.C.; Townsend, H.E. Characterization of iron oxides commonly formed as corrosion products on steel. *Hyperfine Interact.* **1998**, *112*, 59–66. [CrossRef]

67. Legodi, M.A.; de Waal, D. The preparation of magnetite, goethite, hematite and maghemite of pigment quality from mill scale iron waste. *Dyes Pigments* **2007**, *74*, 161–168. [CrossRef]

68. Natkaniec-Nowak, L.; Dumańska-Słowik, M.; Pršek, J.; Lankosz, M.; Wróbel, P.; Gaweł, A.; Kowalczyk, J.; Kocemba, J. Agates from Kerrouchen (the Atlas mountains, Morocco): Textural types and their gemmological characteristics. *Minerals* **2016**, *6*, 77. [CrossRef]

69. Pop, D.; Constantina, C.; Tătar, D.; Kiefer, W. Raman spectroscopy on gem-quality microcrystalline and amorphous silica varieties from Romania. *Stud. UBB Geol.* **2004**, *49*, 41–52. [CrossRef]

70. French, M.W.; Worden, R.H.; Lee, D.R. Electron backscatter diffraction investigation of length-fast chalcedony in agate: Implications for agate genesis and growth mechanisms. *Geofluids* **2013**, *13*, 32–44. [CrossRef]

71. Götze, J.; Plötze, M.; Fuchs, H.; Habermann, D. Defect structure and luminescence behaviour of agate—Results of electron paramagnetic resonance (EPR) and cathodoluminescence (CL) studies. *Mineral. Mag.* **1999**, *63*, 149–163. [CrossRef]

72. Hatipoğlu, M.; Ajò, D.; Kırıkoğlu, M.S. Cathodoluminescence (CL) features of the Anatolian agates, hydrothermally deposited in different volcanic hosts from Turkey. *J. Lumin.* **2011**, *131*, 1131–1139. [CrossRef]

73. Frondel, C. Systematic compositional zoning in the quartz fibers of agates. *Am. Mineral.* **1985**, *70*, 975–979.

74. Heaney, P.J.; Davis, A.M. Observation and origin of self-organized textures in agates. *Science* **1995**, *269*, 1562–1565. [CrossRef] [PubMed]

75. Moxon, T.; Carpenter, M.A. Crystallite growth kinetics in nanocrystalline quartz (agate and chalcedony). *Mineral. Mag.* **2009**, *73*, 551–568. [CrossRef]

76. Zhang, M.; Moxon, T. Infrared absorption spectroscopy of SiO_2-moganite. *Am. Mineral.* **2014**, *99*, 671–680. [CrossRef]

77. Cornell, R.M.; Schwertmann, U. *The Iron Oxides: Structure, Properties, Reactions, Occurences and Uses*, 2nd ed.; Wiley-VCH Verlag GmbH & Co. KGaA: Weinheim, German, 2003; pp. 1–663.

78. Götze, J.; Plötze, M.; Tichomirowa, M.; Fuchs, H.; Pilot, J. Aluminium in quartz as an indicator of the temperature of formation of agate. *Mineral. Mag.* **2001**, *65*, 407–413. [CrossRef]

79. Götze, J.; Möckel, R.; Vennemann, T.; Müller, A. Origin and geochemistry of agates in Permian volcanic rocks of the Sub-Erzgebirge basin, Saxony (Germany). *Chem. Geol.* **2016**, *428*, 77–91. [CrossRef]

80. Harder, H.; Flehmig, W. Quarzsynthese bei tiefen temperaturen. *Geochim. Cosmochim. Acta* **1970**, *34*, 295–305, (In German with English abstract). [CrossRef]
81. Harder, H. Nontronite synthesis at low temperatures. *Chem. Geol.* **1976**, *18*, 169–180. [CrossRef]
82. Dekov, V.M.; Kamenov, G.D.; Stummeyer, J.; Thiry, M.; Savelli, C.; Shanks, W.C.; Fortin, D.; Kuzmann, E.; Vértes, A. Hydrothermal nontronite formation at Eolo Seamount (Aeolian volcanic arc, Tyrrhenian Sea). *Chem. Geol.* **2007**, *245*, 103–119. [CrossRef]
83. Christensen, A.N. Hydrothermal preparation of goethite and haematite from amorphous iron (III) hydroxide. *Acta Chem. Scand.* **1968**, *22*, 1487–1490. [CrossRef]
84. Das, S.; Hendry, M.J.; Essilfie-Dughan, J. Transformation of Two-Line Ferrihydrite to Goethite and Hematite as a Function of pH and Temperature. *Environ. Sci. Technol.* **2011**, *45*, 268–275. [CrossRef] [PubMed]

Article

Mineralogy and Geochemistry of Agates from Paleoproterozoic Volcanic Rocks of the Karelian Craton, Southeast Fennoscandia (Russia)

Evgeniya N. Svetova * and Sergei A. Svetov

Institute of Geology KarRC RAS, Petrozavodsk 185910, Russia; ssvetov@krc.karelia.ru
* Correspondence: enkotova@rambler.ru

Received: 9 November 2020; Accepted: 7 December 2020; Published: 9 December 2020

Abstract: Agates of Paleoproterozoic volcanics (2100–1920 Ma) within the Onega Basin (Karelian Craton, Southeast Fennoscandia) were studied using optical and scanning electron microscopy, X-ray powder diffraction, X-ray fluorescence spectrometry XRF, laser ablation inductively coupled plasma mass spectrometry (LA-ICP-MS), and C-O isotope analysis. Agate mineralization is widespread in the lavas gas vesicles, inter-pillow space of basalts, picrobasalts, basaltic andesites, as well as agglomerate tuffs. Agates are characterized by fine and coarse banding concentric zoning; moss, spotted, veinlet, and poor-fancy texture types were identified. Agate mineralization is represented by silicates, oxides, and hydroxides, carbonates, phosphates, sulfides, and sulfates. Among the silica minerals in agates only chalcedony, quartz and quartzine were found. The parameters of the quartz structure according to the X-ray diffraction data (well-develops reflections (212), (203), (301), large crystallite sizes (C_s 710–1050 Å) and crystallinity index (CI 7.8–10.3) give evidence of multi-stage silica minerals recrystallization due to a metamorphic (thermal) effect. The decreasing trend of trace element concentration in the banded agates from the outer zone to the core suggests a chemical purification process during crystallization. C-O isotope characteristic of agate-associated calcite reflects primary magmatic origin with the influence of hydrothermal activity and/or low-thermal meteoric fluids. Agates were formed under low PT-parameters and related to hydrothermal activity on the first stage of Svecofenian orogeny within 1780–1730 Ma. Thus, it can be believed that the temporal gap between continental flood basalts outflow and agate formation is about 190 Ma.

Keywords: agates; Paleoproterozoic volcanics; Onega Basin; SEM; X-ray diffraction; LA-ICP-MS; C-O isotopes

1. Introduction

Agates are fascinating minerals that have long attracted people's attention with the beauty and uniqueness of the pattern. Agates are found in igneous and sedimentary hosts and questions of agate genesis have been asked for over 250 years [1,2]. However, there have been few answers to key questions. In particular, the agate formation temperature in basic igneous hosts has been proposed in a range of <100 to 1000 °C [3–6]. The nature of the initial silica deposit is unproven and powdered amorphous silica as gel or powder seems the most likely source [7–9]. However, silica glass has also been featured as a possible starting material [10]. Rhythmic banding is not unique to agate but it is this visual property that characterizes the mineral and provoked much discussion [7,11,12]. The question about the temporal gap between the formation of gas vesicles in volcanic rocks and their mineral infilling is under discussion [6,13]. Currently, there is no unified and consistent theory. However new mineralogical, geochemical, and isotopic data obtained over the last 40 years have made a great contribution to our knowledge of the mineralogy and genesis of agates [5,14–25].

The present paper examines the mineralogical and geochemical properties of agates from the famous agate-bearing Paleoproterozoic volcanic complex on the Southeast Fennoscandia. This complex belongs to the Ludicovian Superhorizon (LSH, ca. 2100–1920 Ma) within the Onega Basin [26,27].

Previously, the main research interest in the LSH volcanics of the Onega Basin was focused on the study of mafic and ultramafic rocks (basalts, picrobasalts), as well as volcanic clastic rocks (lava breccias, clastolavas, hyaloclastites, tephroids, and tuffites) [26,28,29]. Hydrothermal mineral systems associated with the postmagmatic processes activity are less well understood [30,31]. Agates are widespread as an infill of fissures, gas vesicles in basalts, and an infill between pillow lavas (inter-pillow lava space).

The agates were initially described by the Russian geologist V. M. Timofeev in 1912 [27]. His thin section micrographs of the agates from Ludicovian basalts of the Suisari Island in Lake Onega were the earliest agate thin sections to be found in the scientific literature. Published scientific works of information about the agates of the Onega Basin are rare [32,33].

The authors started to study Onega Basin agates in 2013. The present investigation aims to highlight both the mineralogical and geochemical features of Onega agates with a reconstruction of the possible conditions of their formation. About 70 specimens of Onega agates and parent volcanic rocks were collected from outcrops for laboratory investigation. Two specimens of agate were donated to the authors by O.B. Lavrov. The collection illustrates the agate varieties available from the Ludicovian basalts within the Onega Basin. The Onega Basin experienced the Svekofennian Orogeny (~1860 Ma) [30] and the agates could provide novel information on the evolution of Paleoproterozoic magmatic complexes in the Karelian Craton.

2. Geological Setting

The Intercraton Onega Basin is located in the southeastern part of the Archaean Karelian Craton, Fennoscandian Shield within an area of more than 35,000 km^2 (Figure 1) [26,34]. Meso-Neoarchaean granite-gneiss and greenstone complexes are dominated within the Craton [35]. These complexes are overlapped with angular unconformity by the Paleoproterozoic volcano-sedimentary sequence (2500–1700 Ma) within the Onega Basin [29].

The LSH within the Onega Basin was initially identified in 1982 by Sokolov and Galdobina [36]. The LSH section is represented by an association of terrigenous, siliceous, carbonate rocks with high organic carbon content (up to 20%), as well as by mafic and ultramafic volcanic complexes related to continental flood basalts. These overlap the volcano-sedimentary succession of the Yatulian Horizon (2300–2100 Ma) [26].

At present, LSH is subdivided into two formations [26,28,29,34]:

(1) Zaonega Formation (ZF) with a total thickness of about 1.8 km that is comprised of basalts, basaltic andesites with lava flows (8–35 m) interbedded with shungite-bearing basaltic tuffs, clay-carbonate rocks, dolomites. The formation was dated with 1980—2050 Ma [26,37,38];
(2) Suisari Formation (SF) has a total thickness up to 1 km and is represented by mafic lava flows interbedded with tuffs, hyaloclastites, tuffites. The subvolcanic complex comprises dolerite and peridotite dykes and sills. A depositional age of the SF by the Sm/Nd isotopic data is 1975 ± 24 Ma [28].

Flows of pillow lava with a terrigenous component is widespread in the inter-pillow space, producing evidence of shallow-water depths conditions during the volcanics formation.

The presence of organic carbon in the inter-pillow lava space is noted. The lavas contain numerous amygdales and gas vesicles in top flows. The pillow cores indicate a high initial gas saturation of melt. A feature of the SF picrobasalts is a specific variolitic structures in massive and pillow lavas [39–41].

The volcano-sedimentary sequence of the LSH was deformed into a system of roughly parallel, northwest-southeast trending folds that overall form a synclinorium [26]. Metamorphism and hydrothermal activity in the Onega Basin are associated with Svecofenian orogeny 1890–1790 Ma [26,30]. The initiation of shear zones (fluid channels controlling ore formation) has occurred during a longer time-lag (>200 Ma). According to the Rb/Sr study of the Onega Basin metasomatites, three peaks of hydrothermal activity have been established (1780–1700, 1600, and 1500 Ma) [31].

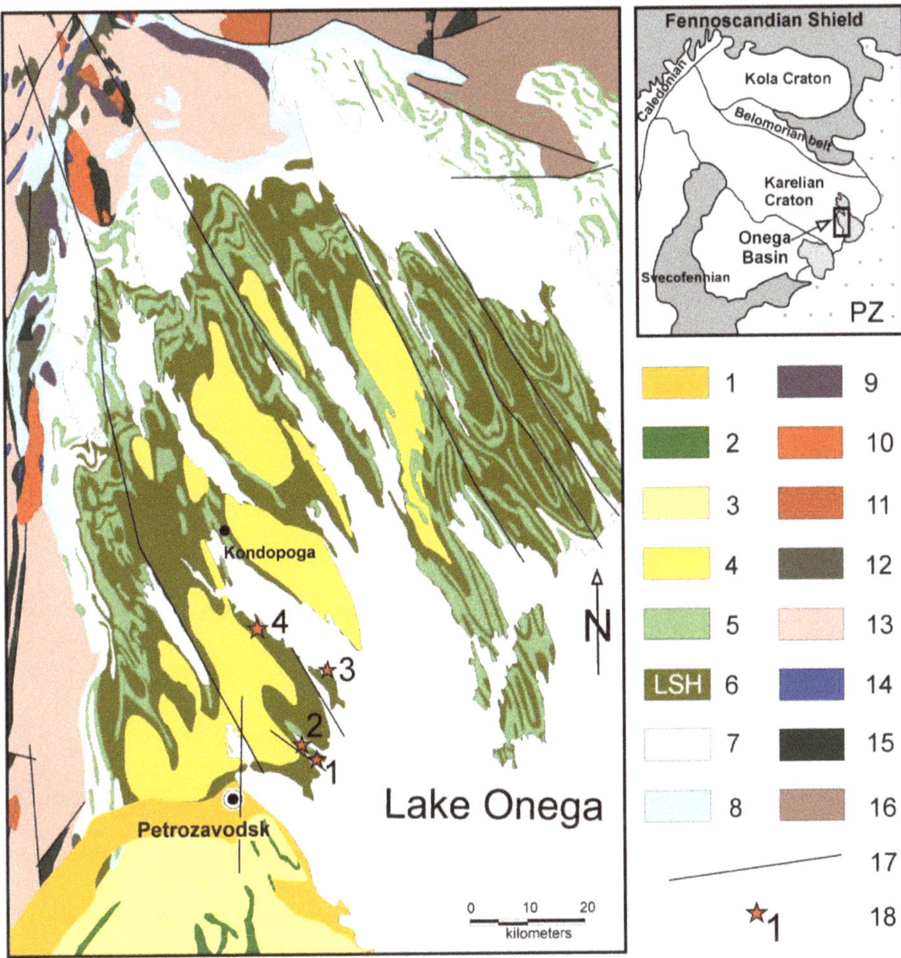

Figure 1. Simplified geological map of the northwestern part of the Onega Basin (based on [34]) and position of studied areas. Legend. Neoproterozoic: 1—ca. 635–542 Ma sandstones, siltstones, argillites, conglomerates; Paleoproterozoic: 2—ca. 1770–1750 Ma gabbro-dolerites; 3—ca. 1800–1650 Ma red quartzitic sandstones, quartzites, conglomerates; 4—ca. 1920–1800 Ma argillites, siltstones, sandstones, conglomerates; 5—ca. 1975-1956 Ma gabbroids; 6—LSH (Ludicovian Superhorizon), ca. 2100–1920 Ma picrites, picrobasalts, tuffs, tuffites, high-carbon rocks (shungites), argillites, siltstones, dolomites, basalts, basaltic andesites; 7—ca. 2300–2100 Ma dolomites, gypsum, salts, anhydrite-magnesite rocks, siltstones, basalts, dolerites; 8—ca. 2300–2100 Ma sandstones, arkoses, conglomerates, basalts; 9—ca. 2500–2300 Ma gritstones, basaltic andesites, conglomerates; Archean: 10—ca. 2740–2650 Ma granites; 11—ca. 2884–2850 Ma granites, plagiogranites; 12—ca. 2900–2800 Ma greenstone belt associations (felsic volcanics and sediments); 13—ca. 2985 Ma tonalities, granodiorites; 14—ca. 2985 Ma peridotites, pyroxenites, gabbro, diorites; 15—ca. 2940–2900 Ma komatiites, basalts, dacites; 16—ca. 3240–3220 Ma tonalities, trondhjemites; 17—faults; 18—sampling occurrences (1—Pinguba, 2—Yalguba Ridge, 3—Suisari Island, 4—Tulguba).

Dolerite sills intruded at the outset of the Svecofenian orogeny led to the initiation of the hydrothermal activity process within the LSH volcanics. For instance, the age of the Ropruchey Sill is estimated at 1770 ± 12 Ma using the U/Pb zircon method [42]. The Shunga gabbrodolerite was dated at 1746 ± 15 Ma, which is in agreement with the age of the hydrothermal zircons from the Onega Basin ore occurrences [43].

It is assumed, that the agates formation period in the Onega Basin is connected with post-magmatic hydrothermal activity initiated by the intrusion of gabbro-dolerite sills (1780–1730 Ma): 190 Ma later than continental flood basalts outflow stage [43]. The LSH volcanics within the Onega Basin were metamorphosed under the prehnite-pumpellyite facies conditions (T: 290–320 °C, P: 1–3 kbar) [32].

3. Agate Occurrences

Agates and parent volcanic rocks were sampled at the following occurrences (Figure 1):

1. Pinguba. Shore outcrops of the northern part of the Pinguba bay of Lake Onega. Here, pillow porphyric basalt lava flows attributed to the upper part of the SF were examined. The lava outcrops cover an area along the water's edge for 1 km at a width of 5–10 m. The pillows have rounded, elongated shapes, shell-like cleavage, and thin chill zones (up to 1 cm), their size ranges from 0.3 × 0.6 to 1.5 × 2.5 m. There are many gas vesicles (1–2 mm in size) filled by chlorite-carbonate-quartz in lavas. Agates form prominent nodules between pillows (Figure 2a) or fill gas cavities in pillow cores. Most agates are cone-shaped; lens-shaped bodies and veinlets are less frequent. The size of such agates ranges from 3–5 to 50 cm across. Agates also occur as rounded fragments on the shore. The frequency of agates occurrence is 3–5 per 10 m^2 of the lava flow surface.

2. Yalguba Ridge. The upper part of the SF section reaches 60 m in depth and is formed by massive and pillow Pl-Cpx porphyric picrobasalts. The section is well exposed in the Yalguba area of Lake Onega. The pillowed and massive lava flow interbed with tuffs with thickness up to 1–3 m. The pillows have compressed shapes, thin chill zones (up to 1 cm), their size ranges from 0.3 × 0.4 to 4 × 5 m. Some pillows have a well-developed sagging tail. Agates are present in the inter-pillow space of lava flows as large (up to 50–60 cm) segregations and veinlets varying in shape (Figure 2b). In some cases, agate mineralization is exposed in the central gas vesicles in the pillow cores. On the lava flow surface, the frequency of agates occurrence is 5–7 per 10 m^2.

3. Suisari Island. In the northeastern part of Lake Onega Island, numerous agates are observed in the Pl-porphyric picrobasalt lava flows. The stratigraphic attribution of volcanic rocks to ZF or SF on Suisari Island is unclear. The pillows are lightly deformed and have chill zones, their diameter ranges from 0.5 to 4 m. Agates fill cavities in the inter-pillow space of basalt flows, as well as in pillow cores (Figure 2e). These are found on the shore and underwater outcrops and can have a size of up to 50 cm. The frequency of agate occurrence is 4–6 per 10 m^2 of the lava flow surface. After storms, agates join the beach pebbles.

4. Tulguba. The studied area was by Lake Onega within an open-pit mine. The middle part of the SF section of the picrobasalt lava shows flows with thickness about 100–120 m. The lavas contain small (from 5 to 30 cm) and large (up to 2.5 m) pillows with chill zones (from 0.5 to 2 cm). Individual pillows have amygdales in the top flow (Figure 2f). Some pillows have flow banding—pahoehoe. The inter-pillow space is composed of a mixture of agglutinate tuff and terrigenous material. Lenses of agglomerate tuffs (with single large volcanic bombs) overlie on the lava top flows. Most of the agates are composed of quartz and calcite. Agates are present in the inter-pillow space, in the pillow cores and tuffs as large (up to 40 cm) segregations. The frequency of agates occurrence is up to 12 per 10 m^2 of the lava flow surface.

Figure 2. Field outcrop photographs illustrating the occurrences of main agate mineralization types in the Paleoproterozoic volcanics of the Onega Basin LSH. (**a**) Agate in the inter-pillow lava space (Pinguba); (**b**) agates in gas vesicle and inter-pillow pycrobasalt lava space (Yalguba Ridge); (**c**) carbonate veins in picrobasalt lavas (Yalguba Ridge); (**d**) scalenohedral calcite crystals from basalt fissure; (**e**) single agate amygdule in the pillow lava (Suisari Island); (**f**) agates in the inter-pillow lava space (Tulguba); (**g**) amygdales in the pycrobasalt top lava flow (Yalguba Ridge); (**h**) elongated pencil-like mineralized channels in pycrobasalt (cross-sectional view from (**g**)); (**i**) quartz-prehnite-carbonate (Qz-Prh-Cal) vein in basalt fissure (Yalguba Ridge). Red arrows point to the agates in gas vesicles, yellow arrows –in inter-pillow lava space.

Thus, agate mineralization in the pillow lavas of the LSH within the Onega Basin has a wide regional distribution and signifies the upper part of the volcano-sedimentary sequence.

4. Methods

Powder X-ray diffraction, XRF and LA-ICP-MS analysis, petrographical and electron microscope studies were carried out at the Institute of Geology, Karelian Research Centre, RAS (IG KRC RAS, Petrozavodsk, Russia). Thin sections were examined using Polam-211 optical microscope. The VEGA II LSH (Tescan, Brno, Czech Republic) scanning electron microscope (SEM) was equipped with energy-dispersive microanalyzer INCA Energy 350. Thin sections of agates were coated with a thin carbon layer for SEM and microprobe investigations and were examined at (W cathode, 20 kV accelerating voltage, 20 mA beam current, 2 μm beam diameter, counting time of 90 s).

Powder XRD used a Thermo Scientific ARL X'TRA diffractometer (CuKα-radiation, voltage 40 kV, current 30 mA, analyst I.S. Inina), Thermo Fisher Scientific, Ecublens, Switzerland. All samples were scanned for review in the 2–156° 2θ range at a scanning step of 0.6° 2θ/min. Diffractograms in the 66–69° and 25–28° 2θ ranges were recorded at a scanning step of 0.2° 2θ/min for more precision measurements of parameters of diffraction reflections. X-ray phase and structural analyses were carried out using by program pack Win XRD, ICCD (DDWiew2008). Agate samples were hand ground to obtained grain sizes <50 μm.

The chemical composition of the basalts host rocks was determined by XRF analyses. For XRF measurements, the powdered sample was mixed and homogenized with Li-tetraborate and analyzed using an ARL ADVANT'X-2331 spectrometer in the IG KSC RAS.

Trace and rare-earth elements in agates were identified by the LA-ICP-MS method on an X-SERIES-2 Thermo scientific quadrupole mass-spectrometer (Thermo Fisher Scientific, Bremen, Germany) with an UP-266 Macro Laser Ablation attachment (New Wave Research) at the IG KSC RAS following the method [44]. A section was cut perpendicular to the agate banding for analysis.

The Nd:Yag laser operates with a wavelength of 266 nm and energy output of 0.133 mJ (scan speed 70 μm/sec and impulse frequency 10 Hz). Measurements were made along with the profile as a series of three laser shots for each analyzed point. During ablation craters with a size of 100–150 × 50 μm and depth up to 70 μm to are formed (Figure 3). Later, crater morphology was controlled using a SEM VEGA II LSH. Crater relief mapping was performed by 3D laser scanning microscope VK-9710 (Keyence, Osaka, Japan) equipped with a surface analysis module. Under laser sampling splashing of matter and formation of high, wide (or excessively deep) crater cones were prevented. The following isotopes were measured: ^{7}Li, ^{9}Be, ^{24}Mg, ^{39}K, ^{45}Sc, ^{47}Ti, ^{51}V, ^{52}Cr, ^{55}Mn, ^{57}Fe, ^{59}Co, ^{60}Ni, ^{65}Cu, ^{66}Zn, ^{69}Ga, ^{72}Ge, ^{75}As, ^{85}Rb, ^{88}Sr, ^{89}Y, ^{90}Zr, ^{93}Nb, ^{95}Mo, ^{107}Ag, ^{115}In, ^{118}Sn, ^{121}Sb, ^{138}Ba, ^{139}La, ^{140}Ce, ^{141}Pr, ^{146}Nd, ^{147}Sm, ^{153}Eu, ^{157}Gd, ^{159}Tb, ^{163}Dy, ^{165}Ho, ^{166}Er, ^{169}Tm, ^{172}Yb, ^{175}Lu, ^{178}Hf, ^{181}Ta, ^{182}W, ^{208}Pb, ^{232}Th, ^{238}U. All measurements were carried out with identical parameters. Standard NIST 612 was used for calibration procedures. Calibration procedures were made at the beginning and the end of the measuring cycle and after analyzing five spots. The measured trace element concentration values are characterized following parameters of relative standard deviation (RSD): for transition metals (Co, Mn, Cr, V) <15%, for LIL (Ba, Sr, Rb) <12%, for HFS (U, Th, Y) <15%, for REE (La, Ce, Eu) <20%, for FME: Li <8%, Be <22%, As <20%. These values are obtained from the results of 50 measurements of the standard and are shown in Table S1.

Isotope studies were performed at the Centre for Collective Usage «Geoscience», Institute of Geology, Komi Scientific Center, Uralian Branch, RAS (IG Komi SC UB RAS, Syktyvkar, Russia). The decomposition of calcite in phosphoric acid and the C-O isotopic composition measurement by flow mass spectrometry (CF-IRMS) were carried out on the Thermo Fisher Scientific analytical complex, which includes a Gas Bench II sample preparation and injection system connected to a mass spectrometer DELTA V Advantage. The δ^{13}C values are given in per mille relative to the PDB standard, δ^{18}O—the SMOW standard. The international standards NBS 18 and NBS 19 were used for calibration. The analytical precision for δ^{13}C and δ^{18}O was ± 0.15 ‰ (1σ).

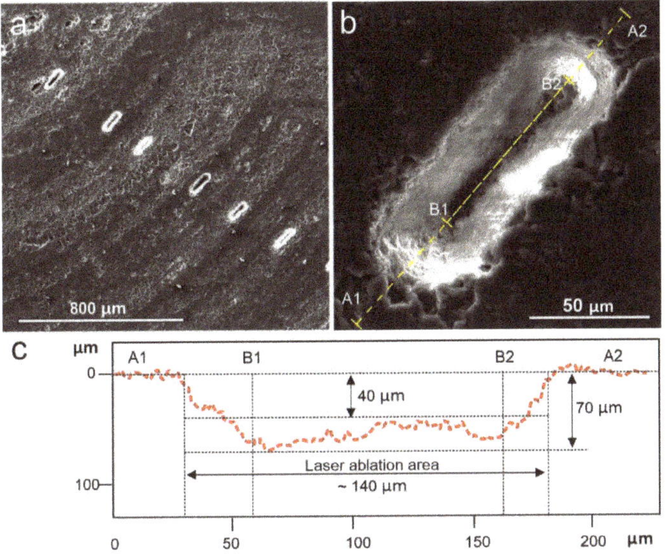

Figure 3. Surface topography of Onega agate samples and laser ablation (LA) crater character: (**a**) BSE image of laser sampling points in banding agate (#SvE-A1, Pinguba); (**b**) BSE image of the typical crater produced with 3 successive laser shots; (**c**) relief profile of ablation crater.

5. Results and Discussion

5.1. Host Rock Characteristic

The examined volcanics are fine-grained, greenish-gray to dark grey. Clinopyroxene (augite), albitized and saussuritized plagioclase (oligoclase-andesine), magnetite are the major rock-forming minerals of LSH volcanics. A secondary assemblage is presented by actinolite, chlorite, epidote, quartz, calcite, sericite, stilpnomelane, biotite. Ilmenite, apatite, titanite, and sulfides are accessory minerals. The primary glassy matrix of lava with a spilitic structure is replaced by a chlorite-actinolite. Plagioclase and pyroxene microliths are replaced by chlorite, actinolite, and calcite-epidotic aggregates. The obtained data on composition and texture of clinopyroxene phenocrysts together with the crystal size distribution analysis suggests that crystallization of the massive lavas mainly involves fractionation in a closed magmatic system, whereas the crystallization of the variolitic lavas is determined by processes in an open magmatic system. The study of pyroxene phenocrysts zoning from SF picrobasalts (Yalguba Ridge) has shown that the evolution of magmatic systems involved complicated processes such as fractional crystallization, magma mixing, and crustal contamination [41]. The vesicular basalts (mandelstones) are locally observed within the flows. Gas vesicles in it are mainly filled with macro crystalline quartz, chalcedony, or calcite. Inside individual pillows, elongated pencil-like mineralized channels are oriented perpendicular to the boundaries (Figure 2h) are observed forming small rounded amygdales on the surface (Figure 2g).

Using the SiO_2 and $Na_2O + K_2O$ content, volcanic rocks relate to picrite, basalt, basaltic andesite group of normal alkalinity and potassic series (Table 1).

Table 1. Chemical composition of agate-bearing volcanics from the LSH, Onega Basin, wt%.

Sample	SiO$_2$	Al$_2$O$_3$	FeO$_{tot}$	Na$_2$O	CaO	K$_2$O	MgO	MnO	TiO$_2$	P$_2$O$_5$	S	LOI	Σ
					Pinguba								
BP-1 *	57.73	12.32	9.05	3.97	3.33	0.39	7.34	0.10	1.55	0.15	0.08	3.87	99.88
BP-2 *	53.87	13.69	9.04	3.94	3.82	0.80	7.96	0.10	1.49	0.15	0.07	5.10	100.03
BP-3 *	49.88	15.01	11.22	4.16	2.62	0.34	8.75	0.12	1.64	0.20	0.07	5.36	99.37
BP-4	50.48	13.92	9.85	4.37	4.00	0.55	9.12	0.11	1.61	0.15	0.08	5.85	100.09
BP-5	49.09	14.33	10.62	4.39	3.22	0.28	9.49	0.10	1.65	0.17	0.07	6.33	99.74
					Yalguba Ridge								
BYa-1 **	39.32	9.30	7.47	0.23	16.01	3.62	7.81	0.15	1.11	0.10	0.07	14.34	99.53
BYa-2	45.91	11.49	14.47	0.13	9.58	0.56	10.82	0.21	1.50	0.14	0.08	4.96	99.85
					Suisari Island								
BC-1	54.78	12.27	8.45	4.59	7.43	0.34	8.14	0.13	1.54	0.17	0.08	1.93	99.85
BC-2	61.08	13.26	8.30	3.66	2.27	1.26	4.89	0.08	1.73	0.23	0.07	2.95	99.78
					Tulguba								
S20-19	47.61	13.19	12.31	1.91	8.00	1.95	9.97	0.14	2.10	0.18	0.00	2.38	99.74
S20-21	48.48	12.44	11.81	2.53	8.78	0.62	10.13	0.17	2.06	0.19	0.00	2.22	99.43

Measurements were obtained using XRF analysis. Samples: BP (1–5)—basalts, basaltic andesites; BYa (1,2)—picrobasalts; BC (1,2)—basaltic andesites; S20 (19,20)—picrobasalts. Vesicular volcanics with gas vesicles filled by (*) silica and (**) calcite.

The geochemical differentiation of the studied volcanic rocks is variable. Yalguba Ridge and Tulguba sections are represented by picrites; Pinguba and Suisare Island—by basalt, basaltic andesite. All volcanic rocks (except for #BC-2) are characterized by high MgO contents (7.34–10.8 wt%), as well as moderate FeO* (7.47–12.31 wt%) and Al$_2$O$_3$ (9.3–15.01 wt%). The Yalguba Ridge vesicular picrite variety is enriched in CaO (9.58–16.01 wt%), which is probably caused by a local hydrothermal alteration of the agate-containing volcanic rocks. This observation is confirmed by the anomalously high LOI values (up to 14.34 wt %, at conventional values 1.93–5.36 wt%).

The geochemical composition of the LSH volcanics gives evidence of the intraplate character of magmatism (continental flood basalts), that took place within the Onega Basin (ca. 2100–1920 Ma). The ZF undergo significant crustal contamination at the initial magmatic stage. Basaltic complexes were formed. This was followed by picrobasaltic complexes of the SF with more primitive compositions. These were comparable with those of modern plume-derived lavas.

5.2. Morphology, Coloration, and Anatomy of Agates

Agates mostly occur in the inter-pillow lavas space as lenses, cone-form segregations, as well as are gas vesicle infill. Agates frequently contain cavities in their geometrical center (initial void relicts). Cavities commonly are incrustated by quartz (Figure 4i,j) or calcite druses, or sinter chalcedony.

Veinlet forms of agates are widespread. They infill tectonic cracks which cut massive lavas (Figure 2c). The thickness of the cracks ranges from a few mm to 10–20 cm. Frequently cracks are composed of scalenohedral crystals of calcite as druses (Figure 2d).

The agates exhibit fine and coarse banded concentric zoning; moss, spotted, poor-fancy varieties are present (Figure 4). No horizontal banded agates were found.

Macro crystalline quartz and chalcedony are the major minerals in Pinguba and Suisari Island agates. Calcite contents prevail over the silica minerals contents in Yalguba Ridge and Tulguba agates.

Silica minerals content in agates is increasing from bottom to top along the Yalguba Ridge section. In the lower section, tectonic cracks and agate segregations are mainly filled with calcite (Figure 2c). In the medium section level where agate mineralization is most widespread, agates have an outer silica layer and massive central calcite cores. Paleocavities and cracks in rocks filled with homogeneous chalcedony are observed in the upper level section. Radial-fibrous crystals of prehnite form kidney-shaped aggregates in agates and layers in veins crosscutting picrobasaltic rocks within the Yalguba Ridge (Figure 2i).

Figure 4. Macrophotographs of Onega agates: (**a**) concentrically zoned agate with green-colored chlorite pigment in the outer layer (Pinguba); (**b**) pseudostalactites of chalcedony in agate amygdule (Suisari Island); (**c**) chalcedony/calcite agate amygdale in variolitic pycrobasalt lava (Yalguba Ridge); (**d,e**) quartz/chalcedony agates with black-and-white graphic ornament (Suisari Island); (**f**) concentric zoned agate with rosy calcite and quartz crystals in the core (Tulguba); (**g**) moss agate with carbon-bearing inclusions (Suisari Island); (**h**) agate with finely dispersed hematite in core in a lava-breccia vesicle (Tulguba); (**i**) amethyst geode from basalt vesicle (Tulguba); (**j**) druse of milky quartz crystals covered Fe-hydroxide film from a fissure in basalt (Tulguba).

Agates of the studied occurrences are characterized by peculiar color due to the composition of minerals involved in their matrix.

The Yalguba Ridge agates are either colorless or smoky with a characteristic mint-green pigment in the outer layers. This is due to mica, epidote, and chlorite inclusions in the silica matrix (Figure 4c). The Tulguba agates macroscopically like those from Yalguba Ridge (Figure 4f). Frequently, they contain rosy and smoky-violet calcite crystals in the inner zones. The amethyst druses occur in the basalt cavities (Figure 4i).

The agates from Pinguba differ from those described above. Pinguba agates are brownish-red and pistachio-green in color. The colors are due to finely dispersed hematite and chlorite in some agate layers (Figure 4a). Separate zones with a homogenous black color or black-and-white pattern cause specific decorative effect of Suisari Island agates (Figure 4d,e).

Moss agates are represented by colorless chalcedony with dispersed carbonaceous black particles (Figure 4g). The mineral composition of Onega agates is listed in Table 2.

Table 2. Mineral composition of Onega agates.

Occurrences	Main Minerals	Accessory Minerals
Pinguba	quartz	calcite, chlorite, epidote, hematite, goethite, pyrite
Yalguba Ridge	calcite, quartz	chlorite, epidote, phengite, prehnite, apatite, titanite, leucoxene, albite, covellite, pentlandite, bornite, chalcopyrite, pyrite, hematite
Suisari island	quartz	calcite, chlorite, epidote, K-feldspar, titanite, barite, carbonaceous phases
Tulguba	calcite, quartz	chlorite, epidote, mica, titanite, chalcopyrite, chalcocite, bornite, hematite, magnetite

5.3. Optical Microscopy

Microscope investigation of fine-banded agates has shown that visible silica layers are present by inequigranular quartz aggregates, fine-flake, and fibrous chalcedony, quartzine (Figure 5a,b). Spherulites composed of fibrous chalcedony or quartzine occurs in the outer zones of agates (Figure 5c). The growth of multiple spheres and semispheres of fibrous chalcedony around detached host rock fragments and the smallest particles is also common. Inclusions of epidote, mica or Fe-oxides developed at the edges of quartz grains emphasize agate zoning (Figure 5d).

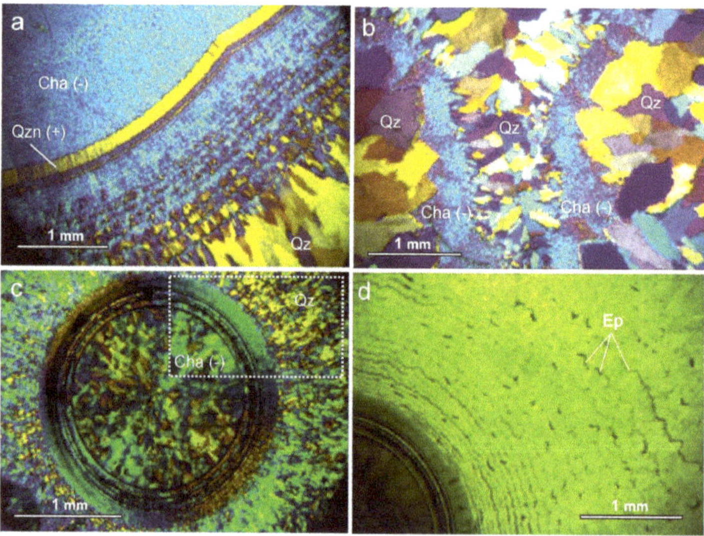

Figure 5. Microphotographs of Onega agates: (**a**) alternating zones of fibrous and flaked chalcedony, quartzine, and inequigranular quartz in concentric zonal agate (Pinguba); (**b**) internal part of the agate formed by euhedral quartz grains and flaked chalcedony (Pinguba); (**c**) spherolite of chalcedony in microcrystalline quartz matrix (Tulguba); (**d**) epidote inclusions developed at the edges of quartz grains (corresponds to dotted field in figure (**c**)). Cha(-): chalcedony, Qzn(+): quartzine, Qz: quartz, Ep: epidote. (**a–c**)—polarized light, crossed nicols, compensator plate; (**d**)—transmitted light, parallel nicols.

Vesicular volcanics contain mineralized vesicles filled mainly with silica. These vesicles often exhibit concentric zoning structures presented as alternating zones of quartz, chalcedony, or quartzine (Figure 6a,c,e). In the Yalguba Ridge occurrences, vesicular volcanics with calcite amygdales are widespread (Figure 6b,d,f). The small amygdales are usually rounded, while large ones have irregular shapes or shapes of coalesced amygdales. In large amygdales, several carbonate generations are distinguished: a fine-grained, fan-shaped, and usually a large single crystal in the center formed possible after quartz crystals (Figure 6f).

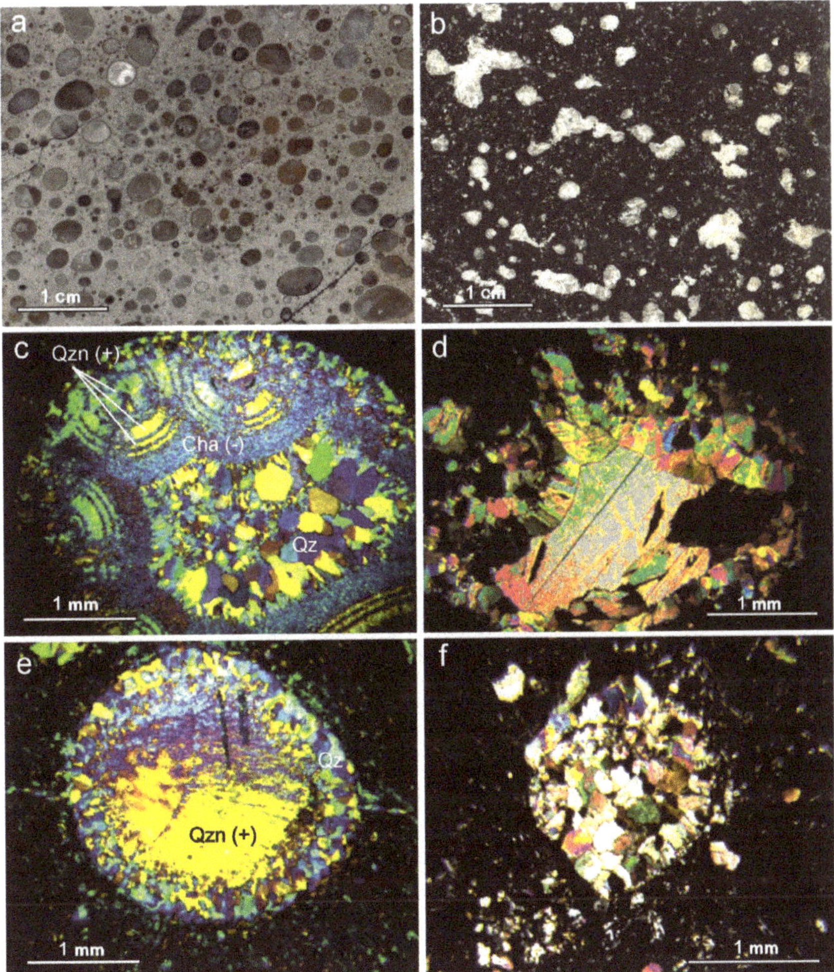

Figure 6. Vesicular basalts (mandelstones) from the LSH of the Onega Basin: Pinguba occurrence (left column, **a,c,e**), Yalguba Ridge occurrence (right column **b,d,f**). Polarized light micrographs (crossed nicols, compensator plate) of concentrically zonal amygdales (**c,e**) filled by fine-flaked and fibrous chalcedony Cha(-), fibrous quartzine Qzn(+), and inequigranular quartz (Qz). Transmitted light micrographs (crossed nicols) of calcite amygdales (**d,f**) filled by fine-grained, fan-shaped, and crystalline calcite.

5.4. X-Ray Powder Diffraction

The phase composition and structural parameters of minerals formed inner agate zones (silicate and carbonate) were examined by the X-ray powder diffraction method. Reflections corresponding to α-quartz only have been identified on the X-ray diffraction patterns of the powder silicate samples. No other silica phases (opal, cristobalite, tridymite, and moganite) have been established. The absence of these phases is typical for agates from old volcanic host rocks [23]. The unit cell parameters of quartz are stable, fluctuations of a and *c* values do not exceed 0.0006 Å (Table 3).

Table 3. Unit cell parameters of quartz and chalcedony from Onega agates.

	Sample	$a \pm \Delta a$, Å	$c \pm \Delta c$, Å	V, Å3	C_s, Å	CI
		Pinguba				
A4x1	Wall-lining chalcedony	4.9131 ± 0.0001	5.4048 ± 0.0002	112.98	1047	8.5
A9x	Quartz from agate core	4.9133 ± 0.0001	5.4051 ± 0.0001	113.00	922	9.1
A10x	Chalcedony from the outer layer of agate	4.9137 ± 0.0001	5.4054 ± 0.0001	113.02	922	8.7
A10k	Quartz from agate core	4.9137 ± 0.0001	5.4054 ± 0.0001	113.02	934	9.1
A11x	Wall-lining chalcedony	4.9137 ± 0.0001	5.4052 ± 0.0001	113.02	848	8.7
		Yalguba Ridge				
QYa-1	Quartz from agate core	4.9133 ± 0.0001	5.4050 ± 0.0002	113.00	908	10.3
QYa-2	Quartz from agate core	4.9136 ± 0.0001	5.4051 ± 0.0002	113.02	817	10.2
		Suisari island				
QC-1	Quartz from agate core	4.9134 ± 0.0001	5.4050 ± 0.0002	113.00	709	9.7
QC-2	Chalcedony from the outer layer of agate	4.9136 ± 0.0001	5.4050 ± 0.0002	113.01	763	7.8
		Tulguba				
QT-2	Quartz from agate core	4.9137 ± 0.0001	5.4050 ± 0.0001	113.02	801	10.1
QT-3	Quartz from agate core	4.9137 ± 0.0001	5.4051 ± 0.0001	113.02	749	10.2
Ref. std.	Quartz crystal	4.9133 ± 0.0001	5.4052 ± 0.0001	113.00	935	10

The mean crystallite size (C_s) of agate samples was determined by the Scherrer equation: $C_s = K\lambda/(\beta\cos\theta)$. The shape factor K was taken as 0.9, λ is the wavelength of Cu-$K_{\alpha1}$ radiation (1.540562 Å), β is the full width at half maximum (FWHM) of the peak (101) in 2θ ≈ 26.6°. C_s (101) is taken to be representative of the average crystallite diameter. Mean crystallite sizes in agate quartz/chalcedony range in 710–1050 Å (Table 3), with much more values (364–567 Å) obtained for agates from young igneous hosts (38–1100 Ma) of various regions [45]. The mean value ~860 Å for Onega agates corresponds to those of agates from close to age Proterozoic and older Archean metamorphosed hosts from western Australia [21]. It is assumed that high crystallite size is the result of the transformation of fibrous to granular α-quartz during late metamorphic alteration [22].

Well-developed reflections [(212), (203), (301) corresponding to macrocrystalline quartz were fixed on the all diffractograms of the Onega agates (Figure 7). The quartz crystallinity index (CI) of agate is defined following Murata and Norman [46] from the peak resolution in the (212) reflection at 2θ≈67.74° as CI = 10Fa/b (Figure 6). A quartz crystal from hydrothermal metamorphogenic veins of the Subpolar Urals (Russia) [47] served as a reference standard with CI = 10. All studied samples of quartz and chalcedony from Onega agates are characterized by high values of the crystallinity index >7.8 (Table 2), which is usually typical for macrocrystalline quartz [Murata and Norman, 1976]. Quartz from the agates cores is characterized by higher CI values (9.1–10.3) as compared to chalcedony from the outer rim CI (7.8–8.7). Thus, high values of crystallinity size and crystallinity index, well-developed 2θ reflections (212), (203), (301), and the absence of moganite evidence suggest a multistage silica minerals recrystallization due to a metamorphic (thermal) effect [22].

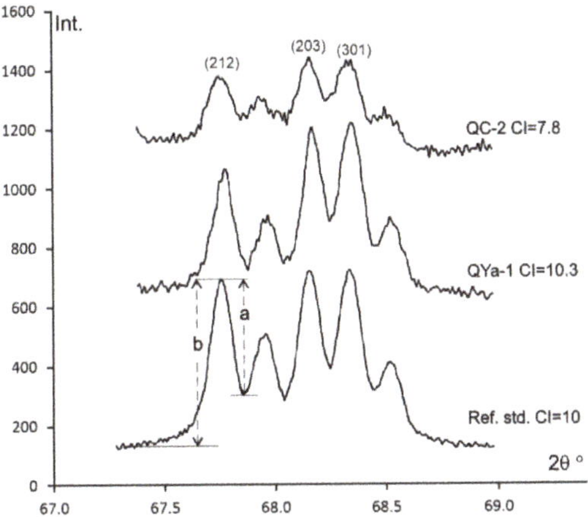

Figure 7. Diffractograms from chalcedony (QC-2, Suisari island), quartz (QYa-1, Yalguba Ridge), and quartz crystal (Ref. std., Subpolar Urals) of the 2θ≈68° quintuplet. a and b parameters are measured for Crystallinity Index (CI) calculation.

Calcite that formed in Yalguba Ridge and Tulguba agates is found as large rhombohedral crystals of varying color (colorless, smoky, and rosy). X-ray diffraction shows that crystalline calcite is free of other mineral phases. The unit cell parameters (Table 4) are close to the reference values. However, are variable depending on color: smoky calcite is characterized by higher values (*a:* 4.9865–4.9907 Å, *c:* 17.056 Å–17.065 Å, *V:* 367.3–368.1 Å^3) in comparison with colorless and rosy (*a:* 4.9810–4.9867 Å, *c:* 17.028 Å–17.057 Å, *V:* 365.9–367.3 Å^3). The lower parameters compared to the reference value probably points to ions with a smaller ionic radius than that of Ca^{2+}, which are incorporated into the calcite structure (examples, Mn^{2+}, Fe^{2+}). In addition, microprobe analysis data indicated the enrichment of colored calcite samples with manganese, magnesium, and iron oxides (up to 1 wt%), which explains the nature of their coloration.

Table 4. Unit cell parameters of calcite from Onega agates.

Sample		$a \pm \Delta a$, Å	$c \pm \Delta c$, Å	V, Å^3
		Tulguba		
CT-1	Colorless crystals	4.9851 ± 0.0003	17.050 ± 0.002	366.9
CT-2	Smoky crystals	4.9872 ± 0.0002	17.057 ± 0.002	367.4
CT-4	Smoky crystals	4.9865 ± 0.0004	17.057 ± 0.002	367.3
CT-5	Smoky crystals	4.9876 ± 0.0004	17.056 ± 0.002	367.4
CSh-2	Rose crystals	4.9865 ± 0.0004	17.057 ± 0.004	367.3
		Yalguba Ridge		
CYa-1	Colorless crystals	4.9867 ± 0.0004	17.041 ± 0.002	367.0
CYa-3	Smoky crystals	4.9907 ± 0.0004	17.065 ± 0.002	368.1
CYa-6	Rose crystals	4.9810 ± 0.0004	17.028 ± 0.002	365.9
CYa-7	Colorless crystals	4.9848 ± 0.0005	17.053 ± 0.002	367.0

5.5. SEM and Microprobe Investigation

Microscopic and microprobe investigation of Onega agates revealed numerous mineral and micromineral inclusions. (Table 2). The most frequent inclusions were identified as chlorite (Mg-Fe variety). The flake chlorite clusters were responsible for the green color (Figure 8a,g). Microinclusions of iron oxides and hydroxides are also characteristic for these agates. The minerals are represented by idiomorphic (up to 200 μm) and needle-like crystals (Figure 8a,b).

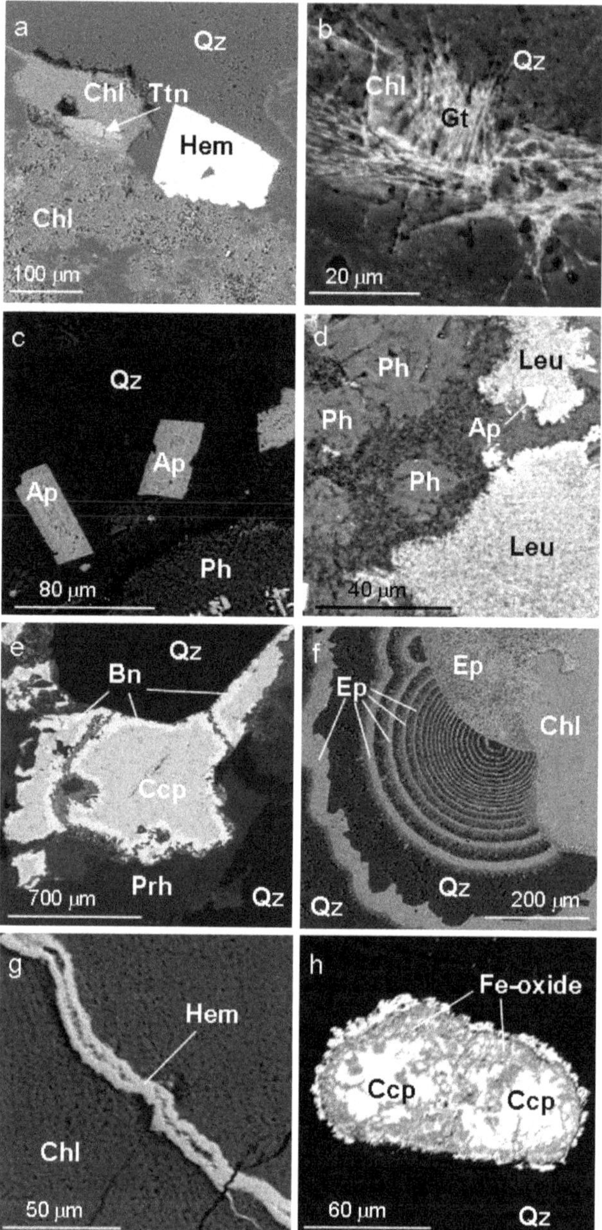

Figure 8. BSE images of microinclusions in Onega agates: (**a**) hematite (Hem) crystal in association with chlorite (Chl), quartz (Qz), and titanite (Ttn); (**b**) needle-like crystals of goethite (Gt) surrounded by silica; (**c**) prismatic apatite (Ap) grains and phengite (Ph) in outer chalcedony zone of agate; (**d**) flacked leucoxene (Leu) and phengite enclosed by chlorite; (**e**) chalcopyrite (Ccp) embedded with bornite (Bn) surrounded by quartz and prehnite (Prh); (**f**) epidote pseudomorphs after banded silica; (**g**) chlorite fissure filled by hematite; (**h**) patchy inclusion of chalcopyrite with Fe-oxide replacement zones.

Multiple epidote after silica pseudomorphs are frequently forming characteristic concentric agate textures are observed at the contact with host rocks (Figure 6f). Fine-flaked mica inclusions that correspond compositionally to phengite are present in the chalcedony and calcite matrixes of the Yalguba Ridge and Tulguba agates (Figure 8d). Various sulfides (chalcopyrite, chalcocite, pyrite, pyrrhotite, bornite) are present in forms of both macroinclusions and microinclusions. Sulfide phases have frequently microheterogeneity. As an example, the inclusions result from, chalcopyrite overgrowth by bornite, or iron oxide substitution of sulfide phase (Figure 8e,h). Hydroxylapatite clusters as flaked aggregates or individual prismatic grains up to 60 μm in size in silica matrix were identified only in Yalguba Ridge agates (Figure 8c). Titanite inclusions from 5 to 100 μm in diameter and leucoxene flakes, which are products of titanate decomposition were revealed in Yalguba Ridge and Tulguba agates (Figure 8a,d). Carbonaceous microinclusions whose presence is confirmed by the previously Raman spectroscopy method were identified in Suisari Island agates only [48]. The determination of carbonaceous matter sources in Onega agates is one of the future challenges of our investigation. For the moment, we speculate that the carbonaceous matter is associated with ZF volcanogenic-sedimentary rocks containing a significant amount of organic carbon [26].

5.6. Geochemistry Investigation

Previous work using trace element analyses has been carried out by the authors on Onega agates and parent host rocks [48].

It is established that agates from Pinguba, Yalguba Ridge, and Suisari Island differ in their chemical composition (Tables S2 and S3). The quartz from Pinguba and Suisari Island agates have high Ti, Cr, Mn, Ni, and Cu (10–120 ppm) and low Li, Co, Ga, Zn, Sr, Zr, Mo, Sn (0.5–10 ppm) concentration. The total rare earth elements (REE) concentration in the quartz from Pinguba agates (3–15 ppm) is higher than that from Suisari agates (0.5–0.6 ppm). A significant difference for other elements is not observed. Heavy REE concentration in all cases is on the detection limit, which is in agreement with data on agates from other localities around the world [19]. The Yalguba Ridge agate calcite is characterized by a high concentration of Mn (1253–6675 ppm), Sn, Ti, Ni, Sr, Y, La, and Nd (5–56 ppm) and Cr, Zn, Sm, Gd, Dy (1–8 ppm). The Cr, Zn, Sm, Gd, and Dy concentrations are lower (1–8 ppm). The contents of other identified trace elements are below 1 ppm. The REE content level in calcite is similar to or slightly higher than that in host volcanics and is higher by 1–2 orders of magnitude than that in agate quartz (Tables S2 and S3). The total REE content ranges from 31–58 ppm in smoky calcite crystals to 158 ppm in colorless ones.

Additionally, we have analyzed the lateral distribution of transition metals (Co, Mn, Cr, V), large ion lithophile (LIL) elements: Rb, Ba, Sr; high field strength (HFS) elements: Y, U, Th; light rare earth (LREE) elements: La, Ce, Eu; fluid mobile elements (FME): As, Be, Li in banded agate from Pinguba occurrences using LA-ICP-MS method of local chemical analysis (Table S4). To obtain valid results, measurements were made using an enhanced field of laser sampling because certain elements occur in very low concentrations (Figure 3). However, even in this case, concentrations of some REE elements are below the detection limit of trace-element analysis. The example of certain trace elements distribution profiles is illustrated in Figure 9.

Transition metals concentrations of clear macrocrystalline quartz area (points 1–2) are low and relatively constant. Their accumulation is observed in outer and colored agate zones due to microinclusions presence (points 3–7). LIL and HFS elements content except for Y are more balanced along with the profile. The highest LREE and Li concentrations are typical for outer agates zones near the contact with host rocks. This effect can probably be resulting from host volcanic rocks alteration. Supposably it is bound with the decomposition of volcanic glass and feldspars, which could be a base source of not only Si, Al, Ca, Na, K but also LIL and HFS elements for agates.

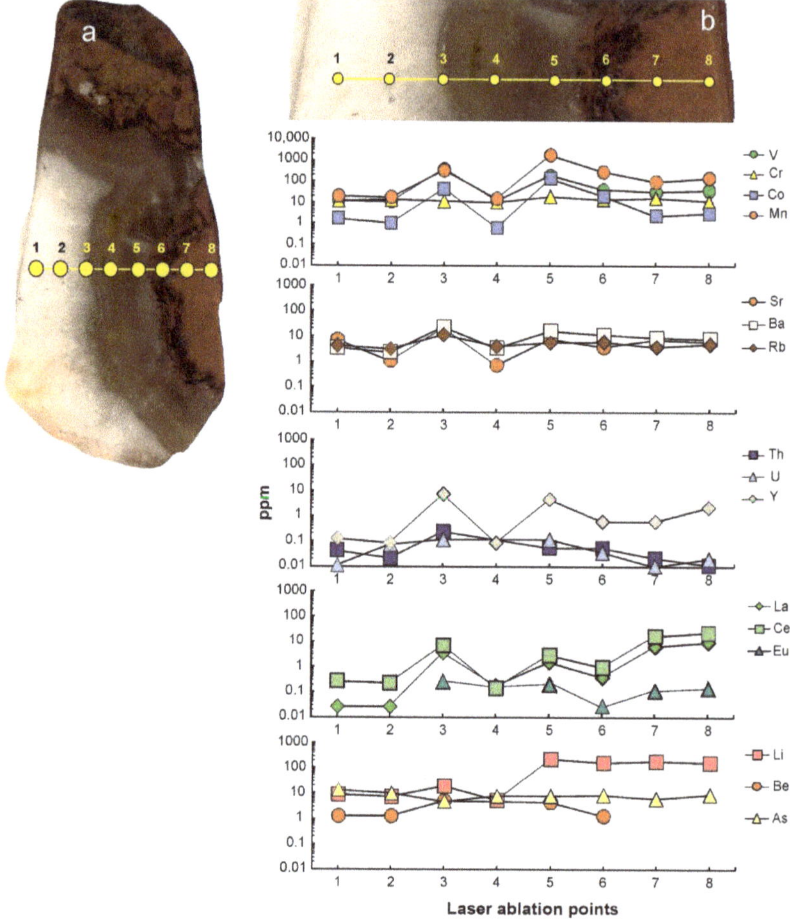

Figure 9. Micrograph of an agate from Pinguba occurrence with a profile showing the analytical points of trace element analysis by LA-ICP-MS (**a**); distribution profiles of certain trace elements in the investigated agate (**b**). Points: 1, 2—macrocrystalline area; 3–5—chalcedony zone; 6–8—outer colored zone.

Möckel et al. [12] established elevated concentrations of V, Co, Mn, Ba, Y, La, Ce, Eu, Li in chalcedonic layers of agates when compared to macrocrystalline quartz areas. This is supported by our investigation. The concentrations of Cr, As, Be, and Rb are only slightly varying along with the profile.

The observed decrease of trace element concentration in banded agates from the agate host rock to core probably reflects the chemical purification process during crystallization [12]. It is established that the chemical composition of agate layers correlate with their colors: the total concentration of trace elements in colored zones is higher than in colorless.

5.7. C-O Isotopic Composition

Preliminary stable isotope studies were carried out on agate-associated calcite samples of Yalguba Ridge occurrence (Table 5).

Table 5. Carbon and oxygen isotope composition of calcite samples from Yalguba Ridge agates.

Sample	$\delta^{13}C$ PDB(‰)	$\delta^{18}O$ SMOW(‰)
CYa-1	−3.31	15.67
CYa-2	−4.98	13.46
CYa-3	−5.12	12.91

Measurements have shown that calcites have a narrow $\delta^{13}C$ (−3.31 … −5.12‰, PDB) and $\delta^{18}O$ (12.91–15.67‰, SMOW) ranges. The $\delta^{13}C$ values are significantly higher than ones for organic carbon in sedimentary rocks (−30 to −10 ‰, [49]) that indicate an inorganic carbon origin. The $\delta^{13}C$ of agate-associated calcite within the range of mantle carbonite [50] and probably reflect primary magmatic origin. In the ratio plot, $\delta^{13}C/\delta^{18}O$ markers are in the transition zone between sedimentary carbonates (normal marine limestone) and magmatic carbonates with the influence of hydrothermal or low-thermal meteoric fluids [50] (Figure 10). In the same plot area, but with a less pronounced hydrothermal trend, there are samples of agate calcite from similar genesis host rocks (Phanerozoic and Precambrian volcanic complexes) from some other locations in the world [19]. The close C–O isotopic characterization of agate calcite from different regions suggests that agates in volcanics were formed with the active participation of post-magmatic hydrothermal fluids.

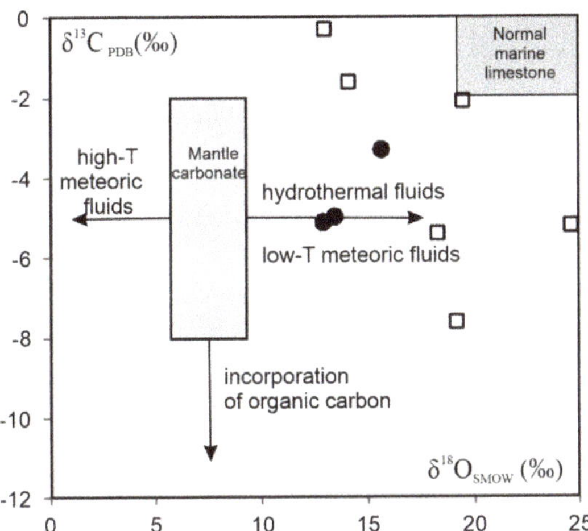

Figure 10. $\delta^{13}C$ and $\delta^{18}O$ diagram for agate-associated calcite samples from: ●—Yalguba Ridge occurrences, □—other localities around the world [19]. Mantle carbonate box and arrows showing how various processes could affect the C and O isotopes of magmatic carbonates are given according to Giuliani et al. [50], normal marine limestone (part of the field)—[51].

6. Conclusions

(1) The present study describes the key occurrences of agate mineralization in volcanic rocks of the LSH within the Onega Basin, Karelian Craton, Southeast Fennoscandia, Russia (2100–1920 Ma). Agates occur in the inter-pillow lava space, fills gas vesicles, and tectonic cracks of volcanics.

(2) Investigations showed that this mineralization is represented by silicates, oxides and hydroxides, carbonates, phosphates, sulfides, and sulfates. Among the silica minerals in agates only chalcedony, quartz, and quartzine are found. The parameters of the quartz structure according to the X-ray diffraction data (well-develops reflections (212), (203), (301), large crystallite sizes (C_s 710–1050 Å) and CI (7.8–10.3) provide evidence for multi-stage silica minerals recrystallization due to a metamorphic (thermal) effect.

(3) The elevated concentrations of certain elements (V, Co, Mn, Ba, Y, La, Ce, Eu, Li) in chalcedonic layers of banded agates in comparison with macro crystalline quartz areas were confirmed. The concentrations of Cr, As, Be, and Rb are only slightly varying. The observed decrease of trace element concentration in banded agates from the agate host rock to core is probably reflecting a chemical purification process during crystallization.

(4) C–O isotope characteristic of agate-associated calcite reflects a primary magmatic origin with the influence of hydrothermal or low-thermal meteoric fluids.

Supplementary Materials: The following are available online at http://www.mdpi.com/2075-163X/10/12/1106/s1, Table S1. The trace elements content in the reference material NIST-612 and average measured values at 50 points by LA-ICP-MS (ppm, RSD %). Table S2: The trace elements contents of agate-bearing volcanics from the LSH, Onega Basin, Russia, on a bulk probes by ICP-MS, ppm; Table S3 The trace elements contents in agates of the LSH, Onega Basin, Russia, on a bulk probes by ICP-MS, ppm; Table S4 The trace elements contents in agates of the LSH, Onega Basin, Russia, by LA-ICP-MS, %, ppm.

Author Contributions: Conceptualization, E.N.S. and S.A.S.; methodology, E.N.S. and S.A.S.; field investigation, E.N.S. and S.A.S.; writing—original draft preparation, E.N.S.; writing—review and editing, E.N.S. and S.A.S.; visualization, E.N.S. and S.A.S. All authors have read and agreed to the published version of the manuscript.

Funding: This research was funded by state assignment to the Institute of Geology Karelian Research Centre RAS programs AAAA-A18-118020290175-2 and AAAA-A18-118020290085-4.

Acknowledgments: We are grateful to A.S. Paramonov, S.V. Bordyukh, I.S. Inina, O.V. Bukchina (IG KRC RAS, Petrozavodsk, Russia), I.V. Smoleva and V.L. Andreichev (IG Komi SC, UB, RAS, Syktyvkar, Russia) for their assistance in analytical investigations. We thank Yu.L. Kyullenen for help in the polished agate sections process. We also thank O.B.Lavrov (IG KRC RAS, Petrozavodsk, Russia) for donating agate specimens (Figure 4b,g). We would like to thank T. Moxon (visiting research worker, Cambridge University) for correcting the English and helpful comments. We are grateful to Guest Editor G.A. Palyanova (Novosibirsk, Russia) for the invitation to write this article for the special issue «Agates: types, mineralogy, deposits, host rocks, ages and genesis» and the advices. We also thank three anonymous reviewers and Academic Editor for their constructive comments, which helped us to improve the manuscript

Conflicts of Interest: The authors declare no conflict of interest.

References

1. Moxon, T. *Agate Microstructure and Possible Origin*; Terra Publications: Doncaster, UK, 1996; p. 106.

2. Moxon, T.; Palyanova, G. Agate Genesis: A Continuing Enigma. *Minerals* **2020**, *10*, 953. [CrossRef]

3. Saunders, J. A Oxygen-isotope zonation of agates from Karoo volcanic of the Skeleton Coast, Namibia: Discussion and Reply. *Am. Mineral.* **1990**, *75*, 1205–1206.

4. Fallick, A.E.; Jocelyn, J.; Donelly, T.; Guy, M.; Behan, C. Origin of agates in the volcanic rocks of Scotland. *Nature* **1985**, *313*, 672–674. [CrossRef]

5. Godovikov, A.A.; Ripinen, O.I.; Motorin, S.G. *Agates*; Nedra: Moscow, Russia, 1987; p. 368. (In Russian)

6. Kigai, I.N. The genesis of agates and amethyst geodes. *Can. Mineral.* **2019**, *57*, 867–883. [CrossRef]

7. Liesegang, R.E. *Die Achate*; T. Verlag von Theodor Steinkopff: Dresden, Germany; Leipzig, Germany, 1915; p. 122.

8. Landmesser, M. Mobility by metastability: Silica transport and accumulation at low temperatures. *Chem. Erde* **1995**, *55*, 149–176.

9. Goncharov, V.I.; Gorodinsky, M.E.; Pavlov, G.F.; Savva, N.E.; Fadeev, A.P.; Vartanov, V.V.; Gunchenko, E.V. *Chalcedony of North-East of the USSR*; Science: Moscow, Russia, 1987; p. 192. (In Russian)

10. Nacken, R. Über die Nachbildung von Chalcedon–Mandeln. *Natur Folk* **1948**, *78*, 2–8.

11. Spiridonov, E.M.; Ladygin, V.M.; Frolova, Y.V. When and how the vesicular lavas are transformed into mandelstones, agate-bearing ones included. In Proceedings of the Conference «Paleovulcanology, Vulcano-Sedimentary Lithogenesis, Hydrothermal Metamorphism and Ore Formation in the Precambrian», Petrozavodsk, Russia, 20–25 August 2001; Svetov, A.P., Ed.; IG KarRC RAS: Petrozavodsk, Russia, 2001; pp. 123–124. (In Russian)

12. Pilipenko, P.P. Zur Frage der Achat genese. *Bul. Soc. Nat. Mosc.* **1934**, *12*, 279–299, (In German and Russian).

13. Mockel, R.; Götze, J.; Sergeev, S.A.; Kapitonov, I.N.; Adamskaya, E.V.; Goltsin, N.A.; Vennemann, T. Trace-Element Analysis by Laser Ablation Inductively Coupled Plasma Mass Spectrometry (LA-ICP-MS): A Case Study for Agates from Nowy Kościoł, Poland. *J. Sib. Fed. Univ. Eng. Technol.* **2009**, *2*, 123–138.

14. Florke, O.W.; Kohler-Herbertz, B.; Langer, K.; Tonges, I. Water in microcrystalline quartz of volcanic origin: Agates. *Contrib. Mineral. Petrol.* **1982**, *80*, 324–333. [CrossRef]

15. Harris, C. Oxygen-isotope zonation of agates from Karoo volcanics of the Skeleton Coast. *Namibia Am. Mineral.* **1989**, *74*, 476–481.

16. Wang, Y.; Merino, E. Self-organizational origin of agates: Banding, fiber twisting, composition, and dynamic crystallization model. *Geochim. Cosmochim. Acta* **1990**, *54*, 1627–1638. [CrossRef]

17. Heaney, P.J. A Proposed Mechanism for the Growth of Chalcedony. *Contrib. Mineral. Petrol.* **1993**, *115*, 66–74. [CrossRef]

18. Götze, J.; Nasdala, L.; Kleeberg, R.; Wenzel, M. Occurrence and distribution of "moganite" in agate/chalcedony: A combined micro-Raman, Rietveld, and cathodoluminescence study. *Contrib. Mineral. Petrol.* **1998**, *133*, 96–105. [CrossRef]

19. Götze, J.; Tichomirowa, M.; Fuchs, H.; Pilot, J.; Sharp, Z.D. Geochemistry of agates: A trace element and stable isotope study. *Chem. Geol.* **2001**, *175*, 523–541. [CrossRef]

20. Götze, J.; Möckel, R.; Vennemann, T.; Muller, A. Origin and geochemistry of agates in Permian volcanic rocks of the Sub-Erzgebirge basin, Saxony (Germany). *Chem. Geol.* **2016**, *428*, 77–91. [CrossRef]

21. Moxon, T.; Nelson, D.R.; Zhang, M. Agate recrystallization: Evidence from samples found in Archaean and Proterozoic host rocks, Western Australia. *Aust. J. Earth Sci.* **2006**, *53*, 235–248. [CrossRef]

22. Moxon, T.; Reed, S.J.B.; Zhang, M. Metamorphic effects on agate found near the Shap granite, Cumbria: As demonstrated by petrography, X-ray diffraction spectroscopic methods. *Miner. Mag.* **2007**, *71*, 461–476. [CrossRef]

23. Moxon, T.; Carpenter, M.A. Crystallite growth kinetics in nanocrystalline quartz (agate and chalcedony). *Miner. Mag.* **2009**, *73*, 551–568. [CrossRef]

24. Moxon, T.; Petrone, C.M.; Reed, S.J.B. Characterization and genesis of horizontal banding in Brazilian agate: An X-ray diffraction, thermogravimetric and microprobe study. *Miner. Mag.* **2013**, *77*, 227–248. [CrossRef]

25. Zenz, J. *Achate*; Bode-Verlag GmbH: Salzhemmendorf, Germany, 2005; p. 656. (In German)

26. Melezhik, V.A.; Medvedev, P.V.; Svetov, S.A. The Onega basin. In *Reading the Archive of Earth's Oxygenation*; Melezhik, V.A., Prave, A.R., Fallick, A.E., Kump, L.R., Strauss, H., Lepland, A., Hanski, E.J., Eds.; Springer: Berlin/Heidelberg, Germany, 2013; pp. 387–490.

27. Timofeev, V.M. Chalcedony of Sujsar Island. *Proc. Soc. St. Petersburg Nat.* **1912**, *35*, 157–174. (In Russian)

28. Puchtel, I.S.; Arndt, N.T.; Hofmann, A.W.; Haase, K.M.; Kröner, A.; Kulikov, V.S.; Kulikova, V.V.; Garbe-Schönberg, C.D.; Nemchin, A.A. Petrology of mafic lavas within the Onega plateau, central Karelia: Evidence for 2.0 Ga plume-related continental crustal growth in the Baltic Shield. *Contrib. Mineral. Petrol.* **1998**, *130*, 134–153. [CrossRef]

29. Kulikov, V.S.; Kulikova, V.V.; Lavrov, B.S.; Pisarevskii, S.A.; Pukhtel, I.S.; Sokolov, S.Y. *The Paleoroterozoic Suisarian Picrite–Basalt Complex in Karelia: Key Section and Petrology*; KNTs RAN: Petrozavodsk, Russia, 1999; p. 96. (In Russian)

30. Glushanin, L.V.; Sharov, N.V.; Shchiptsov, V.V. *Paleoproterozoic Onega Structure: Geology, Tectonics, Structure, and Metallogeny*; Karelian Research Centre, RAS: Petrozavodsk, Russia, 2011; p. 431. (In Russian)

31. Glebovitskii, V.A.; Bushmin, S.A.; Belyatsky, B.V.; Bogomolov, E.S.; Borozdin, A.P.; Savva, E.V.; Lebedeva, Y.M. RB-SR age of metasomatism and ore formation in the low-temperature shear zones of the Fenno-Karelian Craton, Baltic Shield. *Petrology* **2014**, *22*, 184–204. [CrossRef]

32. Spiridonov, E.M.; Putintzeva, E.V.; Lavrov, O.B.; Ladygin, V.M. Kronstedtite, pumpelliite, prehnite and lennilenapeite in the metaagates and metabasalts of the early Proterozoic trap formation in the northern Onega region. In Proceedings of the Conference «Lomonosov Readings», Moscow, Russia, 17–27 April 2017; Moscow State University: Moscow, Russia. Available online: https://conf.msu.ru/file/event/4305/eid4305_attach_b0acc3e7de2cd859225469534617a6272d70ce50.pdf (accessed on 8 December 2020). (In Russian)

33. Spiridonov, E.; Ladygin, V.; Frolova, Y.; Lavrov, O.; Putintseva, E.; Semikolennykh, E.; Sokolov, V.; Chernov, M. Agates in the low-grade metamorphic volcanic rocks: "bearing", "existence", "extinction". In *Book of Abstracts of the International Conference on Clay Science and Technology EUROCLAY 2019, Paris, France, 1–5 July 2019*; Sorbonne University: Paris, France, 2019; p. 563.

34. Kulikov, V.S.; Svetov, S.A.; Slabunov, A.I.; Kulikova, V.V.; Polin, A.K.; Golubev, A.I.; Gorkovets, V.Y.; Ivashchenko, V.I.; Gogolev, M.A. Geological map of Southeastern Fennoscandia (scale 1:750,000): A new approach to map compilation. *Trans. KarRC RAS* **2017**, *2*, 3–41. (In Russian) [CrossRef]

35. Slabunov, A.I.; Lobach-Zhuchenko, S.B.; Bibikova, E.V.; Sorjonen-Ward, P.; Balagansky, V.V.; Volodichev, O.I.; Shchipansky, A.A.; Svetov, S.A.; Chekulaev, V.P.; Arestova, N.A.; et al. The Archaean nucleus of the Fennoscandian (Baltic) Shield. In *European Lithosphere Dynamics*; GEE, D.G., Stephenson, R.A., Eds.; Memoirs, no. 32; Geological Society: London, UK, 2006; pp. 627–644.

36. Sokolov, V.A.; Galdobina, L.P. The Ludicovi—A new stratigraphic subdivision of the lower Proterozoic in Karelia. *Trans. USSR Acad. Sci.* **1982**, *267*, 187–190. (In Russian)

37. Narkisova, V.V. Petrology and geochemistry of igneous rocks in the OPB section. In *Paleoproterozoic Onega Structure: Geology, Tectonics, Structure, and Metallogeny*; Glushanin, L.V., Sharov, N.V., Shchiptsov, V.V., Eds.; KarRC of RAS: Petrozavodsk, Russia, 2011; pp. 195–208. (In Russian)

38. Martin, A.P.; Prave, A.R.; Condon, D.J.; Lepland, A.; Fallick, A.E.; Romashkin, A.E.; Medvedev, P.V.; Rychanchik, D.V. Multiple Palaeoproterozoic carbon burial episodes and excursions. *Earth Planet. Sci. Lett.* **2015**, *424*, 226–236. [CrossRef]

39. Svetov, S.A.; Chazhengina, S.J. Geological Phenomenon of Yalguba Ridge Variolite from F. Yu. Levinson-Lessing's Time until Today: Mineralogical and Geochemical Aspects. *Geol. Ore Depos.* **2018**, *60*, 547–558. [CrossRef]

40. Gudin, A.N.; Dubinina, E.O.; Nosova, A.A. Petrogenesis of Variolitic Lavas of the Onega Structure, Central Karelia. *Petrology* **2012**, *20*, 255–270. [CrossRef]

41. Svetov, S.A.; Chazhengina, S.Y.; Stepanova, A.V. Geochemistry and texture of clinopyroxene phenocrysts from Paleoproterozoic picrobasalts, Onega Basin, Fennoscandian Shield: Records of magma mixing processes. *Minerals* **2020**, *10*, 434. [CrossRef]

42. Bibikova, E.V.; Kirnozova, T.I.; Lazarev, Y.I.; Makarov, V.A.; Nikolaev, A.A. U–Pb isotopic age of the Karelian Vepsian. *Trans. USSR Acad. Sci.* **1990**, *310*, 189–191.

43. Lokhov, K.I.; Goltsin, N.A.; Kapitonov, I.N.; Prasolov, E.M.; Polekhovsky, Y.S.; Bogomolov, E.S.; Akhmedov, A.M.; Sergeyev, S.A. Isotopic dating of Zaonega Formation rocks subjected to stepwise alteration in the Khmelozerskaya syncline. In *Paleoproterozoic Onega Structure: Geology, Tectonics, Structure, and Metallogeny*; Glushanin, L.V., Sharov, N.V., Shchiptsov, V.V., Eds.; KarRC of RAS: Petrozavodsk, Russia, 2011; pp. 297–314. (In Russian)

44. Svetov, S.A.; Stepanova, A.V.; Chazhengina, S.Y.; Svetova, E.N.; Rybnikova, Z.P.; Mikhailova, A.I.; Paramonov, A.S.; Utitsyna, V.L.; Ekhova, M.V.; Kolodey, B.S. Precision geochemical (ICP–MS, LA–ICP–MS) analysis of rock and mineral composition: The method and accuracy estimation in the case study of Early Precambrian mafic complexes. *Tr. KarRC RAS* **2015**, *7*, 54–73. [CrossRef]

45. Moxon, T.; Rios, S. Moganite and water content as a function of age in agate: An XRD and 1497 thermogravimetric study. *Eur. J. Miner.* **2004**, *4*, 693–706.

46. Murata, J.; Norman, M.B. An index of crystallinity for quartz. *Am. J. Sci.* **1976**, *276*, 1120–1130. [CrossRef]

47. Kuznetsov, S.K.; Svetova, E.N.; Shanina, S.N.; Filippov, V.N. Minor Elements in Quartz from Hydrothermal-Metamorphic Veins in the Nether Polar Ural Province. *Geochemistry* **2012**, *50*, 911–925. [CrossRef]

48. Svetova, E.N.; Svetov, S.A. Agates from Paleoproterozoic volcanic rocks of the Onega Structure, Central Karelia. *Geol. Ore Depos.* **2020**, *62*, In Press. [CrossRef]

49. Faure, G. *Principles of Isotope Geology*, 2nd ed.; Wiley: New York, NY, USA, 1986; p. 589.

50. Giuliani, A.; Phillips, D.; Kamenetsky, V.S.; Fiorentini, M.L.; Farquhar, J.; Kendrick, M.A. Stable isotope (C, O, S) compositions of volatile-rich minerals in kimberlites: A review. *Chem. Geol.* **2014**, *374–375*, 61–83. [CrossRef]

51. Valley, J.W. Stable isotope geochemistry of metamorphic rocks. *Rev. Mineral. Geochem.* **1986**, *16*, 445–489.

Publisher's Note: MDPI stays neutral with regard to jurisdictional claims in published maps and institutional affiliations.

 minerals

Article

Black Agates from Paleoproterozoic Pillow Lavas (Onega Basin, Karelian Craton, NW Russia): Mineralogy and Proposed Origin

Evgeniya N. Svetova *, Svetlana Y. Chazhengina , Alexandra V. Stepanova and Sergei A. Svetov

Institute of Geology, Karelian Research Centre of RAS, 185910 Petrozavodsk, Russia; chazhengina@mail.ru (S.Y.C.); stepanov@krc.karelia.ru (A.V.S.); ssvetov@krc.karelia.ru (S.A.S.)
* Correspondence: enkotova@rambler.ru

Abstract: The present study provides the first detailed investigation of black agates occurring in volcanic rocks of the Zaonega Formation within the Onega Basin (Karelian Craton, Fennoscandian Shield). Three characteristic texture types of black agates were identified: monocentric concentrically zoning agates, polycentric spherulitic agates, and moss agates. The silica matrix of black agates is only composed of length-fast and zebraic chalcedony, micro- and macro-crystalline quartz, and quartzine. In addition to silica minerals, calcite, chlorite, feldspar, sulphides, and carbonaceous matter were also recognised. The black colour of agates is related to the presence of disseminated carbonaceous matter (CM) with a bulk content of less than 1 wt.%. Raman spectroscopy revealed that CM from black agates might be attributed to poorly ordered CM. The metamorphic temperature for CM from moss and spherulitic agates was determined to be close to 330 °C, whereas CM from concentrically zoning agates is characterised by a lower temperature, 264 °C. The potential source of CM in moss and spherulitic agates is associated with the hydrothermal fluids enriched in CM incorporated from underlaying carbon-bearing shungite rocks. The concentrically zoning agates contained heterogeneous CM originated both from the inter-pillow matrix and/or hydrothermal fluids.

Keywords: agate; Raman spectroscopy; X-ray powder diffraction; electron scanning microscopy; Paleoproterozoic volcanics; carbonaceous matter

check for updates

Citation: Svetova, E.N.; Chazhengina, S.Y.; Stepanova, A.V.; Svetov, S.A. Black Agates from Paleoproterozoic Pillow Lavas (Onega Basin, Karelian Craton, NW Russia): Mineralogy and Proposed Origin. *Minerals* **2021**, *11*, 918. https://doi.org/10.3390/min11090918

Academic Editors: Galina Palyanova and Frederick Lin Sutherland

Received: 12 July 2021
Accepted: 22 August 2021
Published: 25 August 2021

Publisher's Note: MDPI stays neutral with regard to jurisdictional claims in published maps and institutional affiliations.

1. Introduction

Agate and chalcedony are varieties of silica that are mostly composed of minute crystals of α-quartz. Chalcedony is band-free and characterised by uniform colouration, whereas agate is defined as banded or patterned chalcedony [1]. Agates usually contain significant amounts of other silica phases (opal, cristobalite, tridymite, quartzine, moganite, macro-crystalline quartz) and paragenetic minerals (carbonates, clay minerals, zeolites, iron compounds, etc.) [2]. These impurities may be responsible for the wide diversity in colour and spectacular pattern of agates. White, bluish, and greyish colours of fine-banded agates are mainly related with microstructural features of alternating bands, including porosity and water content [3,4]. The most common agate shades are caused by Fe, Mn, Co, Cr, and some other transition elements [5]. Inclusions of iron oxides/hydroxides are responsible for red, brown, and yellow colours [3,6,7]. Different tints of green colour can be related to chlorite, epidote, or mica inclusions dispersed within the silica matrix [7–9]. Black varieties of agate are rare and have been described from a few occurrences in India, Brazil, South and Western Africa [10], and Bulgaria [3]. The black colouration of agate areas can be caused by micro-inclusions of Fe-, Mn-, and Ti-oxides [3], as well as carbonaceous material [10,11]. The origin of carbonaceous matter (CM) in agates is quite vague. For instance, it is assumed that the origin of carbon-rich inclusions in Mali agates can be related to both the hydrothermal formation of graphite from methane under elevated temperature and graphitisation of organic precursors by secondary hydrothermal or metamorphic overprint [10]. The investigation of the solid hydrocarbon (bitumen) substance in agates from Nowy Kościół (Poland) revealed its algal or algal-humic origin [11].

139

In the Paleoproterozoic (2100–1980 Ma) volcanic rocks within the Onega Basin (Karelian Craton, Fennoscandian Shield, NW Russia) [12], agates are widely distributed and characterised by a diversity of colours and textures. A previous study showed that the light-coloured agates from these volcanic rocks have greenish, brownish-red colour, and fine or coarse banded concentric zoning, with spotted or poor-fancy textures [13]. Black-coloured agates locally occur in this area. Timofeev (1924) was the first to discover black-coloured agates at the north-western coast of Lake Onega [14]. He called them "black agates" and supposed that the black colour is related to the disseminated carbon originating from shungite, typical for the Onega Basin [12].

The present study aims to examine the black agates from Paleoproterozoic pillow lavas within the Onega Basin. To specify the origin of black colouration and the potential source of carbonaceous matter in agates, we studied mineral and chemical compositions of agates as well as characteristics of CM. The results of mineralogical, XRD, and Raman spectroscopy investigations provided novel information on genesis of black-coloured agates from the Paleoproterozoic volcanic rocks within the Onega Basin.

2. Geological Setting

The Onega Basin with an area of more than 35,000 km^2 [15] is located in the south-eastern part of the Archean Karelian Craton in the Fennoscandian Shield (Figure 1A). The Karelian Craton is formed by granite-greenstone and tonalite-trondhjemite-granodiorite complexes of the Meso-Neoarchean [12,16–18]. The Onega Basin has developed in the Paleoproterozoic under the continental rifting regime [12].

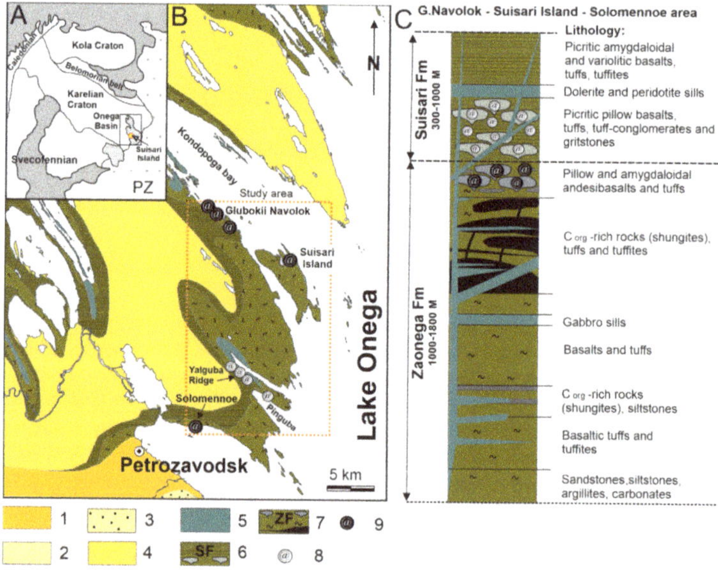

Figure 1. Location of sampling areas (**A**) and simplified geological map of the north-western part of the Onega Basin, Karelian Craton (**B**). Legend: Neoproterozoic: 1—ca. 635–542 Ma sandstones, siltstones, argillites, conglomerates. Paleoproterozoic: 2—ca. 1800–1650 Ma red quartzitic sandstones, quartzites, conglomerates; 3—ca. 1920–1800 Ma grey sandstones, siltstones, argillites, conglomerates; 4—quartzitic sandstones; 5—ca. 1975–1956 Ma gabbroids. Ludicovian Super-horizon, ca. 2100–1920 Ma: 6—picrites, picritic basalts, tuffs, tuffites, tuff-conglomerates, gritstones (Suisari Formation); 7—C_{org}-rich rocks (shungites), sandstones, siltstones, argillites, carbonates, basalts, andesibasalts, tuffs, tuffites (Zaonega Formation); 8—light-coloured agate occurrences; 9—black agate occurrences. (**C**) Simplified geological section of the studied area.

The Paleoproterozoic volcano-sedimentary successions of the Onega Basin are subdivided into six lithostratigraphic units: Sumian (2.5–2.4 Ga), Sariolian (2.4–2.3 Ga), Yatulian (2.3–2.1 Ga), Ludicovian (2.1–1.9 Ga), Kalevian (1.9–1.8 Ga), and Vepsian (1.8–1.7 Ga). The lowermost Sumian Super-horizon unconformably overlies the Archean basement. The Sariolian, Jatulian, Ludicovian, and Kalevian Super-horizons are separated from each other by unconformities and paleo-weathered surfaces [12]. The age boundaries of the formations are determined by the regional chronostratigraphic scheme [18].

Among the Paleoproterozoic volcanic-sedimentary formations of the Onega Basin, two large igneous provinces of plateau-basalts are distinguished: Yatulian dolerite-basaltic (2300–2100 Ma) [15,19] and Ludicovian picrite-basaltic (2050–1970 Ma) [12,20,21]. The agate mineralisation studied in the present work is associated with the Ludicovian volcanic rocks. The Ludicovian Super-horizon (LSH) within the Onega Basin comprises terrigenous, siliceous, carbonate rocks with high organic carbon content, as well as mafic and ultramafic volcanic complexes (Figure 1B). The LSH includes two formations (Figure 1C): the lower Zaonega Formation (ZF) and the upper Suisari Formation (SF) [22].

Zaonega Formation has a total thickness of 1100–1800 m. The base of the sequence (100–200 m) comprises alternating sandstones, siltstones, and mudstones, containing up to 3% organic carbon, and carbonates. Interbedded volcanic and sedimentary rocks prevail in the middle and upper parts of the sequence (800–900 m). Sediments are represented by carbonates and siltstones and contain 20 to 80 wt.% organic carbon [23–28], and were named "shungite rocks" after Inostrantsev [29]. Volcanic rocks include plagio-phyric massive and pillow andesibasalts with a thickness of lava flow ca. 8–35 m, tuffs, and tuffites. The ZF has been dated at 2100–1980 Ma [18,30,31].

Suisari Formation: The sequence has a total thickness of 300–1000 m and is represented by picrite-picrobasaltic massive and pillow lavas interbedded with tuff and tuffite material [12,21]. It is separated from the ZF by a horizon of tuff conglomerates, polymictic conglomerates, and gritstones, overlying the ZF with insignificant angular unconformity. The SF has been dated at 1980–1920 Ma [12,15,21].

The intrusive units of ZF and SF comprise dykes and gabbro-dolerite sills. For example, the age of quartz dolerites from the eastern Zaonezhsky Peninsula (Lebeshchina area) is 1956 ± 5 Ma [30], Konchozero gabbro-dolerite sill is 1975 ± 24 Ma [15], and Paza-Kochkoma gabbro-dolerite sill is 1961.6 ± 5.1 Ma [31].

Pillow lavas formed in shallow underwater environments dominate among the ZF and SF of LSH volcanic rocks. The pillows that range in size from 0.5 to 4 m are differentiated and have thin (<1 cm) chill zones, a fine-grained marginal part, and a medium-grained core. The composition of pillow lavas ranges from basaltic andesite in ZF to picrobasalts in SF. Variolitic structure in the pillows was recognised in the lower part of the SF and is absent in ZF [32–34]. The inter-pillow matrix is represented by lava breccia, tuff, tuffites, or sandstones, sometimes with an admixture of CM. The occurrence of CM is typical for pillow lavas from the upper part of ZF and rarely recorded in the pillow lavas from the lower part of SF. The central parts of the pillows are gas-saturated, which is evidenced by numerous amygdales (0.3–1 cm) and large (0.3–0.6 m) gas vesicles filled by light-coloured and black agates [9].

Available geochronological data indicate that in most parts of the Onega basin, the LSH volcanic, intrusive, and sedimentary rocks, including shungites, were deformed and metamorphosed at ca. 1780–1750 Ma [35,36]. The ZF and SF successions in the studied area experienced prehnite-pumpellyite facies metamorphism under T: 290–320 °C, P: 1–3 kbar [36].

The earliest interval of hydrothermal alteration of the LSH in the ZF is associated with the injection of gabbro-dolerite sills of age 1970–1950 Ma. The sills are triggered hydrothermal fluid flow, redistribution of carbonaceous material, and epigenetic sulphide mineralisation [37]. The Rb-Sr isotope study of altered rocks in the Onega Basin [38] revealed several subsequent peaks of hydrothermal activity at 1780–1700, 1600, and 1500 Ma. The

1780–1700 Ma events were assumed to be related to doleritic sills of an age 1770 ± 12 Ma (U-Pb, zircon [39]) or late Svecofennian orogenic events [38].

3. Agate Occurrence

Agates from pillow lavas along with the inter-pillow matrix were collected from the coastal outcrops of the Petrozavodsk and Kondopoga bay of Lake Onega at the following occurrences: Suisari Island, Glubokii Navolok, and Solomennoe (Figure 1B). Three representative black agate samples with the size varying from 3 to 10 cm with various textures were selected for investigation. Additionally, three samples of inter-pillow matrix from basaltic pillow lavas of Solomennoe occurrences were sampled for comparison.

(1) Suisari Island. In the north-eastern part of the Suisari Island (Keltnavolok cape) (Figure 1B), the ZF plagio-phyric pillow lava flows with a thickness of 25 m are exposed. The pillows are slightly deformed and have chill zones up to 1 cm, fine-grained marginal zones, and a coarse-grained core, and their diameter ranges from 0.5 to 4–5 m. The central and upper parts of the pillows contain gas cavities filled by chalcedony and calcite, sometimes with carbonaceous material. The agates have a black colour. Along with black agates, light-coloured quartz-carbonate-chlorite varieties of agates are widely distributed in the outcrops [9,13]. The inter-pillow matrix is composed of chloritized lava breccia and tuff material. There are numerous cavities in the inter-pillow matrix filled by chalcedony, calcite, and carbonaceous material that form agates with a size up to 70 cm.

(2) Glubokii Navolok cape. The coastal outcrops of Kondopoga bay from Suisari village to Glubokii Navolok (Figure 1B) are formed by the volcano-sedimentary succession that belongs to the lower part of the SF and upper part of the ZF. In this area, the LSH section comprises massive and pillow picrobasalts (SF), plagioclase-pyroxene, and pyroxene-phyric basalts interbedded with lava breccias, agglomerate, and pyroclastic tuffs (ZF). The thickness of lava flows ranges from 5 to 16 m. The agate mineralisation is mainly associated with pillow lava flows and pyroclastic tuffs matrix (ZF, Figure 2A–C). The rounded, elongated pillow-shave chill zones range up to 1 cm and have well-developed sagging tails. The diameter of pillows varies from 0.5 to 3 m. Agates fill cavities in the inter-pillow space, fissures, and gas vesicles in pillow lavas, and are also observed in pyroclastic tuffs. The size of agates ranges from 3 to 25 cm. They are composed of quartz-carbonate material with organic carbon impurities.

(3) Solomennoe. The coastal outcrops near Solomennoe village (Figure 1B) are represented by massive and pillow lavas of plagioclase-phyric basalts (upper parts of ZF). Some pillows have flow banding—pahoehoe. Massive and pillow lavas are often interbedded with lenses of basaltic carbon-rich tuffs and tuffites [15] (Figure 2D,E) that are 20–40 cm in thickness and up to 12 m along the strike. The interbeds are black in colour and often bordered by pillows by thin films. A series of thin picrobasaltic dykes cut the lava flows and overlay sedimentary succession. The hydrothermal alteration is widespread in the pillow lavas at the boundary with dykes, and large carbonate nests (up to 1 m in size) are noted. The agates in this area are rare and were identified exclusively in the cores of large pillows and are represented by light-coloured varieties.

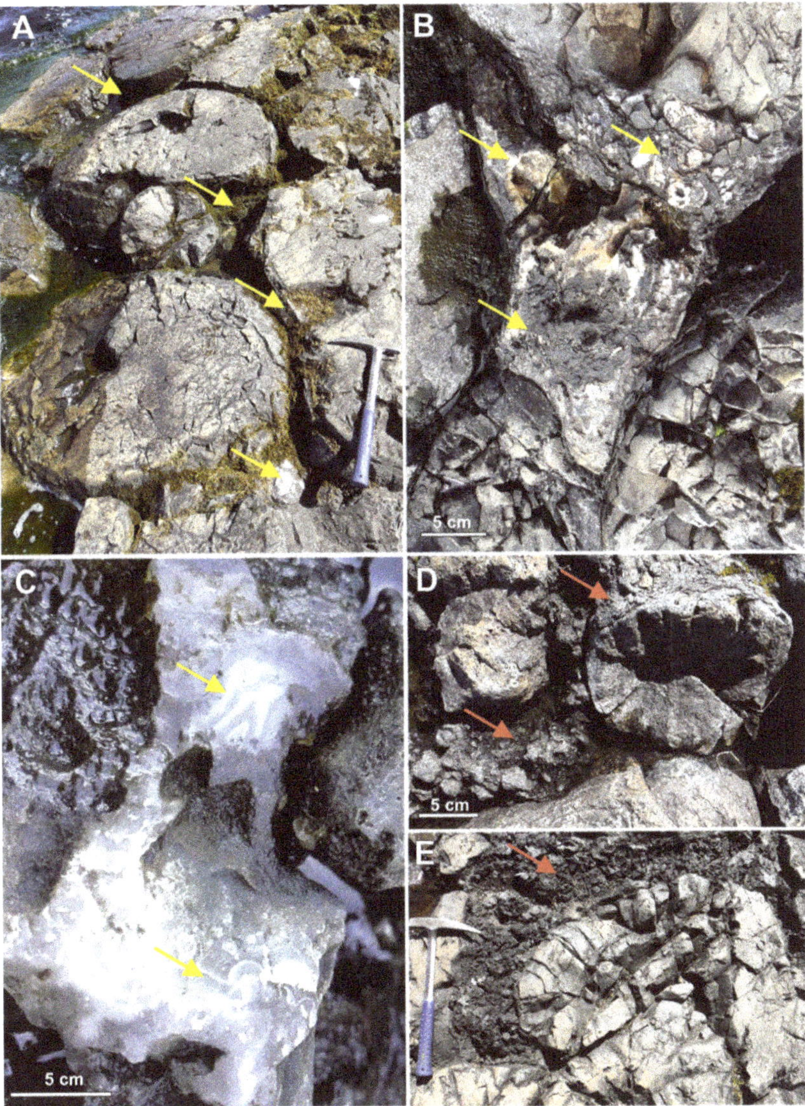

Figure 2. Field outcrop photographs illustrating black agates' location in the Paleoproterozoic volcanics within Onega Basin LSH (ZF). (**A**) Agates in gas vesicles and inter-pillow lava space (Glubokii Navolok), (**B**) agates in the inter-pillow lava space (Glubokii Navolok), (**C**) banded agates in the inter-pillow lava space (covered by water) (Glubokii Navolok), and (**D**,**E**) interbeds of carbon-rich rock ("shungite") in the inter-pillow lava space (Solomennoe). Yellow arrows point to the black agates, and red arrows to carbon-rich rock.

4. Methods

The black agates and inter-pillow matrix were studied using optical and scanning electron microscopy (SEM), electron microprobe analysis, X-ray powder diffraction, X-ray fluorescence spectrometry, thermal analysis, and Raman spectroscopy. Analytical investigations of the agates were carried out at the Institute of Geology, Karelian Research Centre, RAS (IG KRC RAS, Petrozavodsk, Russia).

Standard petrographic analyses were carried out using a Polam-211 optical microscope.

Scanning electron microscopy (SEM) and energy-dispersive X-ray spectrometry (EDS) investigations were applied to study the micromorphological features and chemical composition of individual agate zones. The experiments were performed using a VEGA II LSH (Tescan, Brno, Czech Republic) scanning electron microscope with an INCA Energy 350 energy-dispersive detector (Oxford Instruments, Oxford, UK) at the following parameters: W cathode, 20 kV accelerating voltage, 20 mA beam current, 2 μm beam diameter, and counting time of 90 s. Agate chips were coated by a carbon layer. Polished thin sections of agates were coated with a thin beryllium layer for quantity analysis of carbon content. The following standards were used: calcite, albite, MgO, Al_2O_3, SiO_2, FeS_2, wollastonite, Fe, Zn, InAs, and C (graphite). SEM-EDS quantitative data and determination of the analysis accuracy were acquired and processed using the Microanalysis Suite Issue 12, INCA Suite version 4.01. Standard deviation (SD) for Si: 0.5–0.7%, and for C: 1.0–1.1%. The average composition of the black and white agate zones was determined by the EDS microanalysis.

Powder XRD analysis was carried out using a Thermo Scientific ARL X'TRA (Thermo Fisher Scientific, Ecublens, Switzerland) diffractometer (CuKα-radiation (λ = 0.1790210 nm), voltage 40 kV, current 30 mA). All samples were scanned for review in the 2–80° 2θ range at a scanning step of 0.6° 2θ/min. Diffractograms in the 66–69° 2θ range were recorded at a scanning step of 0.2° 2θ/min for more precise measurements of parameters of diffraction reflections. X-ray phase and structural analyses were carried out using the program pack Win XRD, ICCD (DDWiew2008). Agate samples were hand-ground to obtain grain sizes < 50 μm. The detection limit for XRD phase identification was 3%.

The chemical composition of the agate samples was determined by X-ray fluorescence (XRF) using an ARL ADVANT'X-2331 (Thermo Fisher Scientific, Ecublens, Switzerland) wavelength-dispersive spectrometer with a rhodium tube, working voltage of 60 kV, working current of 50 mA, and resolution of 0.01°. Preliminarily, 2 g of each powdered sample were heated in ceramic crucibles at 1000 °C in a muffle furnace for 30 minutes. The loss of ignition was determined by a change in the mass of the sample after heating. For XRF measurements, 1 g of heated sample was mixed with Li-tetraborate flux and heated in an Au-Pt crucible to 1100 °C to form a fused bead.

Thermal properties of the agates were examined on a NETZSCH STA 449 F1 Jupiter (Selb, Germany) equipment. The powdered samples were placed into a platinum-rhodium crucible in the amount of 10 mg. The analysis was performed over the temperature range of 20–1200 °C in the air atmosphere at a rate of 10 °C/min.

Raman spectroscopy was used to investigate the structural characteristics of the carbonaceous matter recorded in agates. Raman spectroscopy analysis was carried out using a dispersive Nicolet Almega XR Raman spectrometer (Thermo Fisher Scientific, Waltham, MA, USA) with a green laser (532 nm, Nd-YAG) (Thermo Fisher Scientific, Waltham, MA, USA). The spectra were collected at 2 cm^{-1} spectral resolution. The spectrometer was calibrated before each analytical session by 'zero-point' centring and by analysing a Si-standard with a characteristic Si Raman band at 520.4 cm^{-1}. A confocal microscope with a 50× objective was used to focus an excitation laser beam on the sample and to collect a Raman signal from a 2 μm diameter area. Raman spectra were acquired in the 85–4000 cm^{-1} spectral region, with the exposition time between 30 and 100 s for each scan, depending on the signal intensity, and laser power of 2–10 mW to prevent any sample degradation. A total of 75 Raman spectra were acquired from polished cross-sections and thin sections.

A Raman spectrum of CM is composed of two intensive G and D1 bands in the first-order region (1100–1800 cm^{-1}). The graphite band G (~1580 cm^{-1}) corresponds to the in-plane vibration of aromatic carbons in the graphitic structure [40]. The bands D1 (~1350 cm^{-1}) and D2 (~1620 cm^{-1}) appear in disordered CM, and they are assigned to defects in the graphitic structure [40–42]. In poorly ordered CM, at least two additional peaks, D3 (~1500 cm^{-1}) and D4 (~1260 cm^{-1}), can be decomposed in the first-order region of the CM Raman spectrum. Broad D3 band results from out-of-plane defects, related to tetrahedrally coordinated carbons, dangling bonds, and heteroatoms, occur in natural CM [40,41]. In poorly ordered CM, the D4 band appears like a shoulder of the D1 band (~1190–1250 cm^{-1}). It has been tentatively attributed to sp^2–sp^3 bonds or C–C and C=C stretching vibration of polyene-like structures [43,44].

The most intensive bands in the second-order region of CM Raman spectra are located at ~2700 and ~2900 cm^{-1}. The former is attributed to overtone scattering of the D1 band and is absent in poorly ordered CM, while in pure crystalline graphite, it occurs as a doublet of two partially overlapping peaks [41]. The later is attributed to the combination modes (G + D) and appears only in poorly ordered CM or is due to the presence of C–H; bonds [40–42].

For peak decomposition, we used the band protocol described by Kouketsu et al. [45]. Raman spectral data, such as peak position, band area (i.e., integrated area), and band width (i.e., full-width at half maximum (FWHM)), were determined using OMNIC for Dispersive Raman software v.8.2. (v.8.2., Thermo Fisher Scientific, Waltham, MA, USA) with a Gaussian/Lorentzian function. Mean values, estimated for each sample from several spot analyses, and standard deviation, which may characterize the heterogeneity of the carbonaceous matter of a sample, were calculated for each parameter of the Raman spectrum.

5. Results

The agates from ZF with a black and white pattern mostly occur in the inter-pillow lava space as veinlets and prominent nodules. They were also rarely observed within the tuffs matrix. The agates exhibit alternating white and black non-transparent zones, significantly varying in amount and size (Figure 3A–G). The black colour of local agate areas (as shown below) is related to the presence of disseminated CM in the silica matrix. In some cases, agates contain cavities encrusted by quartz druses or sinter chalcedony (Figure 3A). Three texture types of black agates from ZF were identified, including monocentric concentrically zoning agates, polycentric spherulitic agates, and moss agates.

Figure 3. Photographs of Onega black agates in the Paleoproterozoic volcanics within the Onega Basin LSH (ZF). (**A**) Concentrically zoned agate with central cavity covered by sinter chalcedony (Glubokii Navolok), (**B**) uniform opaque agate with quartzine spherulites zone and large calcite inclusions (Suisari Island), (**C**) concentrically zoning agate with black ellipsoid (Suisari Island), (**D**) enlarged fragments of spherulitic zone from (**B**), (**E**) black and white spherulite agate (Glubokii Navolok), (**F**) concentrically zoned agate (Glubokii Navolok), (**G**) moss agate composed by colourless chalcedony with black-rounded segregations of carbonaceous matter (Suisari Island). S. Cha—sinter chalcedony, Cal—calcite. The areas of optical microscopy analysis are marked with orange squares, explanations in Figure 4.

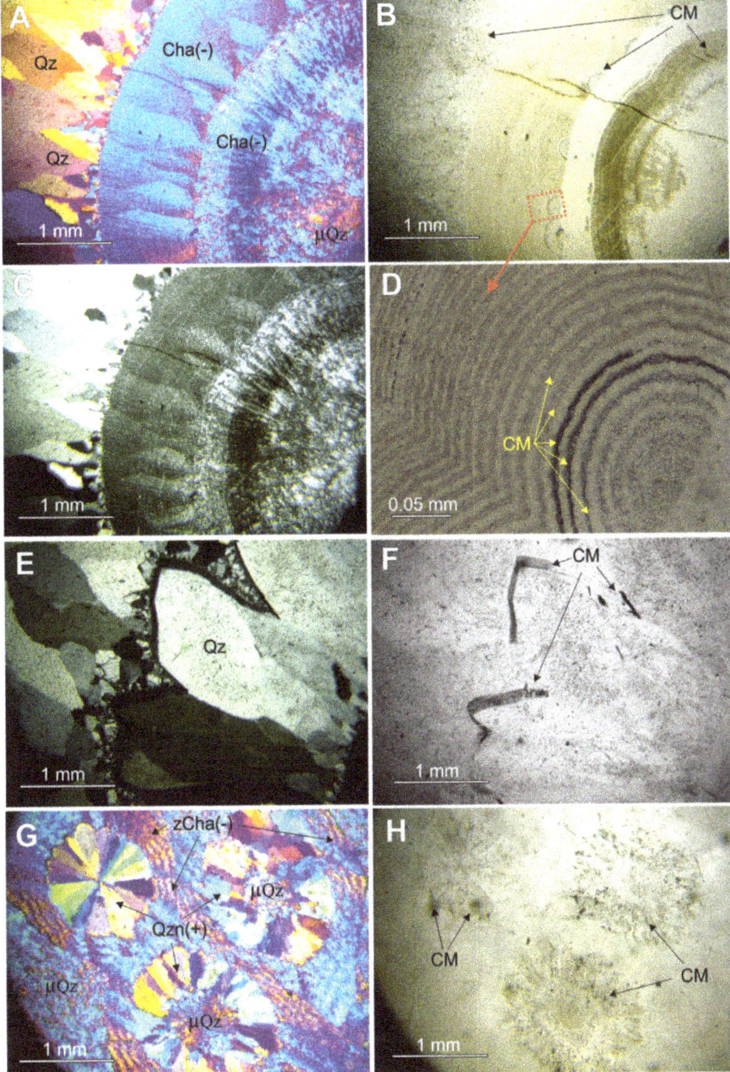

Figure 4. Microphotographs of black Onega agates with disseminated carbonaceous matter (CM): (**A–C**) fragment of the black concentric ellipsoid (from selected area 1, Figure 3C) with alternating zones of length-fast chalcedony, micro- and macro-crystalline quartz in concentrically zoning agate. (**D**) Enlarged fragment (from selected area of (**B**)) with fine concentric bands of chalcedony, between which CM is disseminated. (**E,F**) Fragment of the black concentric ellipsoid (from selected area 2, Figure 3C) with black jagged band containing carbonaceous pigment developed at the edges of euhedral prismatic quartz crystals (Figure 3C, area 2). (**G,H**) Spherolites of radial fibrous quartzine in microcrystalline quartz matrix with zebraic chalcedony zones (Figure 3E). (**A,G**) Polarised light, crossed nicols, compensator plate. (**C,E**) Polarised light, crossed nicols. (**B,D,F,H**) Plane polarised light. (**B,D,F,H**) Photos show that the macroscopic black colour is caused by an accumulation of CM in the local agate zones. Abbreviations: Cha(−): length-fast chalcedony; zCha(−): zebraic chalcedony; Qzn(+): length-slow chalcedony (quartzine); Qz: macrocrystalline quartz; µQtz: micro-crystalline quartz.

5.1. Macro- and Micro-Scopic Characteristics

Concentrically zoning agates (CZ), recognised as nodules, exhibit alternating black and white bands (Figure 3C,F). Black bands are usually localised parallel to the outer shell of an agate nodule close to the contact with the host rock. The thickness of black bands ranges from 0.3 mm to 1 cm. Additionally, black ellipsoids with a size of up to 2 cm were recognised in the inner white part of CZ agates. The petrographic study revealed that the silica matrix of CZ agates is comprised of alternating layers of length-fast chalcedony (with the c-axis situated perpendicular to the fibres), and micro- and macro-crystalline quartz (Figure 4A–C). The thick black bands in the outer part of agates and black ellipsoid inside nodules (Figure 3C) under low magnifications appear to be uniform and completely formed by CM, whereas under high magnification, it shows that they consist of numerous fine concentric bands of microcrystalline quartz and chalcedony, between which carbonaceous matter is disseminated (Figure 4B,D). CM in CZ agates are mostly concentrated in zones of fine-grained quartz and chalcedony. The black jagged bands containing carbonaceous pigment with a thickness up to 0.3 mm occur both in the margins and central parts of the nodules around the ellipsoid (Figure 3C). The optical microscopic study revealed that these bands repeat the morphology of the boundary of the macrocrystalline quartz zone and mark the termination of individual zone crystallisation in banded agate (Figure 4E,F).

Spherulitic agates (SP) occur in the inter-pillow space as nodules. They are represented by black, grey, and white spheroliths with a size of up to 3 mm embedded in a silica matrix (Figure 3D,E), and some of them are coalescent. SP agates contain un-zoned and concentrically zoning spherulites (Figures 3B,D,E and 4G,H). Un-zoned spherulites are composed only of radial fibrous quartzine (length-slow chalcedony with the c-axis situated parallel to the fibres), whereas zoned spherulites comprise a microcrystalline quartz core rounded by a concentric layer of fibrous quartzine (Figure 4G). CM is concentrated within the quartzine fibres and, to a lesser extent, between microcrystalline quartz grains (Figure 4H). Silica matrix-contained spheroliths are represented by microcrystalline quartz and zebraic chalcedony (length-fast with a helical twisting of fibres along the c-axis) (Figure 4G).

Moss agates (MS) are the rarest variety of black agates recognised in ZF. They occur as veinlets in the inter-pillow space. MS agates are composed of transparent colourless chalcedony with numerous black rounded segregations with a size of about 1 mm (Figure 3G). The microscopical study under high magnification in plane polarised light revealed that the black segregations do not have clear contours (Figure 5A,B). The microstructure of the MS agates is presented by microcrystalline quartz (Figure 5C). CM in black segregations is accumulated predominantly between quartz grains (Figure 5D). The micrographs in transmitted light (crossed nicols) of the black segregation area (Figure 5C) of the MS agate show that concentrating of CM is associated with the zones containing finer chalcedony grains.

Inter-pillow matrix is represented by agglomerate tuffs and tuffites of basaltic composition interbedded by the thin layers composed of quartz, calcite, and CM. The inter-pillow matrix has strongly heterogeneous composition within the lava flows. According to the microscopic observation and XRD and Raman analyses, it comprises augite, plagioclase, chlorite, prehnite, calcite, quartz, and CM.

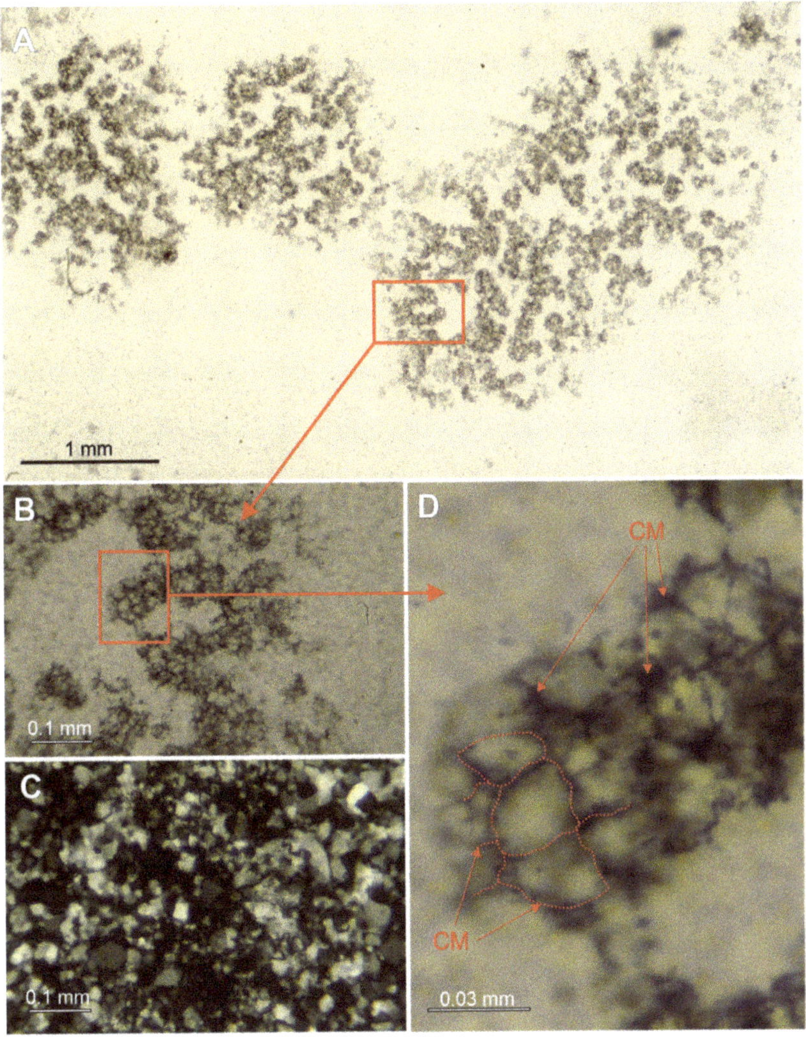

Figure 5. Microphotographs of moss agates (Figure 3G): accumulations of disseminated carbonaceous matter (CM) within the chalcedony matrix in transmitted light (**A**), enlarged fragment from the dark rounded segregation (as indicated in (**A**)) in transmitted light (**B**), and the same fragment between crossed nicols (**C**). (**D**) Enlarged fragment of selected area from (**B**), red dotted lines and arrows shows CM distribution between chalcedony grains.

5.2. Mineral Composition

The phase composition of black agates was studied by the X-ray powder diffraction method. White and black parts of CZ agate, as well as bulk samples of MS and SP agates, were examined separately. Only reflections conforming to α-quartz have been identified on the diffraction patterns (Supplementary Figure S1). No other silica phases (opal, cristobalite, tridymite, and moganite) were found. Raman spectroscopy analysis is in accordance with the X-ray diffraction data showing that agates of all types are composed of α-quartz identified by the characteristic bands at 465, 358, 210, and 130 cm^{-1} (Supplementary Figure S2).

All obtained diffraction patterns are characterised by well-developed reflections, (212), (203), (301), corresponding to macrocrystalline quartz. The quartz crystallinity index (CI) of agate is calculated from the peak resolution in the (212) reflection at $2\theta \approx 67.74°$, as CI = 10Fa/b, following the method in [46] (Figure 6). A colourless quartz crystal from hydrothermal metamorphogenic veins of the Subpolar Urals (Russia) [47] served as a reference standard, with CI = 10. Quartz from the white agate area (CZw) displays higher CI values (10.3) as compared to quartz from the black agate area (CZb) (CI = 9.9), SP (CI = 9.5), and MS (CI = 9.9). No graphite phase or traces of CM were observed on diffractograms (Supplementary Figure S1).

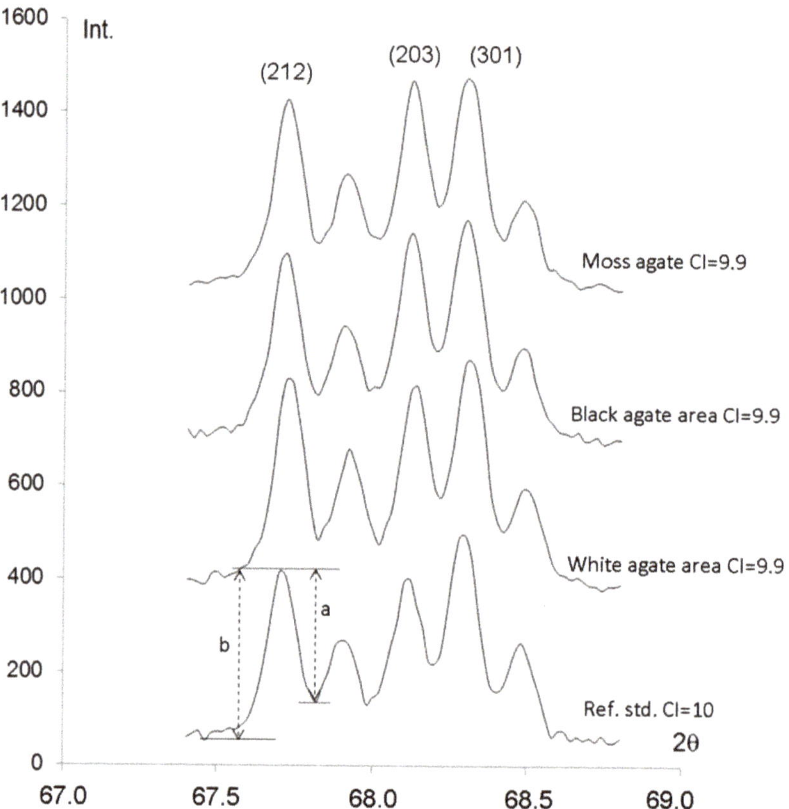

Figure 6. Fragments of X-ray diffractograms of the spherulitic agate, moss agate, black and white agate of concentrically zoning agate, and quartz crystal (Ref. std., Subpolar Urals), with the 2θ 68° quintuplet peak. a and b parameters are measured for Crystallinity Index (CI) calculation.

Scanning electron microscopical and microprobe investigations of agates revealed various mineral micro-inclusions, both in the black and white areas of all agate types. The most frequent inclusions were established as calcite in the form of macro- and micro-inclusions (Figure 7A,B). Micro-inclusions of K-feldspar and chlorite are also characteristic of all types of agates (Figure 7C). Pyrite, sphalerite, and arsenopyrite were recognised in black and white agate areas (Figure 7D–F). Iron oxides/hydroxides inclusions, typical of light-coloured varieties of agates of LSH [9,13], are rare in black agates.

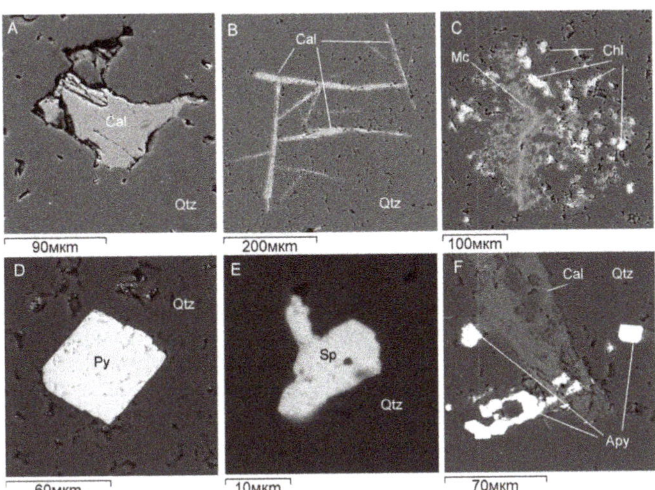

Figure 7. BSE images of micro-inclusions in black Onega agates: (**A**) calcite inclusion in the white area of the concentrically zoning agate, (**B**) calcite (Cal) inclusions in the black marginal area of quartzine spherolithe in spherulitic agate, (**C**) microcline (MC) and chlorite (Chl) in the white area of spherulitic agates, (**D**) prysmatic pyrite (Py) crystals in the colourless area of moss agate, (**E**) sphalerite (Sp) crystal in moss agate, and (**F**) quartz (Qtz) fissure filled by calcite, and accumulations of arsenopyrite (Apy) crystals in spherulitic agate.

Thermal analysis was applied to determine carbon content in white and black areas of CZ agate, as well as in bulk samples of MS and SP agates and the inter-pillow matrix. As a result of high-resolution thermogravimetric analysis and differential thermal analysis, several curves (TG, DTG, DSC, and dDSC) were obtained for each sample (Supplementary Figure S3). On DSC curves of all agate samples, a strong endothermic effect with a maximum at 569 °C was observed. This "quartz peak" is attributed to quartz $\alpha-\beta$ polymorph transition. On DSC curves of CZ, MS, and SP agates, and inter-pillow matrix samples in the temperature range of 500–670 °C, thermal effect attributed to carbon matter combustion [48,49] was detected. The loss in mass resulting from carbon matter combustion is 0.36% for MS agate, 0.18% for SP agate, and 0.69% for black and 0.05% for white areas of CZ agate (Table 1). The maximum carbon content of 1.87 wt.% was determined for the inter-pillow matrix.

Table 1. TG and DSC results for the black agates from the ZF of the Onega Basin.

Sample	Reaction	Temperature Range, °C	Mass Loss, %
CZw	Quartz transition	568.3	-
	Carbonaceous matter combustion	500–611	0.05
CZb	Quartz transition	568.9	-
	Carbonaceous matter combustion	474–669	0.69
MS	Quartz transition	569.2	-
	Carbonaceous matter combustion	534–652	0.36
SP	Quartz transition	568.9	-
	Carbonaceous matter combustion	580–672	0.18
IPM	Sulphide decomposition	460	0.63
	Carbonaceous matter combustion	527	1.87
	Calcite decomposition	746	11.77

Note: The measurements were performed using high-resolution thermal analysis. Samples: white (CZw) and black (CZb) agate areas of concentrically zoning agate, MS: moss agate, SP: spherulitic agate, IPM: inter-pillow matrix with carbonaceous material.

5.3. Chemical Characterisation

X-ray fluorescence spectrometry was used to provide information about the chemical composition of the black and white agate areas of CZ agate, as well as bulk samples of MS and SP agates (Supplementary Table S1). The studied samples are mainly composed of SiO_2 (97.8–98.7 wt.%) and have close contents of other oxides: Al_2O_3 (0.2–0.5%), FeO_{tot} (0–0.7%), CaO (0.5–0.7 wt.%), K_2O (0–0.1 wt.%), MgO (0.2–1.2), and TiO_2 (0.03–0.04 wt.%). The difference in value of losses on ignition (LOI) is established. The dark samples (black agate area of CZ agate and MS agate) exhibited a higher LOI (0.41 and 0.75 wt.%) than the white areas of CZ agate (0.27 wt.%). The lowest value of LOI (0.06%) was determined in SP agate.

The concentrations of the C, Si, and O in the local parts of the black and white (or colourless) areas in CZ, MS, and SP agates revealed differences in the carbon content of the black and white agate areas for all samples (Supplementary Table S2). The carbon content in the MS agate ranges from 10 to 19 wt.% in black to 5–9 wt.% in colourless areas, respectively. The CZ and SP samples are characterised by similar values of carbon content in black (6–12 wt.%, 7–13 wt.%) and white (2–6 wt.%, 4–7 wt.%) agate areas.

5.4. Characteristics of Carbonaceous Matter

The Raman spectra for carbonaceous material from all agate samples and the inter-pillow matrix are consistent with poorly ordered CM [40,41,50–52], because the first-order region of the Raman spectra shows two well-resolved bands: D1 at about 1350 cm^{-1} and G band at about 1600 cm^{-1} (Figure 8). It is noteworthy that the Raman spectra of CM in studied agates always contained the bands of admixture phases, commonly quartz and calcite. The results of spectrum decomposition for CM from agate samples showed the occurrence of two additional disorder-related bands, D3 (~1500 cm^{-1}) and D4 (~1260 cm^{-1}) (Supplementary Table S3). The low-intensity D2 band, which appeared in relatively ordered carbonaceous material [40,41,50–52], was established only in a few Raman spectra of CM from moss agates. In the second-order region, CM from all agate samples is mainly characterised by weak bands located at 2900 and 3200 cm^{-1}.

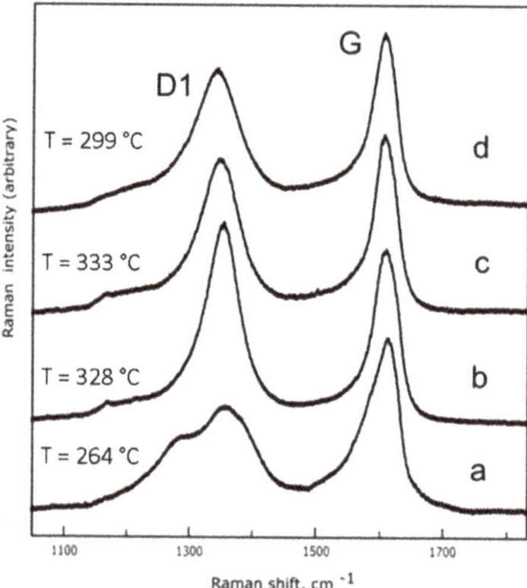

Figure 8. Representative Raman spectra of carbonaceous matter from concentrically zoning (**a**), spherulitic (**b**), and moss (**c**) agates, and inter-pillow matrix (**d**). Calculated metamorphic temperatures refer to the average value (calculated on the basis of 26 measurements for (**a**), 11 for (**b**), 10 for (**c**), and 28 for (**d**)).

However, in spite of the general similarity of CM Raman spectra, there are significant differences in Raman spectroscopic characteristics of CM from various agate types.

CM from CZ agates is characterised by parameter R1 (intensity-based ratio I_D/I_G) = 0.78 ± 0.06 and the high value of the FWHM-D1 = 99 ± 7 cm^{-1}. CMs from CZ agates have the significant scattering of both parameters R1 and FWMH-D1, which corresponds to the high degree of CM heterogeneity. Additionally, the decomposition of Raman spectra of CM from CZ agates shows that the D2 band is absent and the D3 and D4 bands have relatively high intensities; moreover, the latter might appear as a well-pronounced shoulder of the D1 band (Figure 8).

The Raman spectrum of CM from moss (MS) agate is characterised by a higher value of R1 = 0.93 ± 0.05 than the CM from CZ agates. However, the CM from moss agate has a significantly lower FWHM-D1 = 67 ± 0.5, varying in the narrow limits from 66 to 69 cm^{-1}. Unlike the CM from CZ agates, the scattering of the FWHM-D1 parameter for CM from moss agate is low, suggesting its low heterogeneity. The intensities of the D3 and D4 bands in the Raman spectra of CM from moss agates are lower than that of CZ agates (Supplementary Table S3). The D2 band has low intensity or is absent.

CMs from spherulitic (SP) agates are characterised by the higher value of R1 = 1.13 ± 0.5 compared to CZ and MS agates, and a value of FWMH-D1 = 70 ± 4 cm^{-1}, close to MS agates. The degree of scattering of R1 and FWMH-D1 parameters is intermediate between CM from CZ and MS agates. Additionally, the intensities of D3 and D4 bands in the Raman spectra of CM from SP agates have intermediate values between high in CZ agates and low in MS agates. The D2 band is absent similar to the CM from CZ agates.

CM from the inter-pillow matrix is characterised by R1 = 0.90 ± 0.05 and a high value of the FWHM-D1 = 83 ± 4 cm^{-1}. Similar to the CM from CZ agates, CM from the inter-pillow matrix have significant scattering of FWMH-D1, which might correspond to the high degree of CM heterogeneity. Additionally, similar to the CZ agates, the D2 band is absent and the D3 and D4 bands have relatively high intensities; moreover, the latter might appear as a well-pronounced shoulder of the D1 band (Figure 8) (Supplementary Table S3).

6. Discussion

6.1. Agate Varieties in the Onega Basin

Agates are widely distributed in the LSH within the Onega Basin and are mostly associated with the pillow lavas, and to a lesser extent with the massive lavas and pyroclastic tuffs. The previous study of agates [9] revealed that (Figure 1B) only light-coloured agates without carbonaceous material occurred in the lower part of SF (Yalguba Ridge, Pinguba, and Tulguba occurrences), whereas the black agates were not recorded. The light-coloured agates have greenish and brownish-red colours associated with the accessory minerals such as chlorite, phengite, epidote, and iron oxide/hydroxide. The light-coloured agate varieties occur in the inter-pillow lava space as lenses, veinlets, prominent nodules, as well as are gas vesicles' infill. They have fine- or coarse-banded concentric zoning and spotted or poor-fancy textures. Additionally, only a few light-coloured agates have been recorded in the upper part of ZF [9]. Their textural and mineralogical characteristics are similar to that of light-coloured agates from SF (Table 2).

6.2. Origin of CM in Agates

Carbon-bearing agates are rare, and the information on CM composition and origin is scarce in the literature. Apart from calcite, carbon can be presented by organic compounds or carbonaceous matter occurring as fluid or solid inclusions in agates. Various hydrocarbon gases (CH_4, C_2H_4, C_2H_6, C_3H_6, C_3H_8) were recorded in fluid inclusions in agates from the epithermal deposits from the Russian Polar Urals [58], that originated from the oil and gas under-layered sedimentary rocks. Solid hydrocarbon (bitumen) inclusions mostly consisting of asphaltenes have been reported for agates from acidic volcanic rocks (Lower Silesia, Poland) [11]. The stable carbon isotope analysis revealed that the bituminous CM had algal or algal-humic origin. For these rocks, the amorphous carbonaceous matter, probably with low content of hydrocarbons, was also recorded in veinlet and moss agates [59]. The carbonaceous matter with various degrees of structural ordering was recognised in agates from Triassic basalts (Meknés-Tafilalet, Asni and Agouim Regions, Morocco) [6,7] and onyx agates from Mali [10]. The CM from Mali agates was attributed to low-crystalline graphite, which might originate from hydrothermal formation of graphite from methane under elevated temperatures or graphitisation of organic precursors by secondary hydrothermal or metamorphic overprint [10].

Raman spectroscopy analysis shows that CM recorded in all types of black agates from ZF correspond to poorly ordered CM [40,41,50–52], however, the structural ordering is different in various agate types.

The relatively ordered CM was recorded in spherulitic and moss agates, which might be attributed to the medium grade CM according to the classification proposed by Kouketsu et al. [45]. It is characterised by high values of R1 close to 1 (R1 = 0.93–1.13), together with low FWHM-D1 = 67–70 cm^{-1} and low intensive disorder-related bands D3 and D4. The Raman spectroscopic features of CM from CZ agates and inter-pillow matrix, namely the low value of R1 = 0.78 accompanied with high FWHM-D1 = 90 cm^{-1}, and the presence of the intensive disorder-related bands D3 and D4, indicated the relatively low degree of ordering. Raman spectra of CM from CZ agates are constituent to the low-grade CM [45].

The empirical thermometers based on Raman spectroscopic features of CM are widely used to determine the metamorphic temperatures of CM [42,52,60–64]. They are based on the principle that the structural ordering of CM is an irreversible process, so the degree of graphitisation is an indicator of maximum temperature of metamorphic transformation, i.e., the structure and the microstructure of CM are unaffected by retrograde metamorphic events [42,65]. Since CMs from black agates were attributed to low- and medium-grade CM, we applied the Raman spectroscopy thermometer proposed by Kouketsu et al. [45] to determine the metamorphic temperature. This thermometer is based on the linear relation between FWHM-D1 and temperature in the range of T = 150–400 °C. The maximum temperatures of thermal transformation of CM from MS and SP agates are close and were determined to be T = 333 ± 1 °C and T = 328 ± 9 °C, respectively. CM from CZ agates experienced thermal transformations at lower temperatures, T = 264 ± 16 °C. It is noteworthy that temperatures determined for CM from CZ agates show significant scattering, indicating high heterogeneity of CM compared to relatively ordered CM from MS and SP agates (Figure 9). CMs from the inter-pillow matrix have intermediate values of methamorphic temperature, T = 299 ± 9 °C. These temperatures are consistent with prehnite-pumpellyite facies and are in agreement with previous studies [35,36].

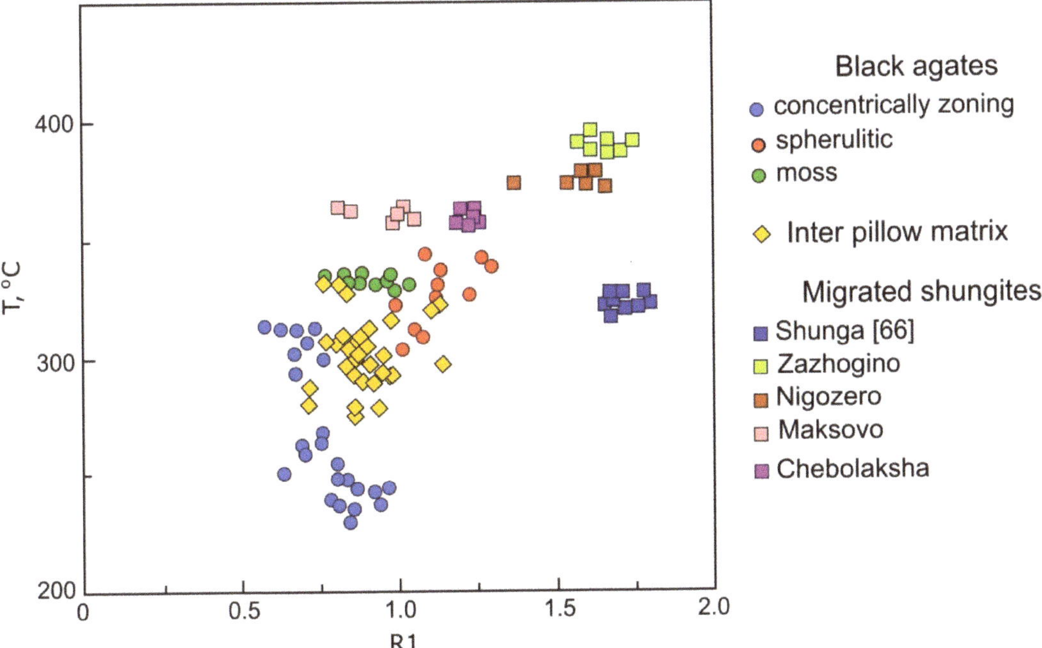

Figure 9. R1 vs. peak metamorphic temperature for the carbonaceous matter from black agates, inter-pillow matrix, and shungites from various deposits. Temperatures were calculated using the thermometer proposed by Kouketsu et al. [45]. Shunga [66].

The following sources of CM in black agates from ZF might be proposed:

1. CM might be incorporated from the inter-pillow matrix considering that initially the inter-pillow matrix might be composed of sedimentary organic matter.
2. CM might be supplied by hydrothermal fluids enriched in CM due to contamination from the host rocks.
3. CM might be formed by the redox reaction between CO_2 and CH_4 in hydrothermal fluids [67,68].

CM from CZ is characterised by high heterogeneity and can be divided into two groups. The first group includes CM with metamorphic temperatures close to 300 °C and similar to that of CM from the inter-pillow matrix (Figure 9), suggesting that CM in CZ agates might be partially incorporated from the inter-pillow matrix. The second group of CM have lower temperatures close to 250 °C and might be produced by the chemical reactions involving CO_2 and/or CH_4 in hydrothermal fluids [69]. Therefore, the data suggest that CM from CZ agates might originate both from the inter-pillow matrix and hydrothermal fluid.

Another possible source of CM in the studied agates might be associated with the underlying carbon-bearing shungite rocks. In order to decipher the model, we compared the metamorphic temperatures determined for migrated shungites from various shungite deposits using the thermometer proposed by Kouketsu et al. [45] with CM from studied agates. Figure 9 shows that CM from MS and SP agates are characterised by the metamorphic temperatures close to migrated shungites (Figure 9). CM from SP agates is characterised by higher heterogeneity than CM from MS agates, and there is a slight overlapping of metamorphic temperatures determined for CM from SP agates and the inter-pillow matrix. Therefore, we cannot exclude that CM from SP agates could partially originate from the CM of the inter-pillow matrix. However, the source of CM in MS and SP agates is mainly associated with the hydrothermal fluids enriched in mechanically derived CM incorporated from underlying shungite rocks.

6.3. Petrological Implications

Summarizing the data on the geology, geochemistry, and texture of black agates from the upper part of ZF obtained in the present study and the results of a previous study of light-coloured agates from the lower part of SF [9], the agate mineralisation in LSH can be generalised into 2 stages, as illustrated in Figure 10. Agate mineralisation in LSH occurred after the formation of volcanic-sedimentary rocks of ZF (2100–1980 Ma), including carbon-bearing mudstones (Figure 10a) and shungite rocks (ca. 2050 Ma) [31] (Figure 10b)—pre-stage "0".

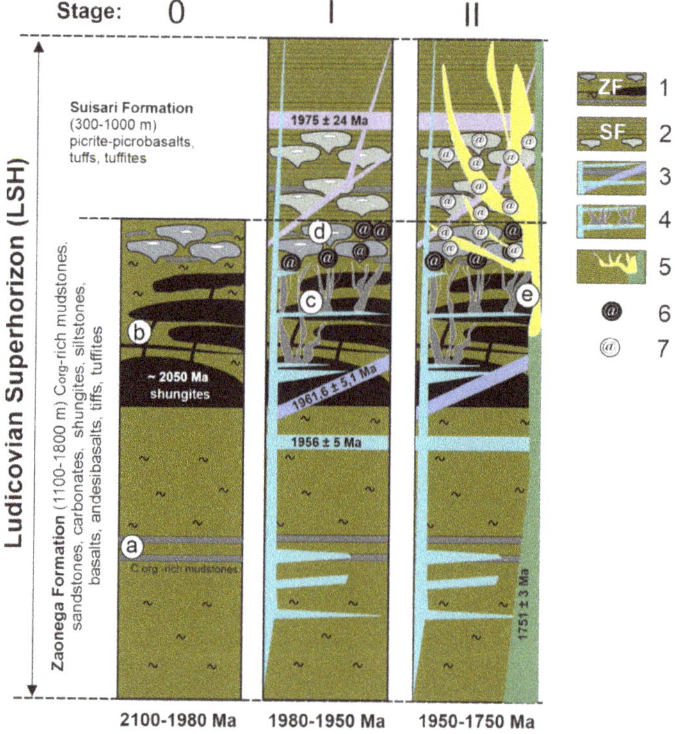

Figure 10. Schematic model of the agate formation stages in the Onega Basin. Compiled using data from [15,18,30,31,70,71]. Legend: Paleoproterozoic, Ludicovian Superhorizon ca. 2100–1920 Ma: 1—C_{org}-rich rocks (shungites), sandstones, siltstones, argillites, carbonates, basalts, andesibasalts, tuffs, tuffites (Zaonega Formation); 2—picrites, picritic basalts, tuffs, tuffites, tuff-conglomerates, gritstones (Suisari Formation); 3—ca. 1975–1956 Ma gabbroids; 4—ca. 1980–1950 Ma hydrothermal fluid flow; 5—ca. 1780–1750 Ma intrusion of gabbro-dolerite sills, hydrothermal fluid flow; 6—black agates; 7—light-coloured agates. Pre-stage 0: formation of carbon-bearing mudstone (**a**) and shungite rocks (**b**) in the volcano-sedimentary succession of the ZF. Stage I: formation of the SF of the volcano-sedimentary succession, multiphase intrusions of peridotite and gabbro-dolerite sills and dikes (**c**), and production of silica-carbonate hydrothermal fluids enriched in CM and formation of black agates in the upper part of ZF (**d**). Stage II: intrusion of gabbro-dolerite sills in the Onega Basin, triggering hydrothermal activity and formation of light-coloured agates in the upper part of ZF and lower part of SF (**e**).

Stage I involves the formation of black agates in the upper part of ZF and is associated with the multiphase intrusions of peridotite and gabbro-dolerite sills and dikes at 1980–1950 Ma [15,30,31] (Figure 10c). This event remobilised the CM from the weakened zones of shungite rocks and produced silica-carbonate hydrothermal fluids enriched in CM. The CM-bearing hydrothermal fluids formed the MS and SP black agates in the upper part of ZF (Figure 10d).

Stage II involves the formation of CZ agates since they contained heterogeneous CM originated both from the inter-pillow matrix and/or hydrothermal fluids. It was associated with a hydrothermal maximum at 1780–1750 Ma, initiated by the intrusion of gabbro-dolerite sills, and occurred at 190 Ma later than the continental flood basalts outflow stage. At this stage, light-coloured agates in both the upper part of ZF and the lower part of SF (Figure 10e) were formed.

7. Conclusions

(1) The present study provided the first detailed investigation of black agates occurring in volcanic rocks of the ZF within the Onega Basin (Karelian Craton, Fennoscandian Shield).

(2) The silica matrix of black agates is only composed by chalcedony, quartz, and quartzine. In addition to silica minerals, calcite, chlorite, feldspar, sulphides, and CM are also identified.

(3) Black colouration of agates is produced by disseminated carbonaceous matter within the silica matrix with a bulk content of less than 1 wt.%.

(4) Three main texture types of black agates were identified: monocentric concentrically zoning, polycentric spherulitic, and moss agates.

(5) The source of CM in moss and spherulitic agates is associated with the hydrothermal fluids enriched in mechanically derived carbonaceous matter incorporated from underlaying shungite rocks. CM from the concentrically zoning agates originate both from the inter-pillow matrix and hydrothermal fluid (i.e., via chemical reactions involving CO_2 and/or CH_4).

The present investigation contributes to understanding possible sources of carbonaceous matter in agates and displays probable mechanisms of redistribution of carbonaceous matter in the hydrothermal systems of the Paleoproterozoic volcanic rocks.

Supplementary Materials: The following are available online at https://www.mdpi.com/article/10.3390/min11090918/s1, Table S1: Chemical composition of black and white agates from the ZF LSH Onega Basin (XRF analysis, wt.%). Table S2: Chemical composition of local (black and white) areas of the agate, EDS microanalysis (wt.%). Table S3: Raman spectral characteristics of CM from black agates and inter-pillow matrix. Figure S1: X-ray diffractograms of the MS and SP agate samples and fragments of X-ray diffractograms in the range of 17–33° 2θ, with the most intense peaks of a-quartz for CZb, CZw, MS, and SP agate samples. Figure S2: Raman spectra of examined agates. (a) Spectrum of black agate area showing characteristic D1 bands at about 1350 cm^{-1} and G bands at about 1600 cm^{-1} of poorly ordered CM, as well as the most intense quartz (Qtz) and calcite (Cal) bands. (b) Spectrum of α-quartz in agate as the only silica phase present. Figure S3: DSC, TG, and DTG curves for examined agate and inter-pillow matrix samples.

Author Contributions: Conceptualisation, E.N.S., S.Y.C. and S.A.S.; investigation, E.N.S., S.Y.C. and S.A.S.; data curation, E.N.S. and S.Y.C.; writing—original draft, E.N.S., S.Y.C. and S.A.S.; writing—review and editing, S.A.S. and A.V.S.; visualisation, E.N.S., S.A.S. and S.Y.C.; supervision, S.A.S. All authors have read and agreed to the published version of the manuscript.

Funding: This research was funded by state assignment to the Institute of Geology Karelian Research Centre RAS.

Data Availability Statement: All data are contained within the article and supplementary materials.

Acknowledgments: We are grateful to S.V. Bordyukh, I.S. Inina, G.S. Ternovaya, and O.V. Bukchina (IG KRC RAS, Petrozavodsk, Russia) for their assistance in analytical investigations. The authors thank O.B. Lavrov (IG KRC RAS, Petrozavodsk, Russia) for consultancy in the field research process. We are grateful to Guest Editor G.A. Palyanova (Novosibirsk, Russia) for the invitation to write this article for the special issue "Agates: types, mineralogy, deposits, host rocks, ages and genesis", the careful analysis of the manuscript, and advice. We would like to deeply acknowledge four anonymous reviewers for their careful revisions and for their interesting suggestions, which helped us to improve the manuscript.

Conflicts of Interest: The authors declare no conflict of interest.

References

1. Moxon, T.; Palyanova, G. Agate Genesis: A Continuing Enigma. *Minerals* **2020**, *10*, 953. [CrossRef]
2. Götze, J.; Möckel, R.; Pan, Y. Mineralogy, geochemistry and genesis of agate—A review. *Minerals* **2020**, *10*, 1037. [CrossRef]
3. Godovikov, A.A.; Ripinen, O.I.; Motorin, S.G. *Agates*; Nedra: Moscow, Russia, 1987; p. 368. (In Russian)
4. Barsanov, G.P.; Yakovleva, V.E. *Mineralogy of Gemstones and Semi-Precious Varieties of Fine-Grained Chalcedony*; Nauka: Moscow, Russia, 1984; 144p. (In Russian)
5. Götze, J.; Tichomirowa, M.; Fuchs, H.; Pilot, J.; Sharp, Z.D. Chemistry of agates: A trace element and stable isotope study. *Chem. Geol.* **2001**, *175*, 523–541. [CrossRef]
6. Natkaniec-Nowak, L.; Dumańska-Słowik, M.; Pršek, J.; Lankosz, M.; Wróbel, P.; Gaweł, A.; Kowalczyk, J.; Kocemba, J. Agates from Kerrouchen (The Atlas Mountains, Morocco): Textural Types and Their Gemmological Characteristics. *Minerals* **2016**, *6*, 77. [CrossRef]
7. Pršek, J.; Dumańska-Słowik, M.; Powolny, T.; Natkaniec-Nowak, L.; Toboła, T.; Zych, D.; Skrepnicka, D. Agates from Western Atlas (Morocco)—Constraints from mineralogical and microtextural characteristics. *Minerals* **2020**, *10*, 198. [CrossRef]
8. Polekhovsky, Y.S.; Punin, Y.O. Agate mineralization in basaltoids of the northeastern Ladoga region, South Karelia. *Geol. Ore Depos.* **2008**, *50*, 642–646. [CrossRef]
9. Svetova, E.; Svetov, S. Mineralogy and Geochemistry of Agates from Paleoproterozoic Volcanic Rocks of the Karelian Craton, Southeast Fennoscandia (Russia). *Minerals* **2020**, *10*, 1106. [CrossRef]
10. Götze, J.; Nasdala, L.; Kempe, U.; Libowitzky, E.; Rericha, A.; Vennemann, T. The origin of black colouration in onyx agate from Mali. *Miner. Mag.* **2012**, *76*, 115–127. [CrossRef]
11. Dumańska-Słowik, M.; Natkaniec-Nowak, L.; Kotarba, M.J.; Sikorska, M.; Rzymełka, J.A.; Łoboda, A.; Gaweł, A. Mineralogical and geochemical characterization of the "bituminous" agates from Nowy Kościoł (Lower Silesia). *J. Mineral. Geochem.* **2008**, *184*, 255–268. [CrossRef]
12. Melezhik, V.A.; Medvedev, P.V.; Svetov, S.A. The Onega basin. In *Reading the Archive of Earth's Oxygenation*; Melezhik, V.A., Prave, A.R., Fallick, A.E., Kump, L.R., Strauss, H., Lepland, A., Hanski, E.J., Eds.; Springer: Berlin/Heidelberg, Germany, 2013; pp. 387–490.
13. Svetova, E.N.; Svetov, S.A. Agates from Paleoproterozoic volcanic rocks of the Onega Structure, Central Karelia. *Geol. Ore Depos.* **2020**, *62*, 669–681. [CrossRef]
14. Timofeev, V.M. To the genesis of Zaonega shungite. *Proc. Soc. St. Petersburg Nat.* **1924**, *39*, 99–122. (In Russian)
15. Puchtel, I.S.; Arndt, N.T.; Hofmann, A.W.; Haase, K.M.; Kröner, A.; Kulikov, V.S.; Kulikova, V.V.; Garbe-Schönberg, C.D.; Nemchin, A.A. Petrology of mafic lavas within the Onega plateau, central Karelia: Evidence for 2.0 Ga plume-related continental crustal growth in the Baltic Shield. *Contrib. Mineral. Petrol.* **1998**, *130*, 134–153. [CrossRef]
16. Slabunov, A.I.; Lobach-Zhuchenko, S.B.; Bibikova, E.V.; Sorjonen-Ward, P.; Balagansky, V.V.; Volodichev, O.I.; Shchipansky, A.A.; Svetov, S.A.; Chekulaev, V.P.; Arestova, N.A.; et al. The Archaean nucleus of the Fennoscandian (Baltic) Shield. In *European Lithosphere Dynamics*; Gee, D.G., Stephenson, R.A., Eds.; Memoirs, No. 32; Geological Society: London, UK, 2006; pp. 627–644.
17. Arestova, N.A.; Chekulaev, V.P.; Lobach-Zhuchenko, S.B.; Kucherovskii, G.A. Formation of the Archean crust of the ancient Vodlozero domain (Baltic shield). *Stratigr. Geol. Correl.* **2015**, *23*, 119–130. [CrossRef]
18. Kulikov, V.S.; Svetov, S.A.; Slabunov, A.I.; Kulikova, V.V.; Polin, A.K.; Golubev, A.I.; Gorkovets, V.Y.; Ivashchenko, V.I.; Gogolev, M.A. Geological map of Southeastern Fennoscandia (scale 1:750,000): A new approach to map compilation. *Trans. KarRC RAS* **2017**, *2*, 3–41. [CrossRef]
19. Stepanova, A.V.; Samsonov, A.V.; Salnikova, E.B.; Puchtel, I.S.; Larionova, Y.O.; Larionov, A.N.; Stepanov, V.S.; Shapovalov, Y.B.; Egorova, S.V. Palaeoproterozoic Continental MORB-type Tholeiites in the Karelian Craton: Petrology, Geochronology, and Tectonic Setting. *J. Petrol.* **2014**, *55*, 1719–1751. [CrossRef]
20. Kulikov, V.S.; Kulikova, V.V.; Lavrov, B.S.; Pisarevskii, S.A.; Pukhtel, I.S.; Sokolov, S.Y. *The Paleoroterozoic Suisarian Picrite–Basalt Complex in Karelia: Key Section and Petrology*; KNTs RAN: Petrozavodsk, Russia, 1999; 96p. (In Russian)
21. Kulikov, V.S.; Rychanchik, D.V.; Golubev, A.I.; Filippov, M.M.; Tarkhanov, G.V.; Frik, M.G.; Svetov, S.A.; Kulikova, V.V.; Sokolov, S.Y.; Romashkin, A.E. Stratigraphy and magmatism. Ludicovian. In *Paleoproterozoic Onega Structure: Geology, Tectonics, Structure, and Metallogeny*; Glushanin, L.V., Sharov, N.V., Shchiptsov, V.V., Eds.; Karelian Research Centre, RAS: Petrozavodsk, Russia, 2011; pp. 67–101. (In Russian)

22. Galdobina, L.P. The Ludicovian Superhorizon. In *Geology of Karelia*; Sokolov, V.A., Ed.; Nauka: Leningrad, Russia, 1987; pp. 59–67. (In Russian)

23. Buseck, P.R.; Galdobina, L.P.; Kovalevski, V.V.; Rozkova, N.N.; Valley, J.W.; Zaidenberg, A.Z. Shungites: The C-rich rocks of Karelia, Russia. *Can. Mineral.* **1997**, *35*, 1363–1378.

24. Melezhik, V.A.; Fallick, A.E.; Filippov, M.M.; Larsen, O. Karelian shungite—An indication of 2.0-Ga-old metamorphosed oil-shale and generation of petroleum: Geology, lithology and geochemistry. *Earth Sci. Rev.* **1999**, *47*, 1–40. [CrossRef]

25. Melezhik, V.A.; Fallick, A.E.; Filippov, M.M.; Lepland, A.; Rychanchik, D.V.; Deines, Y.E.; Medvedev, P.V.; Romashkin, A.E.; Strauss, H. Petroleum surface oil seeps from a Paleoproterozoic petrified giant oilfield. *Terra Nova* **2009**, *21*, 119–126. [CrossRef]

26. Kump, L.R.; Junium, C.; Arthur, M.A.; Brasier, A.; Fallick, A.; Melezhik, V.; Lepland, A.; Črne, A.E.; Luo, G.M. Isotopic evidence for massive oxidation of organic matter following the great oxidation event. *Science* **2011**, *334*, 1694–1696. [CrossRef]

27. Kovalevskii, S.V.; Moshnikov, I.A.; Kovalevski, V.V. Heat-treated nano-structured shungite rocks and electrophysical properties associated. *Nanosyst. Phys. Chem. Math.* **2018**, *9*, 468–472. [CrossRef]

28. Filippov, M.M. *Shungite Rocks of the Onega Structure*; KarNTs RAN: Petrozavodsk, Russia, 2002; 282p. (In Russian)

29. Inostrantsev, A.A. New end member of amorphous carbon. *Gorn. J.* **1897**, *2*, 314–342. (In Russian)

30. Stepanova, A.V.; Samsonov, A.V.; Larionov, A.N. The final episode of the middle proterozoic magmatism in the Onega structure: Data on Trans-Onega dolerites. *Trans. KarRC RAS* **2014**, *1*, 3–16.

31. Martin, A.P.; Prave, A.R.; Condon, D.J.; Lepland, A.; Fallick, A.E.; Romashkin, A.E.; Medvedev, P.V.; Rychanchik, D.V. Multiple Palaeoproterozoic carbon burial episodes and excursions. *Earth Planet. Sci. Lett.* **2015**, *424*, 226–236. [CrossRef]

32. Gudin, A.N.; Dubinina, E.O.; Nosova, A.A. Petrogenesis of Variolitic Lavas of the Onega Structure, Central Karelia. *Petrology* **2012**, *20*, 255–270. [CrossRef]

33. Svetov, S.A.; Chazhengina, S.J. Geological Phenomenon of Yalguba RidgeVariolite from F. Yu. Levinson-Lessing's Time until Today: Mineralogical and Geochemical Aspects. *Geol. Ore Depos.* **2018**, *60*, 547–558. [CrossRef]

34. Svetov, S.A.; Chazhengina, S.Y.; Stepanova, A.V. Geochemistry and texture of clinopyroxene phenocrysts from Paleoproterozoic picrobasalts, Onega Basin, Fennoscandian Shield: Records of magma mixing processes. *Minerals* **2020**, *10*, 434. [CrossRef]

35. Leonov, M.G.; Kulikov, V.S.; Zykov, D.S.; Kolodyazhny, S.Y.; Poleshchuk, A.V. Tectonics. In *Paleoproterozoic Onega Structure: Geology, Tectonics, Structure, and Metallogeny*; Glushanin, L.V., Sharov, N.V., Shchiptsov, V.V., Eds.; Karelian Research Centre, RAS: Petrozavodsk, Russia, 2011; pp. 127–170. (In Russian)

36. Spiridonov, E.M.; Putintzeva, E.V.; Lavrov, O.B.; Ladygin, V.M. Kronstedtite, pumpelliite, prehnite and lennilenapeite in the metaagates and metabasalts of the early Proterozoic trap formation in the northern Onega region. In Proceedings of the Conference Lomonosov Readings, Moscow, Russia, 17–27 April 2017; Moscow State University: Moscow, Russia, 2017. Available online: https://conf.msu.ru/file/event/4305/eid4305_attach_b0acc3e7de2cd859225469534617a6272d70ce50.pdf (accessed on 9 July 2021). (In Russian).

37. Filipov, M.M.; Deynes, Y.E. *Subtabular Type of Shungite Deposits of Karelia*; KarSC RAS: Petrozavodsk, Russia, 2018; 261p. (In Russian)

38. Glebovitskii, V.A.; Bushmin, S.A.; Belyatsky, B.V.; Bogomolov, E.S.; Borozdin, A.P.; Savva, E.V.; Lebedeva, Y.M. RB-SR age of metasomatism and ore formation in the low-temperature shear zones of the Fenno-Karelian Craton, Baltic Shield. *Petrology* **2014**, *22*, 184–204. [CrossRef]

39. Bibikova, E.V.; Kirnozova, T.I.; Lazarev, Y.I.; Makarov, V.A.; Nikolaev, A.A. U–Pb isotopic age of the Karelian Vepsian. *Trans. USSR Acad. Sci.* **1990**, *310*, 189–191.

40. Ferrari, A.C.; Robertson, J. Interpretation of Raman spectra of disordered and amorphous carbon. *Phys. Rev.* **2000**, *61*, 14095–14107. [CrossRef]

41. Wopenka, B.; Pasteris, J.D. Structural characterization of kerogens to granulite-facies graphite: Applicability of Raman microprobe spectroscopy. *Am. Mineral.* **1993**, *78*, 533–557.

42. Beyssac, O.; Goffé, B.; Chopin, C.; Rouzaud, J.N. Raman spectra of carbonaceous material in metasediments: A new geothermometer. *J. Metamorph. Geol.* **2002**, *20*, 859–871. [CrossRef]

43. Dippel, B.; Jander, H.; Heintzenberg, J. NIR FT Raman spectroscopic study of flame soot. *J. Aerosol Sci.* **1999**, *30*, 4707–4712. [CrossRef]

44. Lahfid, A.; Beyssac, O.; Deville, E.; Negro, F.; Chopin, C.; Goffé, B. Evolution of the Raman spectrum of carbonaceous material in low-grade metasediments of the Glarus Alps (Switzerland). *Terra Nova* **2010**, *22*, 354–360. [CrossRef]

45. Kouketsu, Y.; Mizukami, T.; Mori, H.; Endo, S.; Aoya, M.; Hara, H.; Nakamura, D.; Wallis, S. A new approach to develop the Raman carbonaceous material geothermometer for low-grade metamorphism using peak width. *Isl. Arc.* **2014**, *23*, 33–50. [CrossRef]

46. Murata, J.; Norman, M.B. An index of crystallinity for quartz. *Am. J. Sci.* **1976**, *276*, 1120–1130. [CrossRef]

47. Kuznetsov, S.K.; Svetova, E.N.; Shanina, S.N.; Filippov, V.N. Minor Elements in Quartz from Hydrothermal-Metamorphic Veins in the Nether Polar Ural Province. *Geochemistry* **2012**, *50*, 911–925. [CrossRef]

48. Behera, D.; Nandi, B.K.; Bhattacharya, S. Chemical properties and combustion behavior of constituent relative density fraction of a thermal coal. *Energy Sources Part A Recovery Util. Environ. Eff.* **2019**, *41*, 654–664. [CrossRef]

49. Wada, H.; Tomita, T.; Matsuura, K.; Iuchi, K.; Ito, M.; Morikiyo, T. Graphitization of carbonaceous matter during metamorphism with references to carbonate and pelitic rocks of contact and regional metamorphisms, Japan. *Contrib. Mineral. Petrol.* **1994**, *118*, 217–228. [CrossRef]

50. Kholodkevich, S.V.; Berezkin, V.I.; Davydov, V.Y. Specific structural features and thermal resistance of shungite carbon to graphitization. *Phys. Solid State* **1999**, *41*, 1291–1294. [CrossRef]

51. Jehlicka, J.; Urban, O.; Pokorný, J. Raman spectroscopy of carbon and solid bitumens in sedimentary and metamorphic rocks. *Spectrochim. Acta* **2003**, *59*, 2341–2352. [CrossRef]

52. van Zuilen, M.A.; Fliegel, D.; Wirth, R.; Lepland, A.; Qu, Y.; Schreiber, A.; Romashkin, A.E.; Philippot, P. Mineral-templated growth of natural graphite films. *Geochim. Cosmochim. Acta* **2012**, *83*, 252–262. [CrossRef]

53. Moxon, T.; Carpenter, M.A. Crystallite growth kinetics in nanocrystalline quartz (agate and chalcedony). *Miner. Mag.* **2009**, *73*, 551–568. [CrossRef]

54. Moxon, T.; Nelson, D.R.; Zhang, M. Agate recrystallization: Evidence from samples found in Archaean and Proterozoic host rocks, Western Australia. *Austr. J. Earth Sci.* **2006**, *53*, 235–248. [CrossRef]

55. Moxon, T.; Reed, S.J.; Zhang, M. Metamorphic effects on agate found near the Shap granite, Cumbria, England: As demonstrated by petrography, X-ray diffraction and spectroscopic methods. *Miner. Mag.* **2007**, *71*, 461–476. [CrossRef]

56. Götze, J.; Möckel, R.; Kempe, U.; Kapitonov, I.; Vennemann, T. Origin and characteristics of agates in 1479 sedimentary rocks from the Dryhead area, Montana/USA. *Min. Mag.* **2009**, *73*, 673–690.

57. Folk, R.L.; Pittman, J.S. Length-slow chalcedony; A new testament for vanished evaporates. *J. Sediment. Petrol.* **1971**, *41*, 1045–1058.

58. Silaev, V.I.; Shanina, S.N.; Ivanovskii, V.S. Inclusions of oil gases in agate-type secretions: Implications for forecast of the oil- and gas-bearing potential of the Polar Urals. *Dokl. Earth Sci.* **2002**, *383*, 246–252. (In Russian)

59. Powolny, T.; Dumańska-Słowik, M.; Sikorska-Jaworowska, M.; Wójcik-Baniaaet, M. Agate mineralization in spilitized Permian volcanics from "Borówno" quarry (Lower Silesia, Poland)—Microtextural, mineralogical, and geochemical constraints. *J. Ore Geol. Rev.* **2019**, *114*, 103–130. [CrossRef]

60. Sforna, C.; van Zuilen, M.A.; Philippot, P. Structural characterization by Raman hyperspectral mapping of organic carbon in the 3.46 billion-year-old Apex chert, Western Australia. *Geochim. Cosmochim. Acta* **2014**, *124*, 18–33. [CrossRef]

61. Rantitsch, G.; Grogger, W.; Teichert, C.; Ebner, F.; Hofer, C.; Maurer, E.M.; Schaffer, B.; Toth, M. Conversion of carbonaceous material to graphite within the Greywacke Zone of the Eastern Alps. *Int. J. Earth Sci. Geol. Rundsch.* **2004**, *93*, 959. [CrossRef]

62. Rahl, J.M.; Anderson, K.M.; Brandon, M.T.; Fassoulas, C. Raman spectroscopic carbonaceous material thermometry of low-grade metamorphic rocks: Calibration and application to tectonic exhumation in Crete, Greece. *Earth Planet. Sci. Lett.* **2005**, *240*, 339–354. [CrossRef]

63. Aoya, M.; Kouketsu, Y.; Endo, S.; Shimizu, H.; Mizukami, T.; Nakamura, D.; Wallis, S. Extending the applicability of the Raman carbonaceous-material geothermometer using data from contact metamorphic rocks. *J. Metamorph. Geol.* **2010**, *28*, 895–914. [CrossRef]

64. Wiederkehr, M.; Bousquet, R.; Ziemann, M.; Berger, A.; Schmid, S. 3–D assessment of peak–metamorphic conditions by Raman spectroscopy of carbonaceous material: An example from the margin of the Lepontine dome (Swiss Central Alps). *Int. J. Earth Sci.* **2011**, *100*, 1029. [CrossRef]

65. Buseck, P.R.; Beyssac, O. From Organic Matter to Graphite: Graphitization. *Elements* **2014**, *10*, 421–426. [CrossRef]

66. Chazhengina, S.Y.; Kovakevski, V.V. Raman spectroscopy of weathered shungites. *J. Raman Spectrosc.* **2017**, *48*, 1590–1596. [CrossRef]

67. Manning, C.E.; Shock, E.L.; Sverjensky, D.A. The chemistry of carbon in aqueous fluids at crustal and upper-mantle conditions: Experimental and theoretical constraints. *Rev. Mineral. Geochem.* **2013**, *75*, 109–148. [CrossRef]

68. Kríbek, B.; Sykorová, I.; Machovic, V.; Knésl, I.; Laufek, F.; Zacharias, J. The origin and hydrothermal mobilization of carbonaceous matter associated with Paleoproterozoic orogenic-type gold deposits of West Africa. *Precambrian Res.* **2015**, *270*, 300–317. [CrossRef]

69. Foustoukos, D.I. Metastable equilibrium in the C H O system: Graphite deposition in crustal fluids. *Am. Mineral.* **2012**, *97*, 1373–1380. [CrossRef]

70. Hannah, J.L.; Stein, H.J.; Yang, G.; Zimmerman, A.; Melezhik, V.A.; Filippov, M.M.; Turgeon, S.C. Re—Os geochronology of a 2.05 Ga fossil oil field near Shunga, Karelia, NW Russia. In Proceedings of the Abstracts of the 33 International Geological Congress, Oslo, Norway, 5–14 August 2008.

71. Lubnina, N.V.; Pisarevsky, S.A.; Söderlund, U.; Nilsson, M.; Sokolov, S.J.; Khramov, A.N.; Iosifidi, A.G.; Ernst, R.; Romanovskaya, M.A.; Pisakin, B.N. New palaeomagnetic and geochronological data from the Ropruchey sill (Karelia, Russia): Implications for late Paleoproterozoic palaeogeography. In Proceedings of the Supercontinent Symposium 2012—Programm and Abstracts, Espoo, Finland, 25–28 September 2012; University of Helsinki: Espoo, Finland, 2012; pp. 81–82. Available online: https://tupa.gtk.fi/julkaisu/erikoisjulkaisu/ej_084.pdf (accessed on 9 August 2021).

 *minerals*

Article

Copper-Containing Agates of the Avacha Bay (Eastern Kamchatka, Russia)

Galina Palyanova [1,2,*], **Evgeny Sidorov** [3], **Andrey Borovikov** [2] **and Yurii Seryotkin** [1,2]

[1] Sobolev Institute of Geology and Mineralogy, Siberian Branch of Russian Academy of Sciences, Akademika Koptyuga pr. 3, 630090 Novosibirsk, Russia; yuvs@igm.nsc.ru

[2] Department of Geology and Geophysics, Novosibirsk State University, Pirogova str. 2, 630090 Novosibirsk, Russia; borovikov.57@mail.ru

[3] Institute of Volcanology and Seismology, Far East Branch of Russian Academy of Sciences, Piipa Blvd. 9, 683006 Petropavlovsk-Kamchatsky, Russia; mineral@kscnet.ru

* Correspondence: palyan@igm.nsc.ru

Received: 22 October 2020; Accepted: 8 December 2020; Published: 14 December 2020

Abstract: The copper-containing agates of the Avacha Bay (Eastern Kamchatka, Russia) have been investigated in this study. Optical microscopy, scanning electron microscopy, electron microprobe analysis, X-ray powder diffraction, Raman spectroscopy, and fluid inclusions were used to investigate the samples. It was found that copper mineralization in agates is represented by native copper, copper sulphides (chalcocite, djurleite, digenite, anilite, yarrowite, rarely chalcopyrite) and cuprite. In addition to copper minerals, sphalerite and native silver were also found in the agates. Native copper is localized in a siliceous matrix in the form of inclusions usually less than 100 microns in size—rarely up to 1 mm—forming dendrites and crystals of a cubic system. Copper sulphides are found in the interstices of chalcedony often cementing the marginal parts of spherule aggregates of silica. In addition, they fill the micro veins, which occupy a cross-cutting position with respect to the concentric bands of chalcedony. The idiomorphic appearance of native copper crystals and clear boundaries with the silica matrix suggest their simultaneous crystallization. Copper sulphides, cuprite, and barite micro veins indicate a later deposition. Raman spectroscopy and X-ray powder diffraction results demonstrated that the Avacha Bay agates contained cristobalite in addition to quartz and moganite. The fluid inclusions study shows that the crystalline quartz in the center of the nodule in agates was formed with the participation of solutions containing a very low salt concentration (<0.3 wt.% NaCl equivalent) at the temperature range 110–50 °C and below. The main salt components were $CaCl_2$ and NaCl, with a probable admixture of $MgCl_2$. The copper mineralization in the agates of the Avacha Bay established in the volcanic strata can serve as a direct sign of their metallogenic specialization.

Keywords: agate; Avacha Bay; Kamchatka; Russia; copper; SEM; EPMA; XRD; RS; inclusions; genesis

---

1. Introduction

The mineralogy of agate nodules in basalts has been studied and summarized in many papers [1–19]. Agate is a banded chalcedony, which can be intergrown or intercalated with other silica phases (macro-crystalline quartz, quartzine, opal-A, opal-CT, cristobalite, and moganite) [8]. Some common secondary minerals can be found in agate. Oxides of iron add color, but calcite would be regarded as an impurity. Celadonite is often found as an outer coating but can be found in the agate from basic igneous hosts. Nevertheless, copper and copper minerals in agates are rare. The copper-included agates were originally collected in 1952 [20], despite this, the number of papers on them is relatively small [21].

Agates containing native copper and (or) copper sulphides have been identified in some deposits [18–28]. Native copper, cuprite, and zincite are found in the Ayaguz agate deposit (East Kazakhstan) [18]. Here, native copper plates and grains are located in the second cristobalite-chalcedony zone, following a thin first outer zone of an oval-shaped chalcedony. Native copper and copper minerals are often an intergrowth with cristobalite spheroids and cuprite. Additionally, Radko [22] notes the presence of a fine impregnation of native copper and copper oxides in agates of the Norilsk region (Russia). This dispersion of micro grains of copper gives them a reddish color.

Native copper was noted locally, but in considerable concentrations in an inner agate laminae of "bituminous" agates from Nowy Kościół (Lower Silesia, Poland) [23]. Copper-banded agates, in which native copper sometimes fills up to 80% of a nodule's volume, are known from the Wolverine Mine, Wolverine, Houghton County, Michigan (USA) [19,21]. These unusual agates, which range in size from 1 to 4 cm across, occur in a basaltic lava flow that is approximately 1100 Ma. This makes them, along with the regular Lake Superior agates, the oldest agates on the American continent [10]. The Kearsarge copper-bearing amygdaloidal lode in Houghton County, Michigan (USA) formed commercial ore bodies [21]. Nezafati et al. [24] considered the Darhand copper occurrence in Central Iran as an example of Michigan-type native copper deposits.

The genesis of agates with native copper and copper sulphides has been discussed in few papers [18,19,21,24–28]. Agates with native copper were found in paleobasalts from Rudno near Krzeszowice in Lower Silesia (Poland) [25]. The native copper inclusions are irregular with jagged edges and ranged in size from 0.01 to 2 mm. The analyses of sections also showed cuprite, apart from native copper. Cuprite occurs at the contact of native copper and chalcedony or even replaces the former, creating separate pockets. The relation of copper mineralization to other minerals of the agate paragenesis suggests its secondary origin and formation in the zone of oxidation.

Copper sulphides without native copper are found in agates from Sidi Rahal and Kerrouchen (Khénifra Province, Meknés-Tafilalet Region), in the Atlas mountains of Morocco [26,27]. Moroccan agates occur within Triassic basaltoids as lens-shaped specimens in sizes up to 25–30 cm. Tiny single crystals of copper sulphides (unkown compostion) have been found in the central part of agate nodules from Sidi Rahal and were accompanied by idiomorphic quartz [26]. Copper sulphides, calcite, and an organic substance were incorporated during post-magmatic, hydrothermal or hypergenic conditions. The aggregates of copper sulphides and titanium oxides (rutile) in agate nodules from Kerrouchen (Khénifra Province, Meknés-Tafilalet Region) in Morocco [27] were most likely deposited during post-magmatic processes. It was also proposed that the origin of solid bitumen in these agates was the result of low temperature hydrothermal or hypergenic processes.

Powolny et al. [28] studied vein agates and moss agates, which were encountered for the first time within spilitized (albite-rich) alkali-basalts from the upper parts of the Borówno quarry (Intra-Sudetic Basin, Lower Silesia, Poland). Vein agates consist predominately of length-fast chalcedony evolving into strongly zoned prismatic quartz with numerous solid inclusions such as Fe-Cu sulphides that included chalcocite and chalcopyrite.

Agate-bearing strata have been detected in the vicinity of Petropavlovsk-Kamchatsky, located along the eastern shore of Avacha Bay (Eastern Kamchatka, Russia). Sidorov et al. [29] were the first to report native copper inclusions in agates from the Shlyupochnaya Bay in the Avacha Bay. However, to date, detailed studies of agates with copper have not been made. Taking this into account, the authors in the present paper undertook to carry out detailed researches of the copper containing agates, using a set of various analytical methods: optical microscopy, scanning electron microscopy (SEM), electron microprobe analysis (EPMA), X-ray powder diffraction (XRD), Raman spectroscopy (RS), and fluid inclusions studies.

The prime objective of this study is to examine the copper and other minerals in agates of the Eastern Kamchatka, Russia. Of particular interest was the relationship between copper and the supporting silica minerals. The article provides new information on the mineralogical features of agates from the coastal outcrops of Shlyupochnaya Bay in the Avacha Bay and substantiates the conditions for the genesis of copper mineralization in agates. The copper content of the Avacha Bay agates is low compared to similar agates from Michigan, USA. Nevertheless, the Avacha Bay agates have sufficient copper and compounds to allow comment on its possible genesis.

2. Geological Setting and Description of Agates

2.1. Geological Setting

Agate-containing strata are found along the eastern coast of Avacha Bay (Figure 1). The most ancient formations developed along this territory are the volcanic-sedimentary deposits of the Nikolskaya strata (K_2nk) [30]. The strata is divided into two substructures: the lower strata is a layer of green clay schist and siliceous schist with weakly metamorphosed sandstone, siltstone and mudstone. The upper substructure is composed of volcanogenic-terrigenous rocks: siliceous rocks, metamorphosed siltstone and sandstone, greenstone altered tuffs, and spherical basalt.

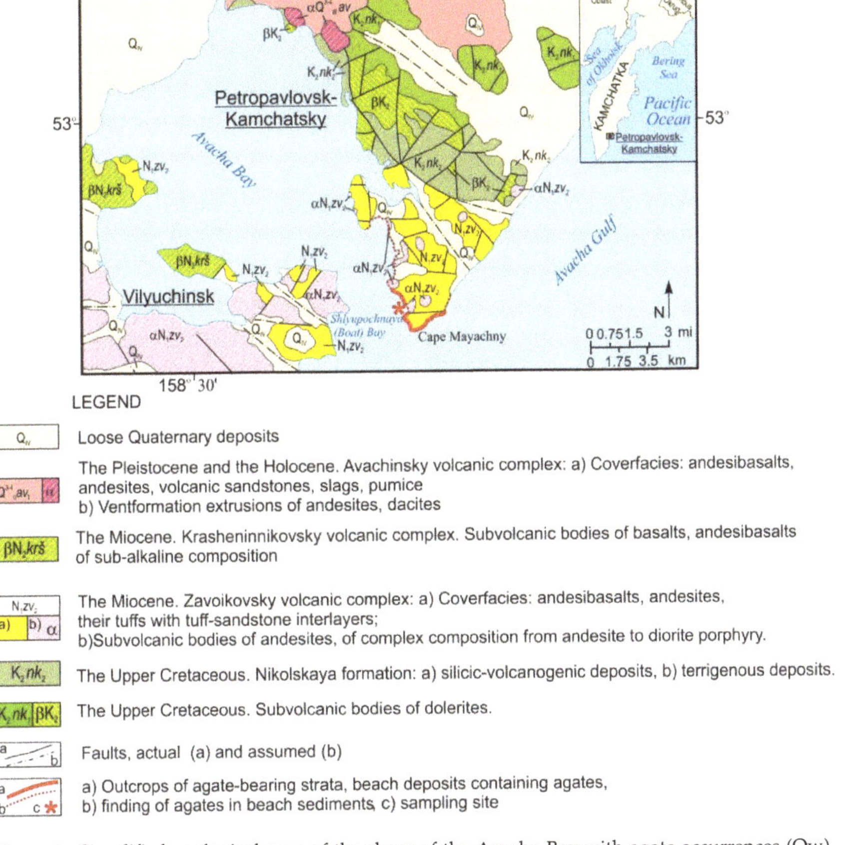

Q$_{IV}$	Loose Quaternary deposits
Q³⁴$_{II}$av$_1$	The Pleistocene and the Holocene. Avachinsky volcanic complex: a) Coverfacies: andesibasalts, andesites, volcanic sandstones, slags, pumice b) Ventformation extrusions of andesites, dacites
βN$_1$krš	The Miocene. Krasheninnikovsky volcanic complex. Subvolcanic bodies of basalts, andesibasalts of sub-alkaline composition
N$_1$zv$_2$ a) b) α	The Miocene. Zavoikovsky volcanic complex: a) Coverfacies: andesibasalts, andesites, their tuffs with tuff-sandstone interlayers; b)Subvolcanic bodies of andesites, of complex composition from andesite to diorite porphyry.
K$_2$nk$_2$	The Upper Cretaceous. Nikolskaya formation: a) silicic-volcanogenic deposits, b) terrigenous deposits.
K$_2$nk βK$_2$	The Upper Cretaceous. Subvolcanic bodies of dolerites.
a ‒ ‒ b	Faults, actual (a) and assumed (b)
a b c ✳	a) Outcrops of agate-bearing strata, beach deposits containing agates, b) finding of agates in beach sediments, c) sampling site

Figure 1. Simplified geological map of the shore of the Avacha Bay with agate occurrences (Qw), after Sheymovich [30].

The Late Cretaceous age of the Nikolskaya formation was determined from the discovered remains of fauna, spores, and pollen in the northern part of the city area [30,31]. Cretaceous rocks are overlain by tuff of the early phase of the Avacha volcanic complex of Pleistocene-Holocene age (Figure 1).

Agate-bearing effusive-pyroclastic formations in the vicinity of Petropavlovsk-Kamchatsky exposed in the cliffs along the shores of the Avacha Bay and on the ocean coast are united in the Zavoykovsky complex. This complex combines cover and subvolcanic intrusive formations developed in the coastal zone of the Avacha Bay. The cover phase unites lava flows, lavobreccia, and the tuff of predominantly andesite/basalt composition. Effusive-pyroclastic formations are common in the shore cliffs of the Avacha Bay [31].

Tuffs, mainly agglomerate, rarely psephite, occur as separate layers with a thickness of 10 to 40 m—rarely up to 150 m. Among the lavas, porphyritic rocks prevail andesites and basalts. Subvolcanic formations are represented by stocks, which are dominated by andesite and bodies of complex structure, the composition of which varies from andesite to diorite porphyrite. Morphologically, subvolcanic bodies are often distinguished in the form of separate dome-shaped peaks; in plain view they have an oval shape. The area of their outcrops ranges from 0.5 to 40 km^2.

The effusive-pyroclastic facies of the volcanic complex contain organogenic remains that indicate the Miocene age of the enclosing sediments. This dating was later confirmed by the K-Ar age determinations of subvolcanic intrusions of andesite composition. Their ages are in the range of 12.5 to 18.4 Ma [30].

Low grade metamorphism of the agate-containing Miocene volcanic rocks is characterized by the zeolite facies and the uneven processing of the rocks. Secondary minerals are dominated by zeolites, hematite, goethite, calcite, and corrensite. Celadonite stains the rock bright green [32].

Agates and chalcedony are detected among the lava flows and breccia of the late phase of the Zavoykovsky complex. The agates are to be found in veins, veinlets, and rounded openings. The latter can be 20 cm in diameter.

At the foot of rocky coastal cliffs are exposures of volcanites with an almond-stone texture. These were formed by contractions and veinlets of agate and chalcedony in the host volcanites (Figure 2a,b). Well-rounded beach agates show that agates are being released from the host rock.

Figure 2. (**a**) Coastal outcrops of Shlyupochnaya Bay (Avacha Bay, Eastern Kamchatka, Russia) (photo G. Palyanova); (**b**) agate nodules (1–10 cm) in basalt (photo by D. Bukhanova).

2.2. Description of Agates

The agates from the coastal zone of the Avacha Bay are grey but can show a bluish tint that can reach a sapphire blue color in intensity [29,33]. The agate bands range in size and generally show alternating zones of a bluish color. Moss agates can be found but are less common. One unusual feature of agates and chalcedony from the Zavoykovsky volcanites is the frequent presence of native copper. These copper inclusions can range in size from fractions of 10 µm up to 1 mm. Native copper in these agates shows a variety of forms: filamentary aggregates, thin plates, dendrites, and individual perfectly faceted cubic crystals, octahedra and their twins [29]. Figure 3 shows a fragment of rock with dendrites of native copper in the cubic habit, mainly octahedrons.

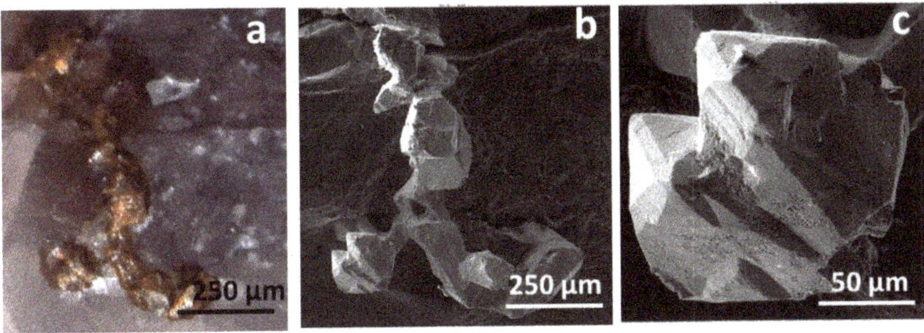

Figure 3. The dendrites of native copper on the surface of a fragment of rock. (**a**) Photo by G. Palyanova, (**b,c**) SEM micrographs.

3. Materials and Methods

The research was based on the collection of agates and agate-bearing basalts from the Avacha Bay (Eastern Kamchatka, Russia). Agates were taken from the coastal outcrop and coastal strip of Shlyupochnaya Bay (Figure 2). The agates were examined using optical microscopy, SEM, EPMA, XRD, RS, and a fluid inclusions study. Analytical studies were carried out in the Analytical Centre for Multielemental and Isotope Research SB RAS (Novosibirsk, Russia). Preliminary results of electron microprobe analysis of inclusions in agates were obtained in the Institute of Volcanology and Seismology, Far East Branch of Russian Academy of Sciences (Petropavlovsk-Kamchatsky, Russia).

The four typical samples of agates Nos. 4, 5, 7, 8 shown in Figures 4a–c and 8a were prepared by being polished and cut into thin sections. The thickness of the thin sections was 0.09 mm. Agate No. 8 with macrocrystalline quartz in the center of the nodule (Figure 8a) was selected for the study of fluid inclusions.

Polished and thin sections of all agate samples were examined with an Olympus BX 51 (Olympus NDT, Inc.", USA) polarizing microscope with a magnification range from 40× to 400×. The agates have been examined using SEM with EDS and XRD techniques. Chemical analyses of mineral phases were carried out using a MIRA LMU electron scanning microscope (TESCAN, Czech) with an INCA Energy 450þ X-Max energy-dispersion spectrometer (Oxford Instruments, UK). The operation conditions were: an accelerating voltage of 20 kV, a probe current of 1 nA and a spectrum recording time of 15 to 20 s.

The XRD experiments were performed using X-ray powder diffractometer ARL X'Tra (Thermo Scientific) (CuKα radiation, 40 kV, 25 mA) equipped with a linear detector. Powder diffraction patterns were collected over the 2θ angular range of 5 to 60 with a speed rotation of 5° 2θ and 17 to 30 with a speed rotation of 2° 2θ. Phase analysis was carried out using the PDF-4 database (The Powder Diffraction File PDF-4þ) [34].

Raman spectra of silica phases were recorded using a Ramanor U–1000 spectrometer (Horiba Scientific, Japan), with detector JobinYvon LabRAM HR800 (Jobyn Yvon Instruments S.A. Inc., France) and a laser MillenniaPro (Spectra–Physics, USA) with a nominal wavelength of $\lambda = 532$ nm. An upright microscope Olympus BX-51 WI with a 100 magnifying objective was used to direct the laser beam onto the sample and to collect the Raman signal under the following parameters: spectral resolution of 2.09 cm^{-1}, exposure time of 10 s, five repetitions and filtration D1. The spectra were calibrated against the emission lines of a standard neon lamp, and the peak positions were accurate to within 0.2 cm^{-1}.

Cryo- and thermometry methods were used to determine the temperatures of phase transitions in fluid inclusions in quartz (a THMSG-600 microthermal chamber from "Linkam" with a measurement range of −196/+600 °C). The total concentrations of salts in solutions of fluid inclusions and their belonging to one or another water-salt system were determined using the cryometry method [35–37].

4. Results

4.1. Macro- and Microscopic Observations

The nodule agates were mostly oval (Figure 4a,b, agates No. 5, 4; Figure 8a, agate No. 8) or a deformed oval form (Figure 4c, agate No. 7). The size of agates was up to 10 cm. The colors of these agates range from light blue to ash gray tones, often with a reddish tint. Numerous reddish native copper inclusions are concentrated in the margins of agates and include black or beige rock fragments (basalts and altered basalts) (Figure 4e,f).

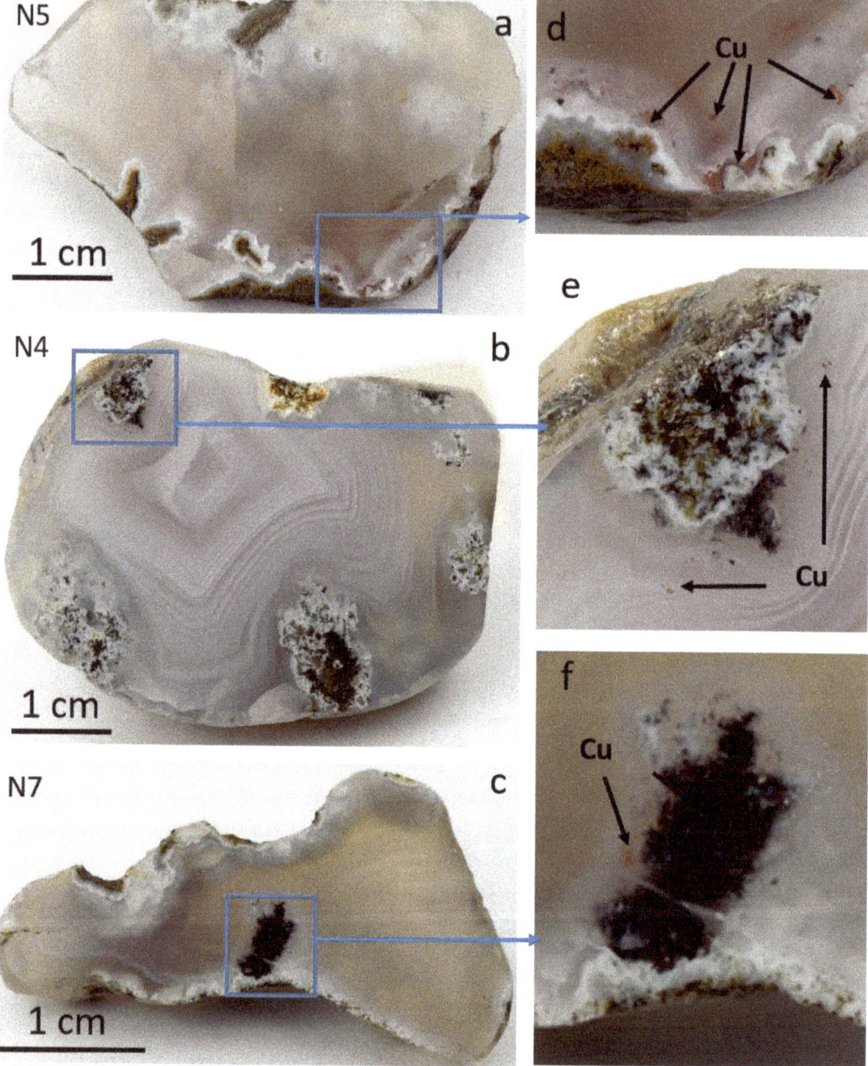

Figure 4. Macrophotographs of studied agates (**a–c**) and enlarged fragments (**d–f**) with microinclusions of native copper (red) and other minerals. The Avacha Bay (East Kamchatka, Russia). Polish sections of agates. Photos by A. Vishnevskii.

The tested agate samples differ in morphology. In Figure 4a, the silica matrix of agate No. 5 seems to be almost homogeneous. In this agate, the concentric layers are only in the outer region and absent in the center. Agate No. 4 (Figure 4b,e) belongs to the mono-centric type with elements of "moss agate" on the edge of the nodule. Agate No. 7 (Figure 4c,d) has a more irregular shape and is polycentric, also with fragments of "moss agate" in the marginal parts of the nodule [38]. Agate No. 8 contains the fragments of "moss agate" and crystalline quartz in the center of the nodule (Figure 8a).

From the periphery to the center of the tested agate nodule, the white shell is replaced by alternating layers of different colors and transparency. The outer hardened white shell, characteristic of agates, was fully or partially preserved in the studied agates 5 and 7. It repeats the shape of agate almonds and fragments of host rocks in agates, which act as seeds, around which a white shell is noted.

The microstructure of agates (Nos. 4, 5, 7) is shown in Figures 5–7. For the outer part of agate No. 5 and agates No. 4 and 7, a fibrous and twisted-fibrous texture (wavy fibers) is typical. Microscopically, intergrowth of length-slow and length-fast chalcedony (Figures 5c, 6b and 7d) is clearly visible. The micrographs in transmitted light (crossed polars) of the edge parts of agate, shows the presence of chalcedony spherulites (Figure 9c).

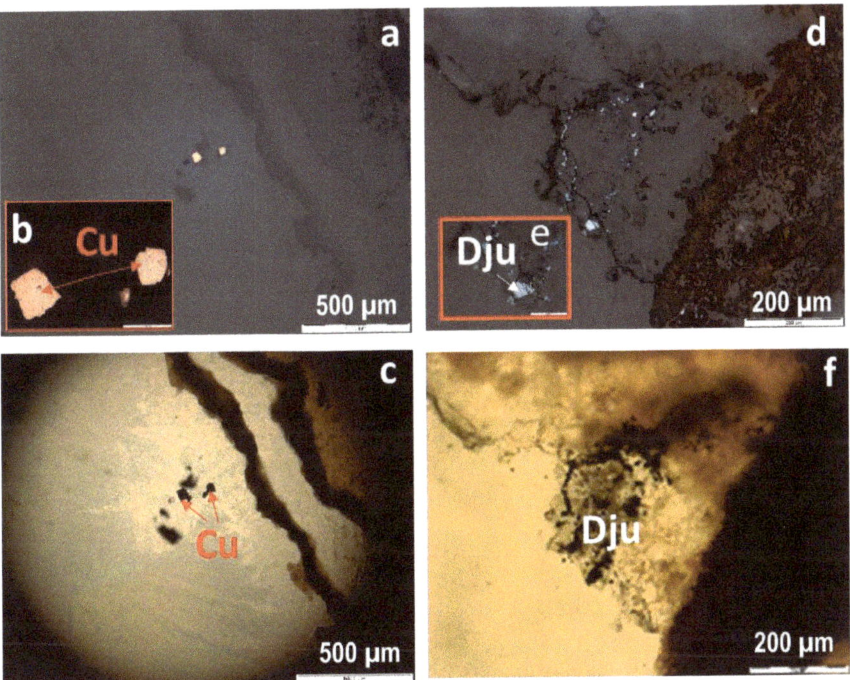

Figure 5. Inclusions of native copper (Cu) (**a–c**) and djurleite (Dju) (**d–f**) surrounded by chalcedony in the outer part of agate No. 5 (Figure 4a): (**a,b,d,e**), in reflected light; (**c,f**), in transmitted light.

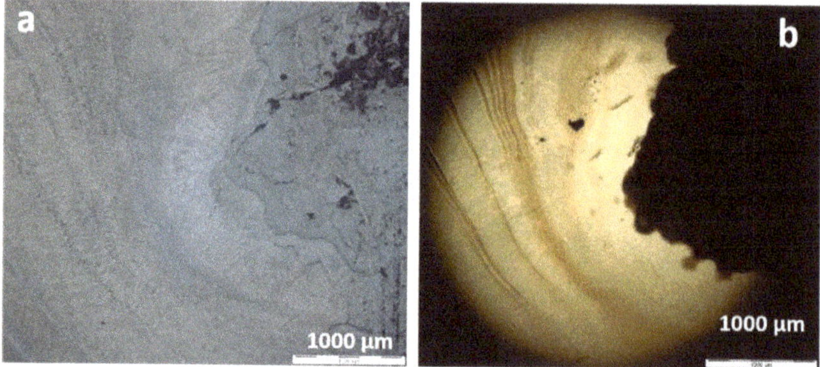

Figure 6. Microstructure of agate No. 4 (Figure 4b) near rock fragments: (**a**) in reflected light; (**b**) in transmitted light. A thin section of the agate.

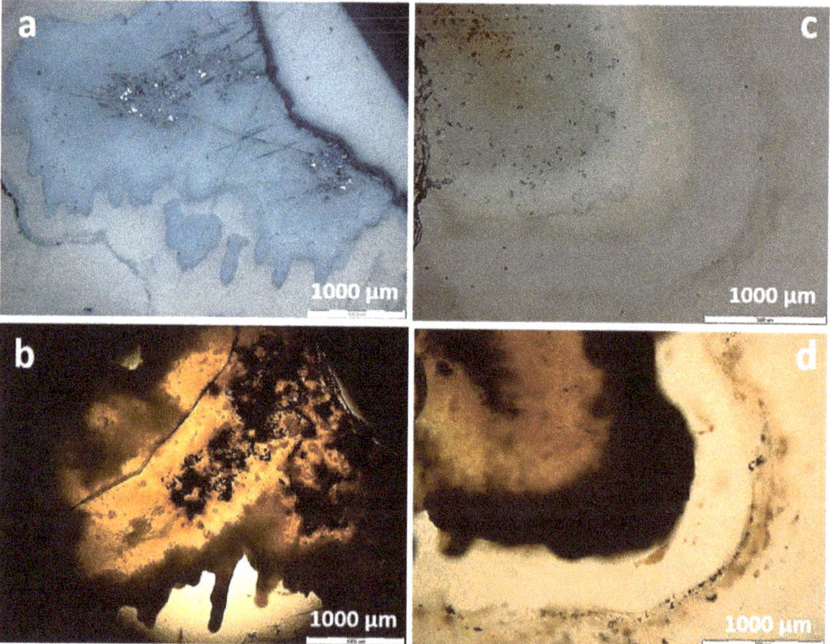

Figure 7. Microstructure of agate No. 7 (Figure 4c) with near inclusions of copper sulphides: (**a**,**c**) in reflected light; (**b**,**d**) in transmitted light. A thin section of the agate.

The form of the copper minerals is varied. The idiomorphic crystals of a cubic habit indicate the presence of native copper (10–500 μm in size) or their dendrites that are intergrown with silica (Figure 5a, Figures 8a and 9c). A significant part of the copper dendrites is located in the outer marginal zones of the agate nodules near rock fragments (Figure 4d–f). There are clear boundaries of crystals of native copper within a silica matrix.

Copper sulphides form clusters of individual grains in the interstices of chalcedony (Figure 5d,f and Figure 7). The grains range in size from 0.1 mm to 1 micron. Inclusions of copper sulphides are often concentrated in the marginal parts of chalcedony spherulites and are in the form of hypidiomorphic grains. Native copper is sometimes located in the center of spherulites formed by copper sulphide dendrites and chalcedony (Figures 13 and 14, see Section 4.3).

Native copper, cuprite, chalcopyrite, sphalerite, and native silver (Figure 8b–f) were determined in agate No. 8 (Figure 8a), with a macrocrystalline quartz in the center of the nodule. The diagnostics of ore minerals were based on their optical properties, established by microscopic examination in reflected light. Later, their compositions were confirmed by EPMA.

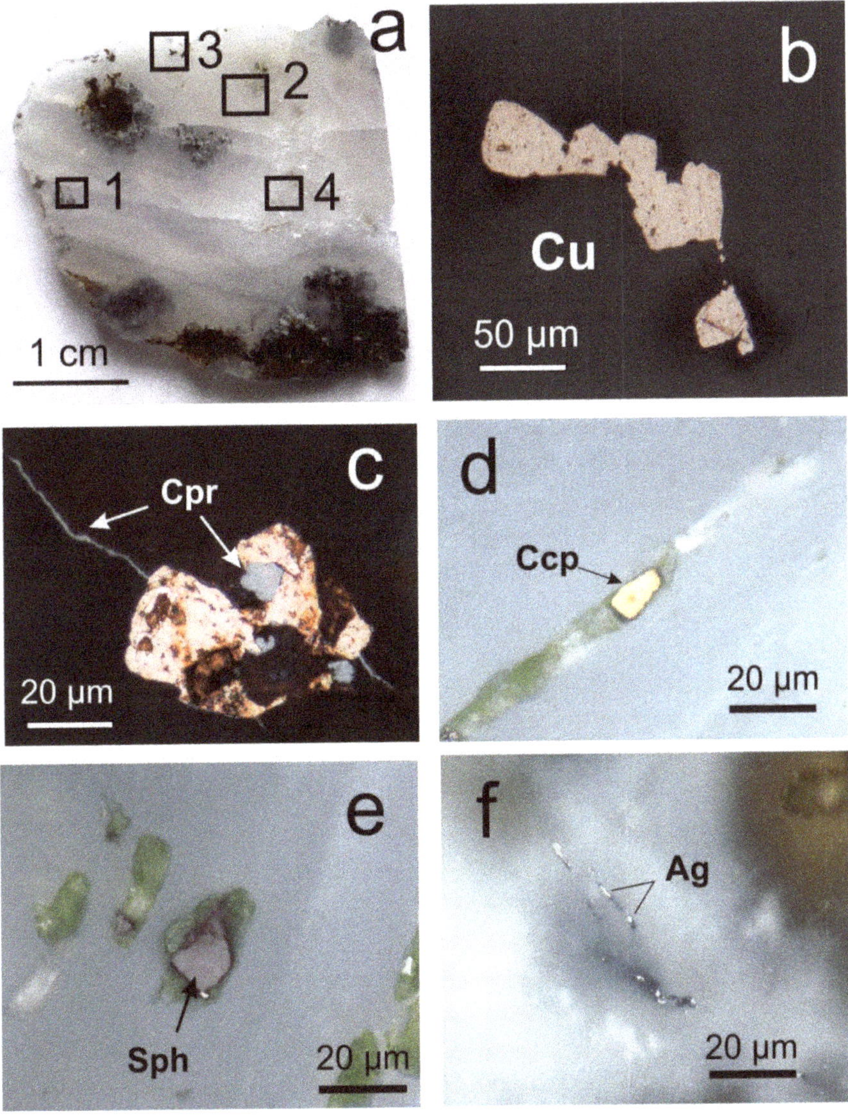

Figure 8. Macrophotograph of the agate No. 8 (**a**) and ore minerals in it (**b–f**): (**b**) dendrites of native copper; (**c**) cuprite (Cpr) (veins and inclusions) in intergrowth with native copper (Cu) among fine-grained quartz; (**d**) chalcopyrite (Ccp) in a cavity in a pseudomorphs of quartz (Figure 9d); (**e**) sphalerite (Sph); (**f**) native silver (Ag). (**b–f**) micrographs in reflected light. Squares 1–4 are RS research areas.

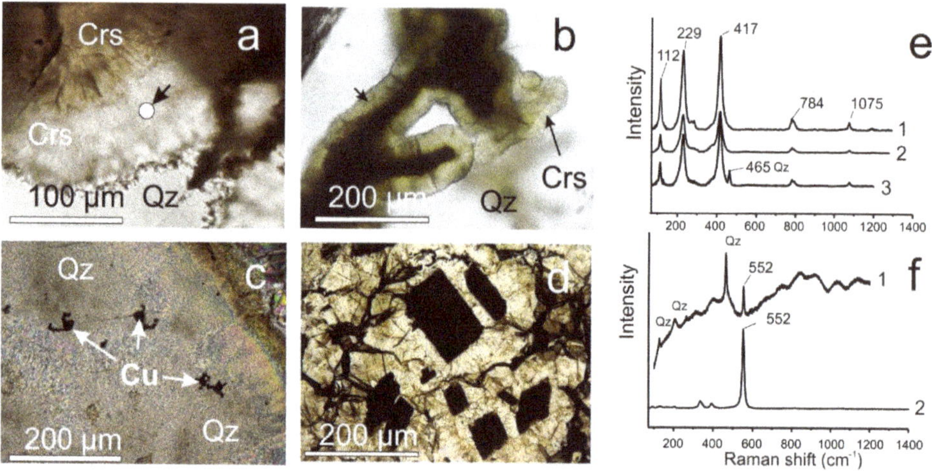

Figure 9. Photographs in the transmitted light of agate No. 8 with native copper: (**a**) cristobalite, the arrow indicates one of the points of exposure of its Raman spectrum (Figure 9e, spectrum 3); (**b**) cristobalite among a fine-grained aggregate of spherulite quartz; (**c**) grains of native copper in quartz at the contacts of intergrown quartz spherulites (crossed polars); (**d**) voids of leaching among quartz. The voids are partially filled with quartz aggregate. Raman spectra of the agate: (**e**) spectra of cristobalite (1 and 2) from the RRUFF database [39], respectively, cristobalite spectra X050046 and X050047; the colloform SiO_2 structures spectrum (3) in the point of exposure Figure 9a; (**f**) spectrum of SiO_2 and cryolite on the surface of the void (1) among quartz and the spectrum of cryolite (2) from the RRUFF database, Cryolite R050287 [39].

4.2. Raman Spectroscopy Results

Raman spectroscopy was used to study agate No. 8 with crystalline quartz in the center of the nodule (Figure 8a) and zonal agate No. 4 (Figure 4b).

Figure 9 shows the varieties of SiO_2 in agate No. 8 (Figure 8a). Cristobalite is present among the fine grained quartz spherulite aggregate (Figure 9a,b). Microinclusions of native copper are located near accrete sintered buds of quartz aggregate (Figure 9c). The coarse quartz of the geode contains quartz pseudomorphs after colloform silica aggregates, which have filled the leaching voids (Figure 9d). They sometimes form complete pseudomorphs. The identification of varieties of SiO_2 were based on their Raman spectrum from the RRUFF database [39] (Figure 9e,f).

The regions of zonal agate No. 4 with different crystallization time and texture were studied. The earliest formations inside the nodule are dark areas with a "moss" texture that is confined to its periphery (Figure 4b,e). These areas are wrapped in alternating concentric layers of transparent and translucent silica, which fills the central portions of the nodule (Figure 10a). According to Raman spectroscopy data [39], the substance of the "moss" formations is represented by an aggregate of cristobalite and fluorophlogopite $KMg_3(Si_3Al)O_{10}F_2$ (Figure 10a,b).

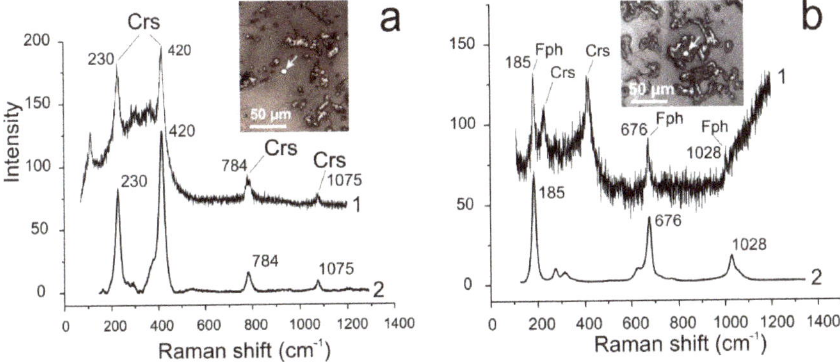

Figure 10. Raman spectra of "moss" aggregates in semitransparent (**a**) and cristobalite areas (**b**) in the thin section fragment of agate No. 4 (Figure 4b,e). The exposure sites are taken in reflected light. (**a**) a spectrum of SiO_2 in a semitransparent area (1) together with an cristobalite spectrum (2) from the RRUFF database the spectrum R060648 [39] (https://rruff.info/); (**b**) the combined spectrum of cristobalite and fluorophlogopite (Fph) (1) in the cristobalite area, compared with the spectrum of fluorophlogopite (2) from the RRUFF database, the spectrum R040075 [39].

The silica that forms the later concentric textures in the center of the nodule is mostly quartz. At the same time, in all the obtained silica spectra of this part of the nodule, the Raman band of moganite is at 503 cm^{-1}, which suggests the presence of this phase in the studied sample. This is also confirmed by similar Raman spectra of numerous agate layers of different transparency near the central part of the agate, where the "moss" formations are absent (Figure 11).

"Moss" white formations in agate No. 4 consist of cristobalite, which is easily identified by the 229 and 418 cm^{-1} lines, according to Raman standards from the RRUFF database [39]. The α-quartz with admixture of moganite folds up light concentric zones (Figure 11a).

The most intense quartz and moganite bands are shown at 465 and 503 cm^{-1}, respectively (Figure 11). The ratio of the main quartz and moganite bands provides information about the relative abundances of these two silica phases. The value of this ratio is much less than unity, which lies outside the scope of the formula based on the data from Goetze [40] and does not allow quantitative determination of the percentage of content of moganite. However, it can be said that quartz prevails over moganite.

4.3. Scanning Electron Microscopy and Electron Microprobe Analysis Results

The EPMA results of native copper in the agates from the Avacha Bay showed the absence of impurities. Figure 12 shows the distribution of elements over the scanning area from the thin section of agate No. 5 containing the inclusion of a cubic crystal of native copper in a silica matrix. Attention should be paid to the clear boundaries and contacts of native copper with quartz.

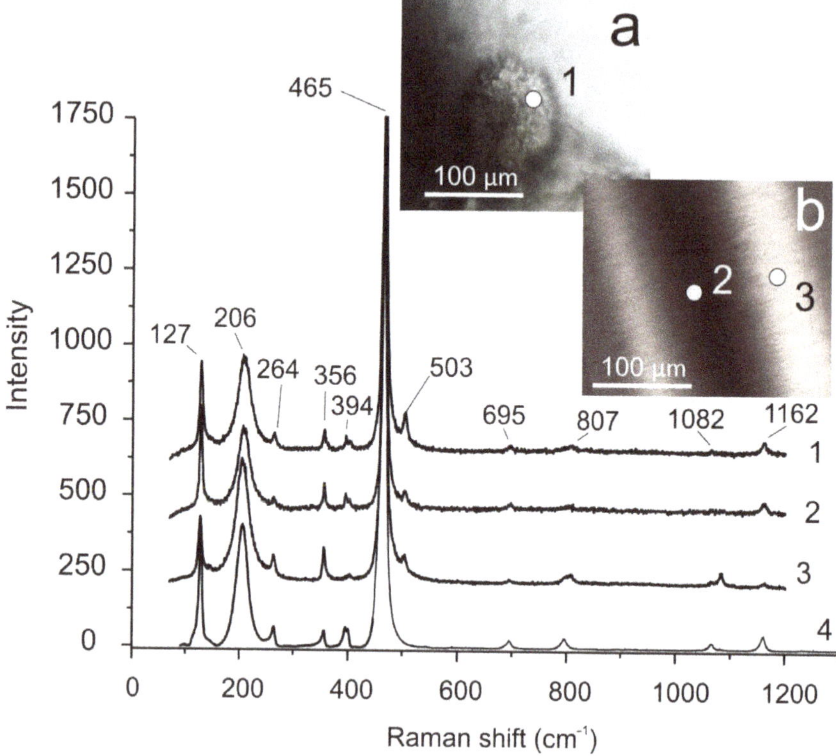

Figure 11. Raman spectra of agate No. 4 (Figure 4b,e) have been taken at various exposure points (1–3) shown in the micrographs (**a,b**) that were taken in transmitted light. Spectrum 1 is from a semitransparent globule (**a**) located on the periphery of "moss" aggregates. The micrograph b and spectras 2 and 3 are the respective translucent and transparent layers at the center of the nodule. The most intense quartz and moganite bands are shown at 465 and 503 cm^{-1}, respectively. All other bands are quartz signals. Spectrum 4 is quartz (X080016) from the RRUFF database [39].

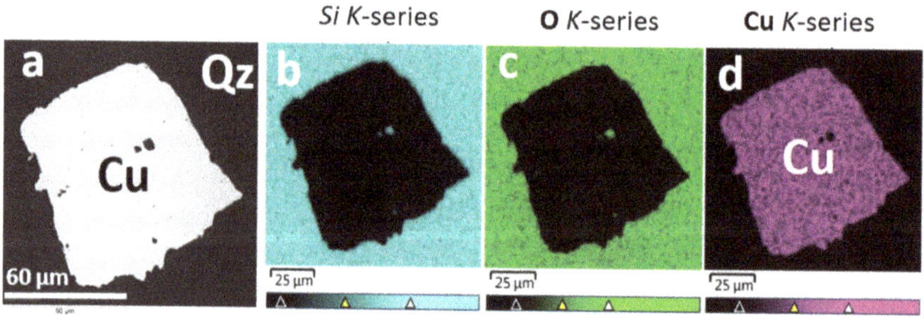

Figure 12. Native copper in a silica matrix: (**a**) SE micrograph, (**b–d**) distribution of elements over the scanning area from the thin section of agate No. 5 (Figure 4a,d).

The inclusions of copper sulphides—djurleite ($Cu_{1.96}S$), anilite ($Cu_{1.75}S$), and yarrowite ($Cu_{1.1}S$)—were located in the interstices of a fine-grained spherulitic aggregate of quartz (Figure 13a,b). Djurleite and barite ($BaSO_4$) filled the micro veins (Figure 13c,d).

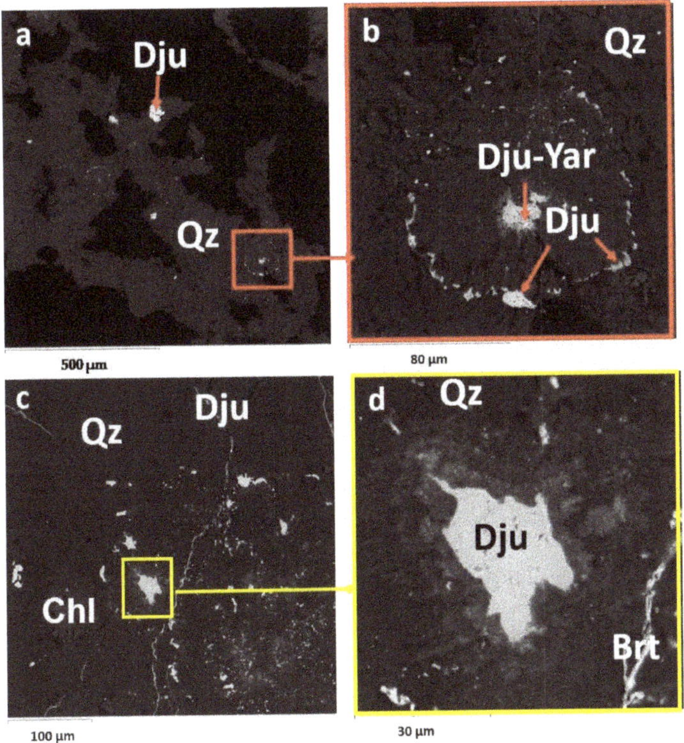

Figure 13. SE micrographs of fragments of agate No. 5 (Figure 4d) with inclusions of copper sulphides: (**a**,**b**) djurleite-anilite-yarrowite (Dju-Yar) were present in the interstices of fine-grained spherulite aggregate of quartz; (**c**,**d**) djurleite (Dju) and barite (Brt) micro veins were present in the quartz or quartz-chlorite (Chl) aggregate.

Figure 14 shows the absolute concentrations of the elements Si, O, Cu, and S along the scanning line of an agate fragment containing native copper in the center of ovoids, which is replaced by growing copper sulphide dendrites. The composition of copper sulphides varies from $Cu_{1.94}S$ to $Cu_{1.75}S$ and $Cu_{1.1}S$, which corresponds to the stoichiometry of such minerals as djurleite, anilite and yarrowite [39].

Table 1 shows the EPMA results of the Cu-S minerals in agates No. 4, 5, 7. The amount of copper and sulphur in sulphides varies, which corresponds to chalcocite (Cu_2S), djurleite ($Cu_{1.96}S$), digenite ($Cu_{1.75-1.78}S$), anilite ($Cu_{1.75}S$), and yarrowite ($Cu_{1.1}S$) [39].

Table 1. Representative composition of Cu-S minerals (EPMA data in wt.% and formula) in agates from the Avacha Bay (Eastern Kamchatka, Russia).

No	O	Si	S	Fe	Cu	Total	Formula	Mineral
5/1-1	1.01	-	31.35	0.19	67.48	100.03	$Cu_{1.1}S$	yarrowite
5/1-2	1.19	0.43	31.48	-	66.19	99.3	$Cu_{1.1}S$	yarrowite
5/1-3	2.28	0.29	21.68	0.26	74.03	98.54	$Cu_{1.73}S$	anilite
5/1-4	1.92	0.36	21.99	0.38	75.07	99.72	$Cu_{1.73}S$	anilite
5/3-1	1.5	0.49	29.93	0.59	66.32	98.83	$Cu_{1.12}S$	yarrowite
5/3-1	3.21		19.98		74.65	97.84	$Cu_{1.89}S$	djurleite
5/3-2	2.79	0.37	20.18		76.04	99.54	$Cu_{1.90}S$	djurleite
4/1			20.24		79.1	99.34	$Cu_{1.98}S$	chalcocite-djurleite
4/26			20.43		78.62	99.05	$Cu_{1.94}S$	djurleite
4/27			20.24		78.06	98.30	$Cu_{1.95}S$	djurleite
7/2-1	2.11	0.28	20.5		76.17	99.05	$Cu_{1.88}S$	djurleite-digenite
7/2-2	2.72	0.64	20.7		74.83	98.89	$Cu_{1.83}S$	digenite
7/2-3	1.18	0.28	22.18		75.09	98.73	$Cu_{1.71}S$	anilite

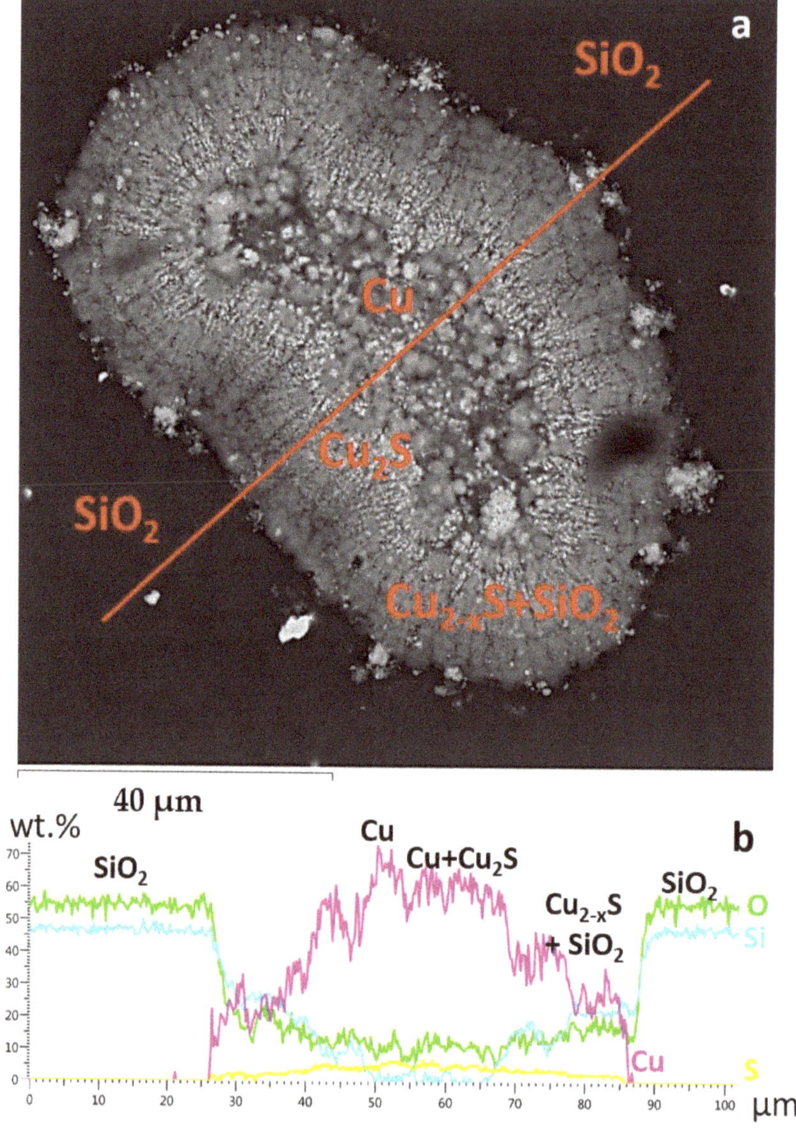

Figure 14. Native copper and dendrites of copper sulphides are present in quartz ovoid (**a**). Concentrations of the elements Si, O, Cu, and S (**b**) (in wt.%) are shown along the scanning line of the fragment agate No. 7 (Figure 4f). (**a**) SEM micrograph.

In the marginal part of agate No. 7 (Figure 4c,f) calcite intergrown with quartz is shown on Figure 15. It contains quartz inclusions and djurleite microinclusions. Some djurleite microinclusions are located at the contact between quartz and calcite or in calcite. Calcite contains Mn impurities up to 0.5 wt.%. The dark zone of quartz around calcite and quartz inclusions in calcite contain Al impurities (0.2–0.3 wt.%).

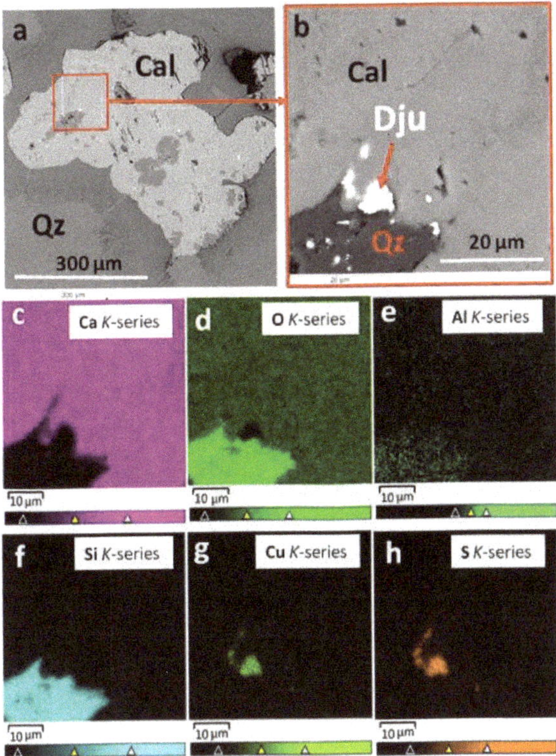

Figure 15. Calcite containing SiO$_2$ inclusions and djurleite (Dju) micro inclusions in a quartz matrix (**a**,**b**) from agate No 7 (Figure 4c,f). Element distribution in a fragment of calcite (Cal) (red square on a) (**b**) was determined using characteristic rays (**c**–**h**). (**a**,**b**) SE micrographs.

Microinclusions of djurleite, phlogopite, and K-feldspar are located in quartz matrix (Figure 16a) at the edge of agate No. 4 (Figure 4b,e). Figure 16b–i shows the distribution of Si, O, K, Al, S, Cu, Mg, and Fe over the scanned area.

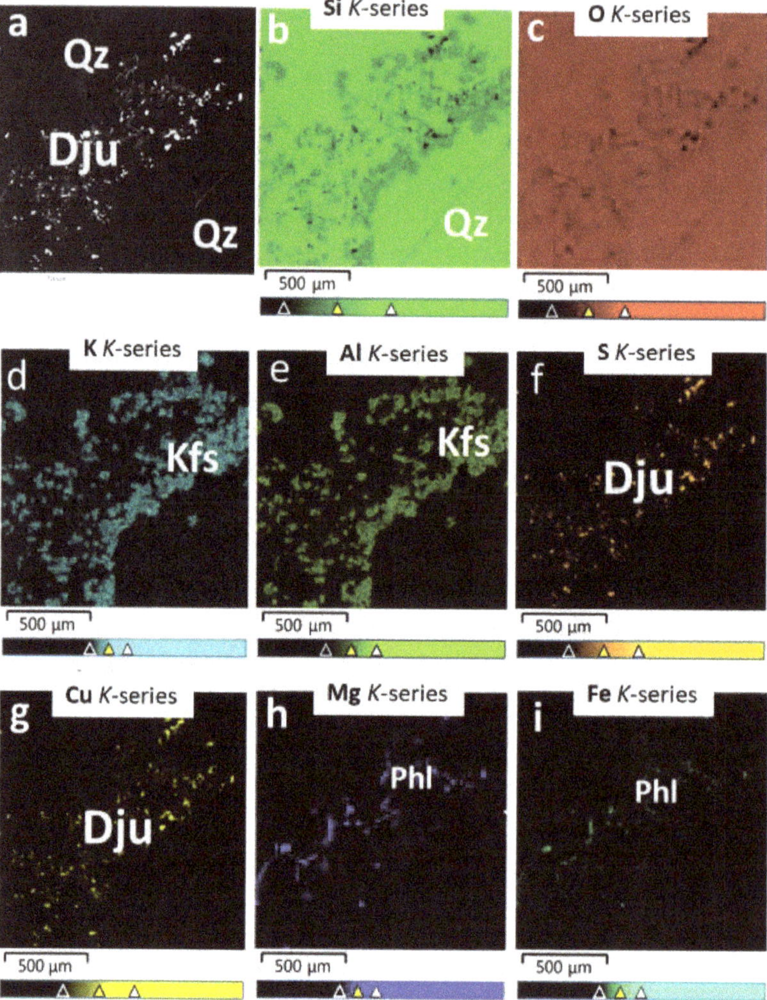

Figure 16. SE micrograph (**a**) at the edge of agate No 4 (Figure 4b,e) containing micro inclusions of djurleite (Dju), K-feldspar (Kfs), and phlogopite (Phl) in quartz, and maps showing the distribution of the elements over the scanned area (**b–i**).

In silica interlayers, impurities of elements such as Na, K, Ca, Mg, Al, and Fe are often noted. The concentrations of Fe and Mg vary from 0.2 to 10 wt.%, Na up to 8.7 wt.%, K and Ca up to 4 wt.%, Al up to 7 wt.%. There are sometimes also impurities Cl (up to 2 wt.%), P (up to 0.2 wt.%), Ba (up to 1.4 wt.%), Ti (up to 3 wt.%), Sr (up to 3 wt.%), Mn (up to 0.4 wt.%), V (up to 0.9 wt.%), Cu (up to 3 wt.%), and S (up to 2.6 wt.%). The presence of these impurities is most likely to be associated with the minerals of ore-hosting rocks.

4.4. Results of The Fluid Inclusions Study

Fluid inclusions were found in agate with macrocrystalline quartz in the central part under the faces of individual quartz crystals (Figure 8a). The inclusions trace fragments of crystal growth zones (Figure 17a) and are primary or pseudo-secondary according to the Roedder's classification [36]. In terms of phase filling, the inclusions are divided into gas single-phase, liquid single-phase, and two-phase, containing a gas bubble and a liquid (Figure 17b–d). In quantitative terms, gas and liquid single-phase inclusions predominate. Two-phase inclusions are rare.

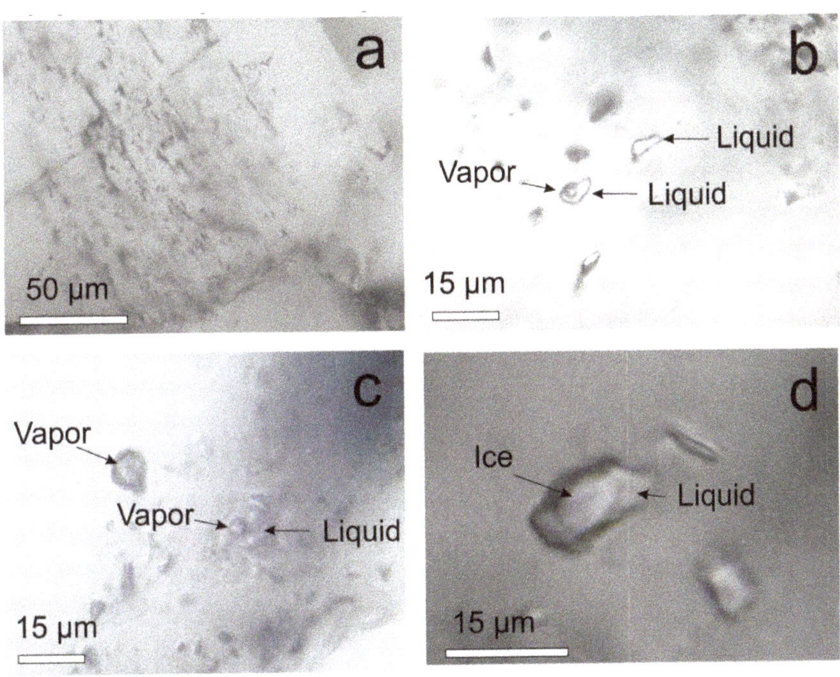

Figure 17. Fluid inclusions in crystalline quartz: (**a**) a fragment of the growth zone traced by fluid inclusions, (**b**) two-phase and single-phase water-salt inclusions, (**c**) single-phase gas and two-phase water-salt inclusions, (**d**) single-phase water-salt inclusions at a temperature of −10 °C, the frozen-out phase of ice is visible, which finally melts at −0.3 °C.

No liquefaction or freezing of gas was observed with the deep cooling of gas inclusions to a temperature of −190 °C, which demonstrates that its density is low. Phases of eutectic and ice appeared in single-phase and two-phase water-salt inclusions with deep cooling. Melting of the eutectic (the beginning of a noticeable melting of the ice phase) was observed in the temperature range of −55 to −52 °C. The ice phase finally disappeared at a temperature of −0.3 °C. According to the melting point of the eutectic, the composition of solutions of inclusions can be attributed to the water-salt system $CaCl_2$-$NaCl$-H_2O, with a probable admixture of $MgCl_2$ [35]. After several cycles of cooling/heating in the temperature range from +30 to −120 °C, a small gas bubble appeared in some single-phase water-salt inclusions, which dissolved in the liquid in the range from 110 to 78 °C and lower. The salts concentrations in solution are very low (less than 0.3 wt.% NaCl eqv.).

Gas-filled two-phase inclusions with a large bubble are homogenized at high temperatures of 250 °C and more, but such temperatures are unrealistic, since sulphide mineralization is confined to amorphous SiO_2 modifications.

The presence of primary gas and two-phase and single-phase liquid fluid inclusions in macrocrystalline quartz indicates boiling of the mineral-forming fluids [36,41,42]. Investigation of the homogenization temperature of two-phase fluid inclusions shows that this quartz could have been formed at temperatures from 110 to 78 °C. At the same time, it should be taken into account that agate specimens have undergone many natural freeze-thaw cycles, at least during the Quaternary period. According to Hardie et al. [43], the force of ice crystallization during the cooling of single-phase inclusions leads to an increase in their volume as a result of ice pressure on the vacuole walls, and then to the appearance of a gas bubble during thawing. It is also possible to assume complete or partial decrepitation of single-phase liquid fluid inclusions during the many cycles of natural freezing-thawing and transformation of single-phase inclusions into two-phase ones. Thus, some of the detected gas and two-phase fluid inclusions in crystalline quartz could be decrepitated, and the measurements of the temperature of homogenization of fluid inclusions could be overestimated relative to the true temperature of their capture. In this case, given the presence of primary single-phase liquid inclusions in crystalline quartz, its crystallization could occur from a homogeneous solution at a temperature lower than 50 °C. Taking into account all the data, the crystallization temperature at the final stages of agate formation could be 110 to 50 °C and lower.

4.5. XRD Results

The content of individual minerals (in wt.%) in the tested samples (No. 4, 5, 7) is automatically calculated by the diffractometer. For example, the X-ray pattern of sample No. 4 is shown in Figure 18. The main phase of the analyzed samples is α-quartz (content 80–90 wt.%). The presence of moganite and cristobalite was also identified. The most intense peaks of these minerals occur in the range of the angles 19–28° 2θ (Figure 19). Cristobalite is reliably diagnosed by a broad diffraction peak at 21.8° 2θ [39]. (https://rruff.info/cristobalite/R060648). Its content (1–3 wt.%) was determined using the program MATCH and the PDF-4 database based on the calculated intensities of the phases present. The position of the cristobalite peak suggests that this is its low-temperature tetragonal modification. As for moganite, quartz peaks overlap its most intense reflections. Its presence can be judged only by the small shoulders at the base of the quartz reflections. Obviously, the assessment of the content of moganite in the studied samples (6–17 wt.%) has a large error.

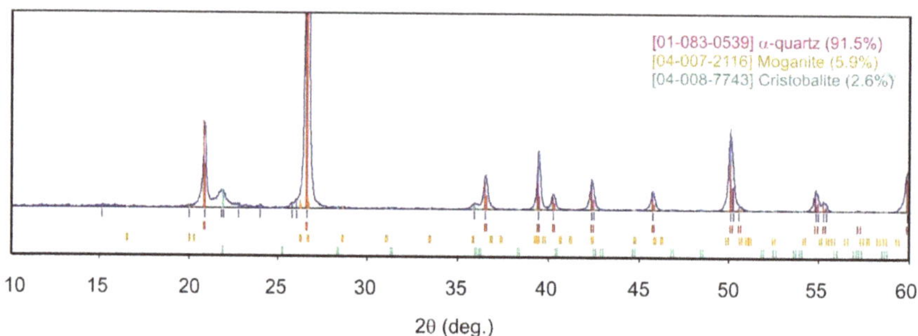

Figure 18. X-ray diffraction pattern of the studied sample (agate No. 4) (CuKα radiation).

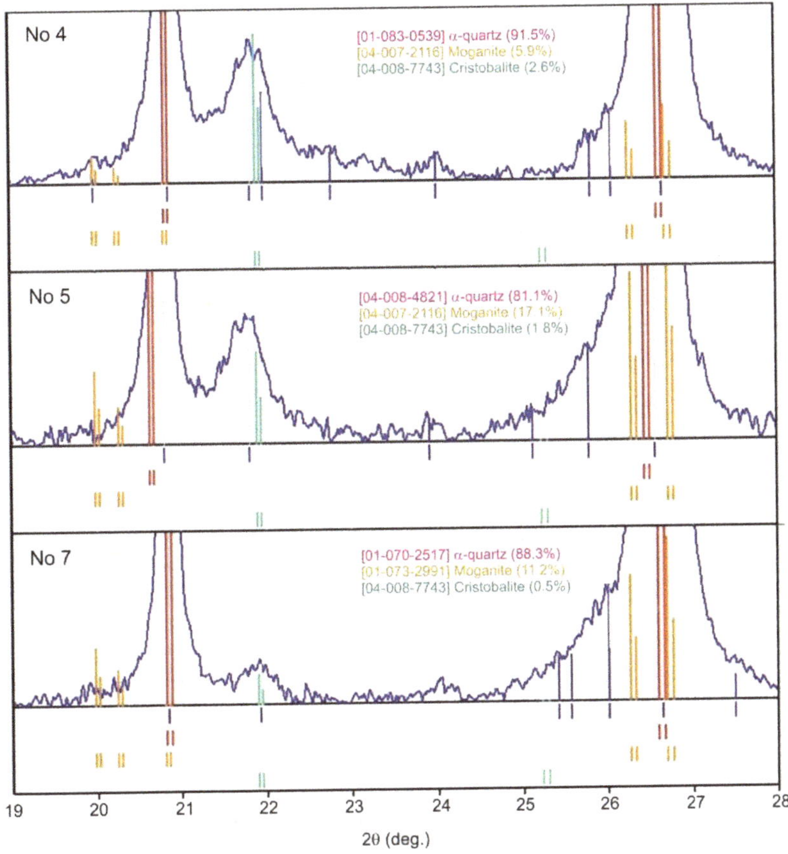

Figure 19. Fragments of X-ray diffraction patterns of the studied samples No. 4, 5, 7, in the range of angles 19–28° 2θ (CuKα radiation) with the most intense peaks of cristobalite and moganite.

5. Discussion

Agates often contain considerable amounts of mineral inclusions. The mineralogical complexity of agate is a result of its complex formation history [8]. Raman spectroscopy, obtained in the present study, showed that the Avacha Bay agates contained cristobalite in addition to quartz and moganite. XRD results demonstrated that the silica matrix in the agates consists of predominantly α-quartz, low moganite (from 6 to 17 vol.%), and minor cristobalite (1–3 vol.%). The moganite content in agates usually range from 0 to 20% [44,45]. Our data are within this specified interval. This silica polymorph is nearly absent in agates found within relatively old volcanos, but could still be present in relatively young host basaltoids. The preliminary rough estimate of agate age on the examination of moganite content in samples 4, 5, 7 shows a wide variation, from 18 to 48 Ma.

In addition to the silica polymorphs, the native copper, copper sulphides (chalcocite, djurleite, digenite, anilite, yarrowite and chalcopyrite), sphalerite, native silver, barite and cuprite have been identified in the Avacha Bay agates. Native copper occurs as cubic crystals or dendrites in a silica matrix (Figure 5a–c, Figures 8b and 12). Sulphides often cement the marginal parts of spherule aggregates of silica and fill the microveins, which occupy a cross-cutting position with respect to the concentric bands of quartz (Figure 13b–d). Cuprite occurs in intergrowth with native copper and forms micro veins in quartz (Figure 8c). The textural relationship between silica and ore minerals suggests that the deposition of native copper occurred simultaneously with the early stage of agate formation.

The deposition of sulphides (copper sulphides and sphalerite), barite and cuprite occurred later and is apparently associated with the second stage of agate formation.

In the agates of the Avacha Bay, copper mineralization is characterized as a two-stage character. The idiomorphic crystals of native copper and clear boundaries with the silica matrix makes it possible to make judgments about the possible simultaneous crystallization. Native copper in the absence of copper sulphides and cuprite indicates a slight deficiency of sulfur and reducing conditions in an early stage. The replacement of native copper with copper sulphides in the pore and interstitial space of spheroidal-layered silica aggregates occurs with the participation of later solutions containing $H_2S(aq)$ or Cu-ions, which leads to the sulphurisation of native copper and later the deposition of Cu-S minerals. The presence of barite and copper sulphides filling the cracks indicates redox conditions close to sulfide-sulfate H_2S/HSO_4^- equilibrium [46]. The presence of cuprite indicates their later deposition and the oxidizing conditions. Cuprite and copper sulphides are a typical secondary mineral, which accompanies native copper in the oxidized zone of copper sulphide deposits [47].

According to the Cu-S phase diagram [46], djurleite is stable at <93 °C and anilite <75 °C. Our results of the fluid inclusions study show that the crystallization temperature at the final stages of agate formation could be 110 to 50 °C and below. The absence of fluid inclusions in the entire volume of agate nodules does not allow one to estimate the entire temperature range of the formation of the studied agates. The minimum homogenization temperatures for agates from volcanic rocks of some German occurrences vary between 95 and 186 °C [8]. Analogic data for deposits in the north-east of Russia show 120–170 °C [4]. The P,T-parameters of the zeolite facies are spread in the sequence of volcanic rocks, containing agates: 90–290 °C, 1–5 kb [12]. There is a general consensus that agates in basic igneous hosts form at <100 °C [48].

According to Götze et al. [8], in agates from basic and intermediate rocks, quartz mostly contains inclusions having low salinity (<4 wt.% NaCl eqv.). The results obtained in this study are in good agreement with these data. The concentrations of salts in the composition of fluid inclusions in agates from Avacha Bay are a very low (less than 0.3 wt.% NaCl eqv.). The main salt components were $CaCl_2$ and NaCl with a probable admixture of $MgCl_2$.

The presence of calcite in the intergrowth with quartz in the marginal parts of some agates (No. 7), containing microinclusions of djurleite at the contact of quartz (Figure 15), indicates the participation of Ca^{2+}-, CO_3^{2-}-ions in low hydrothermal solution at early stage. Geochemical investigations of calcite in agates provide a temperature of crystallization in the range between 20 and 230 °C [7,49].

The relationships between native copper, copper sulphides and quartz in the Avacha Bay agates differ from other locations—the Ayaguz agate deposit (East Kazakhstan) [18], the Wolverine Mine, Wolverine, Houghton County, Michigan (USA) [19,21], Rudno near Krzeszowice in Lower Silesia (Poland) [25], Sidi Rahal, and Kerrouchen (Khénifra Province, Meknés-Tafilalet Region) in the Atlas mountains of Morocco [26,27]. For example, native copper found in agates from Rudno near Krzeszowice in Lower Silesia (Poland) [25] suggests its secondary origin. The formation of vein agates within spilitized (albite-rich) alkali-basalts from the upper parts of Borówno quarry (Intra-Sudetic Basin) (Lower Silesia, Poland) was marked by recrystallization of metastable silica phases (i.e., moganite), coupled with the local replacement of pre-existing sulphates and/or carbonates by silica during boiling-related conditions [28].

Rosemeyer [19,21] shows the Kearsarge copper-bearing agate photos confirming a partial or almost complete replacement of the chalcedony banding by native copper. This replacement mechanism does not explain the relationships of quartz with native copper in the agates from Avacha Bay. The replacement mechanism is a more appropriate explanation for the formation of a late-stage copper sulfide with the participation of H_2S solutions (Figures 13 and 14). The genesis of the native copper lodes on the Keweenaw Peninsula of northern Michigan is most commonly explained by a metamorphogenic model [50]. A meteoric and metamorphic model was also developed by Brown [51,52].

The identification of copper in the Avacha Bay agates suggests a potential copper source in the surrounding basalts. According to our RFA analysis data, the amount of copper in them is 160 ppm, which is more than Clarke content, containing 55 ppm [53]. More than 30 areas of basaltic laves worldwide are known for containing native copper, some of which occur in deposits of potential commercial importance [47]. In 15 of these areas, certain flows contain moderate amounts of finely disseminated primary native copper, whereas the remaining flows contain little or no visible copper. The native copper dendrites in rock fragments in coastal outcrops of the Avacha Bay were found by E.Sidorov, one of the authors of the first paper on agates with native copper in Kamchatka [29]. The photo of this sample is shown on Figure 3. The low-temperature waters could be responsible for the transport of copper. The source of fluid is a task for future research.

It should be noted that agates containing native copper were found in areas where large deposits of native copper are known; Turin mines (Ural) [54], Dzhezkazgan (Kazakhstan) [18], and Michigan (Keweenaw Peninsula, USA) [19,21,50–52]. Preliminary data were obtained on the presence of native silver in some agates from Avacha Bay. The presence of Cu, Ag, Zn, Sr, and V-bearing ore occurrences is predicted.

6. Conclusions

(1) Copper mineralization in agates from the Avacha Bay (Eastern Kamchatka, Russia) is represented by native copper, as well as copper sulphides (chalcocite, djurleite, digenite, anilite, yarrowite, rarely chalcopyrite) and cuprite. In addition to copper minerals, sphalerite and native silver are also found in agates from this location.

(2) Raman spectroscopy and XRD results demonstrated that the Avacha Bay agates contained cristobalite in addition to quartz and moganite. The substance of "moss" formations is represented by an aggregate of cristobalite and fluorophlogopite $KMg_3(Si_3Al)O_{10}F_2$.

(3) The native copper crystallized simultaneously with early silica. Copper sulphides, sphalerite, native silver, cuprite, and barite were deposited later apparently with the participation of low temperature hydrotherms with H_2S, which were replaced by sulphate solutions caused by a change in redox change in hypergenic conditions.

(4) Macrocrystalline quartz in the center of agate nodules could be formed at a temperature from 110 to 50°C and below. The main salt components of the fluid inclusions in macrocrystalline quartz were $CaCl_2$ and $NaCl$ with a probable admixture of $MgCl_2$. The salt concentration of solutions is lower than 0.3 wt.% NaCl equivalent.

(5) The surrounding basalts could be the source of copper for agates of the Avacha Bay. The presence of copper minerals and other ore elements in agates of volcanogenic strata of Eastern Kamchatka can serve as a direct indicator of the high ore potential in this territory.

Author Contributions: Conceptualization, methodology, G.P. and E.S.; investigation, A.B. and Y.S.; writing and editing, G.P.; visualization G.P., E.S., A.B. and Y.S.; supervision G.P. All authors have read and agreed to the published version of the manuscript.

Funding: The studies were carried out within the framework of the state assignment of the Sobolev Institute of Geology and Mineralogy of Siberian Branch of Russian Academy of Sciences and by the grant no. 13.1902.21.0018 "Fundamental Problems of the Development of the Mineral Resource Base of the High-Tech Industry and Energy in Russia" from Ministry of Science and Higher Education of the Russian Federation.

Acknowledgments: The authors thank N. Karmanov, M. Khlestov (VS Sobolev Institute of Geology and Mineralogy, Siberian Branch, Russian Academy of Sciences) and V. Chubarov (Institute of Volcanology and Seismology, Far East Branch of Russian Academy of Sciences) for help in carrying out a large amount of work on a scanning electron microscope and an X-ray spectral microanalyzer. We are grateful to A. Vishnevskii for macrophotographs of agates (Figure 4) and D. Bukhanova for photograph of agate amydgales in basalt (Figure 2b). The authors also thanks V. Zinina and T. Zhuravkova. We would like to thank T. Moxon (visiting research worker, Cambridge University) for the invaluable advices. We are very grateful to Lucyna Natkaniec-Nowak (AGH University of Science and Technology, Kraków, Poland) and three anonymous reviewers for their comments, which helped to improve the quality of the paper. Appreciation is given to the Academic Editor for careful reading of the manuscript and comments.

Conflicts of Interest: The authors declare no conflict of interest.

References

1. Tripp, R.B. The mineralogy of Warsaw Formation geodes. *Iowa Acad. Sci. Proc.* **1959**, *66*, 350–356.
2. Barsanov, G.P.; Yakovleva, M.E. Mineralogy, macro- and micromorphological features of agates. *New Data Miner.* **1982**, *30*, 3–26. (In Russian)
3. Godovikov, A.A.; Ripinen, O.I.; Motorin, S.G. *Agates*; Nedra: Moscow, Russia, 1987; p. 368. (In Russian)
4. Goncharov, V.I.; Gorodinsky, M.E.; Pavlov, G.F.; Savva, N.E.; Fadeev, A.P.; Vartanov, V.V.; Gunchenko, E.V. *Chalcedony of North-East of the USSR*; Science: Moscow, Russia, 1987; p. 192. (In Russian)
5. Heaney, P.J. A proposed mechanism for the growth of chalcedony. *Am. Min.* **1993**, *115*, 66–74. [CrossRef]
6. Graetsch, H. Structural characteristics of opaline and microcrystalline silica minerals. In Silica. *Rev. Mineral.* **1994**, *29*, 209–232.
7. Götze, J.; Tichomirow, M.; Fuchs, H.; Pilot, J.; Sharp, Z.D. Chemistry of agates: A trace element and stable isotope study. *Chem. Geol.* **2001**, 523–541. [CrossRef]
8. Götze, J.; Möckel, R.; Pan, Y. Mineralogy, geochemistry and genesis of agate—A review. *Minerals* **2020**, *10*, 1037. [CrossRef]
9. Moxon, T.; Ríos, S. Moganite and water content as a function of age in agate: An XRD and thermogravimetric study. *Eur. J. Mineral* **2004**, *16*, 269–278. [CrossRef]
10. Moxon, T. *Studies on Agate: Microscopy, Spectroscopy, Growth, High Temperature and Possible Origin*; Terra Publications: Doncaster, UK, 2009; p. 96.
11. Lyashenko, E.A. Agates of Russia. *Mineral. Alm.* **2010**, *15*, 6–27. (In Russian)
12. Spiridonov, E.M.; Ladygin, V.M.; Yanakieva, D.Y.; Frolova, J.V.; Semikolennykh, E.S. Agates in metavolcanics. Bulletin of the Russian Federal Property Fund. 2014. Available online: https://www.rfbr.ru/rffi/ru/bulletin/o_1923809#8 (accessed on 22 October 2020).
13. Ottens, B.; Götze, J.; Schuster, R.; Krenn, K.; Hauzenberger, C.; Zsolt, B.; Vennemann, T. Exceptional multi-stage mineralization of secondary minerals in cavities of flood basalts from the Deccan Volcanic Province, India. *Minerals* **2019**, *1019*, 351. [CrossRef]
14. Gliozzo, E. Variations on the silica theme: Classification and provenance from Pliny to current supplies. *EMU Notes Mineral.* **2019**, *2*, 13–85.
15. Pršek, J.; Dumańska-Słowik, M.; Powolny, T.; Natkaniec-Nowak, L.; Toboła, T.; Zych, D.; Skrepnicka, D. Agates from Western Atlas (Morocco)—Constraints from mineralogical and microtextural characteristics. *Minerals* **2020**, *10*, 198. [CrossRef]
16. Moxon, T. A re-examination of water in agate and its bearing on the agate genesis enigma. *Min. Mag.* **2017**, *81*, 1223–1244. [CrossRef]
17. Zhang, X.; Ji, L.; He, X. Gemological characteristics and origin of the Zhanguohong agate from Beipiao, Liaoning province, China: A combined microscopic, X-ray diffraction, and Raman spectroscopic study. *Minerals* **2020**, *10*, 401. [CrossRef]
18. Yusupov, S.S. Thermobarogeochemical Conditions for the Formation of Agate Deposits in the Urals and Kazakhstan. In *Questions of Mineralogy, Geochemistry and Genesis of Minerals of the Southern Urals*; Bashk. Prince Publishing House: Ufa, Russia, 1982; pp. 92–99.
19. Rosemeyer, T. The Kearsarge copper-bearing amygdaloidal lode, Houghton and Keweenaw countries, Michigan. *Rocks Miner.* **2007**, *82*, 276–297. [CrossRef]
20. Rosemeyer, T. A spectacular find of amygdaloidal agates with native copper inclusions from Michigan's Copper Country. *Rocks Miner.* **2001**, *76*, 403. [CrossRef]
21. Rosemeyer, T. Copper-banded Agates from the Kearsarge Copper-bearing Amygdaloidal Lode, Houghton County, Michigan. *Rocks Miner.* **2012**, *87*, 352–365. [CrossRef]
22. Radko, V.A. *Agates, Carnelian, Jasperoids of Norilsk*; GeoKniga: St. Petersburg, Russia, 2013; p. 128.
23. Dumańska-Słowik, M.; Natkaniec-Nowak, L.; Kotarba, M.J.; Sikorska, M.; Rzymełka, J.A.; Łoboda, A.; Gaweł, A. Mineralogical and geochemical characterization of the "bituminous" agates from Nowy Kościol Lower Silesia. *N. Jb. Miner Mh.* **2008**, *184*, 255–268. [CrossRef]
24. Nezafati, N.; Momenzadeh, M.; Pernicka, E. Darhand Copper Occurrence: An Example of Michigan-Type Native Copper Deposits in Central Iran. In *Mineral Deposit Research: Meeting the Global Challenge*; Mao, J., Bierlein, F.P., Eds.; Springer: Berlin/Heidelberg, Germany; New York, NY, USA, 2005; Volume 1, pp. 165–166.
25. Krawczyński, W. Native copper in agates from Rudno near Krzeszowice. *Mineral. Pol.* **1995**, *26*, 27–32.

26. Dumańska-Słowik, M.; Natkaniec-Nowak, L.; Weselucha-Birczyńska, A.; Gaweł, A.; Lankosz, M.; Wróbel, P. Agates from Sidi Rahal, in the Atlas Mountains of Morocco: Gemmological characteristics and proposed origin. *Gems Gemol.* **2013**, *49*, 148–159. [CrossRef]

27. Natkaniec-Nowak, L.; Dumańska-Słowik, M.; Pršek, J.; Lankosz, M.; Wróbel, P.; Gaweł, A.; Kowalczyk, J.; Kocemba, J. Agates from Kerrouchen (the Atlas Mountains, Morocco): Textural types and their gemological characteristics. *Minerals* **2016**, *6*, 77. [CrossRef]

28. Powolny, T.; Dumańska-Słowik, M.; Sikorska-Jaworowska, M.; Wójcik-Bania, M. Agate mineralization in spilitized Permian volcanics from "Borówno" quarry (Lower Silesia, Poland)—Microtextural, mineralogical, and geochemical constraints. *Ore Geol. Rev.* **2019**, *114*, 103–130. [CrossRef]

29. Sidorov, E.G.; Kutyev, F.T.; Anikin, P.P. Native Copper Agates of the Kuril-Kamchatka Province. In *Native Metals in Postmagmatic Formations*; Yakutsk Publishing House: Yakutsk, Russia, 1985; pp. 72–73. (In Russian)

30. Sheymovich, V.S. *The State Geological Map of the Russian Federation, Scale 1:200,000*; South-Kamchatka Series. Sheets N_57_XXI (Northern Koryaks), N_57_XXVII (Petropavlovsk-Kamchatsky), N_57_XXXIII (Mutnovskaya hill); Explanatory Note: Moscow, Russia, 2000; p. 302. (In Russian)

31. Savelyev, D.P.; Palechek, T.N.; Portnyagin, M.V. Campanian oceanic siliceous-volcanogenic deposits in the basement of the Eastern Kamchatka volcanic belt. *Pac. Geol.* **2005**, *24*, 46–54. (In Russian)

32. Frolova, Y.V.; Blyumkina, M.E.; Bolshakov, I.E.; Ermolinsky, A.B. *Comparative Petrophysical Characteristics of Volcanic Rocks of the Cretaceous and Miocene Age of Avacha Bay. Volcanism and Related Processes*; Materials of the XXIII Annual Scientific Conference Dedicated to the Volcanologist's Day; IViS: Petropavlovsk-Kamchatsky, Russia, 2020; pp. 68–71.

33. Saveliev, D.P. A scattering of agates at Cape Vertikalny, Eastern Kamchatka. *Bull. Kamchatka Reg. Assoc. Educ. Sci. Cent. Ser. Earth Sci.* **2020**, *47*, 3. [CrossRef]

34. *The Powder Diffraction File PDF-4p*; International Centre for Diffraction Data: Newtown Square, PA, USA, 2009.

35. Borisenko, A.S. Analysis of the Salt Composition of solutions of gas-liquid inclusions in minerals by cryometry. In *The Use of Methods of Thermobarogeochemistry in the Search and Study of Ore Deposits*; Laverova, N.P., Ed.; Nedra: Moscow, Russia, 1982; pp. 37–47. (In Russian)

36. Roedder, E. Fluid inclusions. *Rev. Mineral.* **1984**, *12*, 79–108.

37. Bodnar, R.J.; Vityk, M.O. Interpretation of microthermometric data for NaCl–H_2O fluid inclusions. In *Fluid Inclusions in Minerals: Methods and Applications*; De Vivo, B., Frezzotti, M.L., Eds.; Virginia Polytechnic Inst State Univ: Blacksburg, VA, USA, 1994; pp. 117–131.

38. Dong, G.; Morrison, G.; Jaireth, S. Quartz textures in epithermal veins, Queensland; classification, origin, and implication. *Econ. Geol.* **1995**, *90*, 1841–1856. [CrossRef]

39. Lafuente, B.; Downs, R.T.; Yang, H.; Stone, N. The power of databases: The RRUFF project. In *Highlights in Mineralogical Crystallography*; Armbruster, T., Danisi, R.M., Eds.; W. De Gruyter: Berlin, Germany, 2015; pp. 1–30.

40. Götze, J.; Nasdala, L.; Kleeberg, R.; Wenzel, M. Occurrence and distribution of "moganite" in agate/chalcedony: A combined micro-Raman, Rietveld, and cathodoluminescence study. *Contrib. Mineral. Petrol.* **1998**, *133*, 96–105. [CrossRef]

41. Bodnar, R.J. Interpretation of data from aqueous-electrolyte fluid inclusions. In *Fluid Inclusions: Analysis and Interpretation*; Samson, I., Anderson, A., Marshall, D., Eds.; Short Course Series; Mineralogical Association of Canada: Ottawa, ON, Canada, 2003; pp. 81–100.

42. Goldstein, R.H.; Reynolds, T.J. *Systematics of Fluid Inclusions in Diagenetic Minerals*; SEPM Short Course: Tulsa, OK, USA, 1994; Volume 31, p. 199.

43. Hardie, L.A. Origin of $CaCl_2$ brines by basalt-seawater interactions insights provided by some simple mass balance calculations. *Contrib. Mineral. Petrol.* **1983**, *82*, 205–213. [CrossRef]

44. Heaney, P.J.; Post, J.E. The Widespread Distribution of a Noel Silica Polymorph in Microcrystalline Quartz Varieties. *Science* **1992**, *255*, 441–443. [CrossRef]

45. Moxon, T.; Carpenter, M.A. Crystallite growth kinetics in nanocrystalline quartz (agate and chalcedony). *Miner. Mag.* **2009**, *73*, 551–568. [CrossRef]

46. Barton, P.B., Jr.; Skinner, R.J. Sulphide mineral stabilities. In *Geochemistry of Hydrothermal Ore Deposits*; Barnes, H.L., Ed.; Wiley: New York, NY, USA, 1979; pp. 278–403.

47. Cornwall, H.R. A summary of ideas on the origin of native copper deposits. *Econ. Geol.* **1956**, *51*, 615–631. [CrossRef]

48. Moxon, T.; Palyanova, G. Agate genesis: A continuing enigma. *Minerals* **2020**, *10*, 953. [CrossRef]

49. Gilg, H.A.; Morteani, G.; Kostitsyn, Y.; Preinfalk, C.; Gatter, I.; Strieder, A.J. Genesis of amethyst geodes in basaltic rocks of the Serra Geral Formation (Ametista do Sul, Rio Grande do Sul, Brazil): A fluid inclusion, REE, oxygen, carbon, and Sr isotope study on basalt, quartz, and calcite. *Miner. Depos.* **2003**, *38*, 1009–1025. [CrossRef]

50. Stoiber, R.E.; Davison, E.S. Amygdule mineral zoning in the Portage Lake lava series, Michigan copper district. *Econ. Geol.* **1959**, *54*, 1444–1460. [CrossRef]

51. Bornhorst, T.J. Tectonic context of native copper deposits of the North American Midcontinent rift system. *Geol. Soc. Am. Spec. Pap.* **1997**, *312*, 127–136.

52. Brown, A.C. Genesis of native copper lodes in the Keweenaw district, northern Michigan: A hybrid evolved meteoric and metamorphic model. *Econ. Geol.* **2006**, *101*, 1437–1444. [CrossRef]

53. Taylor, S.R. Abundance of chemical elements in the continental crust: A new table. *Geochim. Cosmochim. Acta* **1964**, *28*, 1273–1285. [CrossRef]

54. Savchuk, Y.S.; Volkov, A.V.; Aristov, V.V. Cupriferous basalts of the Northern Urals. *Litosfera* **2017**, *17*, 133–144.

Publisher's Note: MDPI stays neutral with regard to jurisdictional claims in published maps and institutional affiliations.

Article

Occurrence and Distribution of Moganite and Opal-CT in Agates from Paleocene/Eocene Tuffs, El Picado (Cuba)

Jens Götze [1,*], Klaus Stanek [2], Gerardo Orozco [3], Moritz Liesegang [4] and Tanja Mohr-Westheide [5]

[1] Institute for Mineralogy, TU Bergakademie Freiberg, Brennhausgasse 14, 09599 Freiberg, Germany

[2] Institute for Geology, TU Bergakademie Freiberg, Cottastraße 4, 09599 Freiberg, Germany; klaus-peter.stanek@geo.tu-freiberg.de

[3] Departamento de Geología, Universidad de Moa, Avenida Calixto Garcia Iñiguez 15, Moa, Holguín 83330, Cuba; orozcog49@gmail.com

[4] Institute for Geological Sciences, FU Berlin, Malteserstraße 74–100, 12249 Berlin, Germany; m.liesegang@fu-berlin.de

[5] Institute for Evolution and Biodiversity Science, Museum für Naturkunde, Invalidenstraße 43, 10115 Berlin, Germany; Tanja.Mohr-Westheide@mfn-berlin.de

* Correspondence: jens.goetze@mineral.tu-freiberg.de

Abstract: Agates in Paleocene/Eocene tuffs from El Picado/Los Indios, Cuba were investigated to characterize the mineral composition of the agates and to provide data for the reconstruction of agate forming processes. The volcanic host rocks are strongly altered and fractured and contain numerous fissures and veins mineralized by quartz and chalcedony. These features indicate secondary alteration and silicification processes during tectonic activities that may have also resulted in the formation of massive agates. Local accumulation of manganese oxides/hydroxides, as well as uranium (uranyl-silicate complexes), in the agates confirm their contemporaneous supply with SiO_2 and the origin of the silica-bearing solutions from the alteration processes. The mineral composition of the agates is characterized by abnormal high bulk contents of opal-CT (>6 wt%) and moganite (>16 wt%) besides alpha-quartz. The presence of these elevated amounts of "immature" silica phases emphasize that agate formation runs through several structural states of SiO_2 with amorphous silica as the first solid phase. A remarkable feature of the agates is a heterogeneous distribution of moganite within the silica matrix revealed by micro-Raman mapping. The intensity ratio of the main symmetric stretching-bending vibrations (A_1 modes) of alpha-quartz at 465 cm^{-1} and moganite at 502 cm^{-1}, respectively, was used to depict the abundance of moganite in the silica matrix. The zoned distribution of moganite and variations in the microtexture and porosity of the agates indicate a multi-phase deposition of SiO_2 under varying physico-chemical conditions and a discontinuous silica supply.

Keywords: agate; alpha-quartz; chalcedony; moganite; opal-CT; Raman spectroscopy

Citation: Götze, J.; Stanek, K.; Orozco, G.; Liesegang, M.; Mohr-Westheide, T. Occurrence and Distribution of Moganite and Opal-CT in Agates from Paleocene/Eocene Tuffs, El Picado (Cuba). *Minerals* **2021**, *11*, 531. https://doi.org/10.3390/min11050531

Academic Editor: Galina Palyanova

Received: 27 April 2021
Accepted: 12 May 2021
Published: 18 May 2021

Publisher's Note: MDPI stays neutral with regard to jurisdictional claims in published maps and institutional affiliations.

1. Introduction

Cuba is a country that is particularly famous for its enormous deposits of Ni ores. Huge surficial laterite horizons in the eastern part of the Caribbean island belong to the most important nickel deposits in the world and contain considerable amounts of other valuable chemical compounds of, e.g., Sc, Co, or rare earth elements (REE) [1]. However, occurrences of other mineral deposits are rare and restricted to a few locations.

For instance, different gemstones were exploited between 1989 and 1998 in Cuba, especially silica minerals and rocks such as chalcedony, opal, jasper or silicites [2]. Remarkable are chalcedony from Palmira near Cienfuegos, opal from Loma de los Ópalos and Pontezuela (near Camagüey) as well as silicites from Corralillo on the northern coast of Cuba (Figure 1). Silicified wood occurs in Upper Cretaceous volcanic sequences of the Sierra de Najasa SE of Camagüey (Sibanicu) and above ophiolites in the Sierra de Cubitas north of Camagüey (Miocene) [2,3].

187

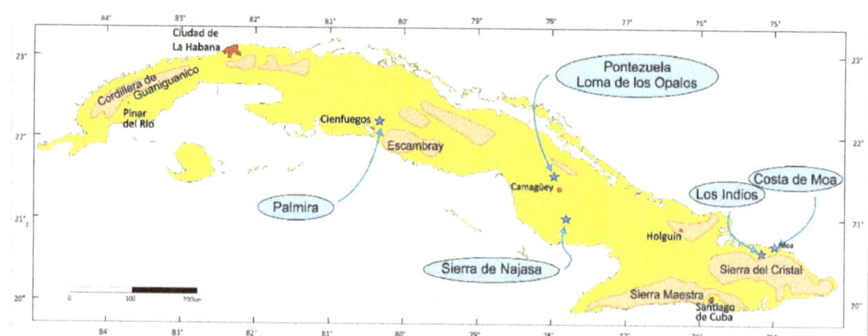

Figure 1. A geomorphological map of Cuba showing occurrences of different SiO_2 mineralization.

Agates were sporadically found in Upper Cretaceous rocks overlaying ophiolites of the Camagüey fault zone NE of Camagüey. In addition, chalcedony, jasper and silicified corals are reported from the Rio Yumuri region near Matanzas [2]. In the eastern part of Cuba, silica mineralization occurs in particular in the surrounding area of Moa. Jasper was found in altered serpentinites of the Costa de Moa region, whereas agates/chalcedony occur in altered volcanic rocks in the region of El Picado, western Moa as well as near Baracoa on the river Toa, eastern Moa (Figure 1).

Because of the lack of mineralogical data, the agates of the Moa region were in the focus of the present study. Mineralogical and geochemical investigations were aimed to characterize the silica minerals in the agates and to provide data for the reconstruction of agate formation processes in Moa. On the other hand, the results of the study should provide additional general arguments for the ongoing discussion concerning the enigma of agate genesis [3,4]. Emphasis was placed on the detection and distribution of the monoclinic silica polymorph moganite and opal-CT, which both occur in considerable amounts in the agates from the surroundings of Moa.

2. Geological Background and Sample Material

The geology of Cuba is dominated by metamorphic and sedimentary rocks. The geographic position in a tectonically active region, with plate movements in Mesozoic to Cenozoic times along NE trending fault systems, resulted in a geologic trisection [5]. The western part of the island is mainly characterized by carbonate sediments and widespread karst formation. The central part (Escambray) consists of ophiolites, lifted metamorphic rocks and carbonate sediments. The eastern part of Cuba is dominated by the largest ophiolite complexes of the Caribbean, K-poor volcanic rocks (and associated intrusive rocks) of the Paleogene to Eocene island arc volcanism, as well as overlaying calcareous sediments [5].

The investigated agate material derives from El Picado near the village of Los Indios, ca. 15 km west of the city of Moa (Figure 1). The host rocks are strongly altered Paleocene/Eocene tuffs (ca. 55 Ma) locally intercalated with limestones. The altered volcanic rocks are fractured and contain numerous fissures and veins mineralized by quartz and chalcedony, indicating a secondary silicification of the pyroclastic material during tectonic activities (Figure 2).

Figure 2. (**a**) A layered and strongly altered tuff and (**b**) a brecciated volcanic host rock with secondary quartz and chalcedony veins (arrows) from El Picado, Cuba.

XRD studies of the surrounding host rocks provided an average mineral composition of 8–10 wt% plagioclase (andesine), ca. 20 wt% dioctahedral illite-smectite mixed-layer minerals, 11–14 wt% zeolite (clinoptilolite), 6–8 wt% quartz, ca. 40 wt% opal-CT, and 10–12 wt% calcite. These results document that the primary mineral composition of the volcanic rocks is strongly modified during the alteration processes, resulting in the release of remarkable amounts of SiO_2.

The occurrence of agate in the investigation area around El Picado/Los Indios is variable. Agates appear both within the outcropping volcanic host rocks and as loose material surficially distributed in the field (Figure 3). Chalcedony shows different colors (white, grey, blue, yellow) and often collomorph or botryoidal habit (Figures 3a and 4c). Agates may occur in remarkable sizes but often exhibit only weak banding (Figures 3b and 4a,b).

Figure 3. (**a**) Chalcedony with botryoidal texture and (**b**) large yellow agate in weathered volcanic bedrock from El Picado, Cuba.

The collected sample material was documented and macroscopically described. Several agates were cut and polished to better visualize the agate textures. Finally, ten samples of different characteristics and textures were selected and prepared for detailed mineralogical investigations representing three main types (ACu1–3; Figure 4).

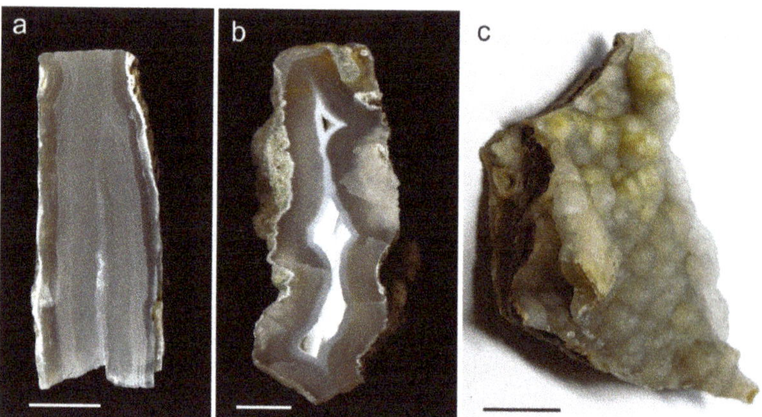

Figure 4. Selected samples of the investigated agate/chalcedony material from El Picado/Los Indios, Cuba representing three main types: (**a**) finely banded, vein-like agate (sample ACu1); (**b**) agate with a white center (ACu2); (**c**) botryoidal chalcedony (ACu3); the scale bar is 2 cm.

3. Analytical Methods

The mineral composition of the agates and volcanic host rocks was analyzed by X-ray diffraction (XRD) measurements on selected and prepared (<20 µm) sample material. The qualitative and quantitative phase compositions were analyzed using an URD 6 (Seifert/Freiberger Präzisionsmechanik) and a METEOR 0D Si drift detector (GE Sensing & Inspection Technologies GmbH, Ahrensburg, Germany) with Co Kα-radiation in the range of 5–80° (2θ). Analytical conditions included a detector slit of 0.25 mm, 0.03° step width and 5 s measuring time per step. Data evaluation was realized using the Analyse RayfleX v.2.352 software and subsequent Rietveld refinement with Autoquan v.2.7.00 [6].

The identification and characterization of different SiO_2 phases was performed by a combination of X-ray diffraction and Raman spectroscopy. Raman spectroscopy analyses were conducted on a Horiba Jobin Yvon LabRAM HR 800 instrument coupled to an Olympus BX41 microscope. For mappings, unpolarized spectra were collected with the LabSpec 6 software over a range from 90 to 600 cm^{-1} at a step size of 8 µm. A 633 nm laser was used to excite the sample with a 50× objective, at a spectral integration time of 5 s, and two accumulations. Scattered Raman light was collected in backscattering geometry and dispersed by a grating of 600 grooves/mm after passing through a 100 µm entrance slit. The confocal hole size was set to 1000 µm. An internal intensity correction (ICS, Horiba, Kyoto, Japan) was used to correct detector intensities. The instrument was calibrated using the Raman band of a silicon wafer at 520.7 cm^{-1}.

Microscopic investigations comprised an integrated analysis by polarizing, cathodoluminescence (CL) and scanning electron microscopy (SEM) on polished thin sections (30 µm). Conventional polarizing microscopy in transmitted light was conducted with a Zeiss Axio Imager A1m microscope. Additional Nomarski differential interference contrast (DIC) imaging in reflected light was performed using specific Nomarski DIC prisms to visualize variations of the surface topography.

These investigations were completed by SEM studies as well as CL analyses with an optical CL microscope HC1-LM on carbon-coated, polished thin sections [7]. The SEM measurements (SE, BSE), including local chemical analyses (EDX), were performed using a JEOL JSM-7001F (20 kV, 2.64 nA) with a BRUKER Quantax 800 EDX system. CL microscopy and spectroscopy were operated at 14 kV accelerating voltage and a current of 0.2 mA with a Peltier cooled digital video camera (OLYMPUS DP72) and an Acton Research SP-2356 digital triple-grating spectrograph with a Princeton Spec-10 CCD detector, respectively. CL spectra in the wavelength range from 380 to 900 nm were measured under standardized conditions (wavelength calibration by a Hg-halogen lamp, spot width 30 μm, measuring time 5 s).

4. Results

4.1. Mineralogy and Microstructure of the Agates

The mineralogical composition of the agates from El Picado comprises different phases. In addition to microcrystalline (chalcedony) and macrocrystalline quartz, the agates contain elevated contents of moganite and opal-CT, as well as minor amounts of calcite (Table 1, Figure 5).

Table 1. The mineral composition (wt%) of agate samples from El Picado (Cuba) determined by X-ray diffraction measurements with Rietveld refinement.

	ACu1	ACu2	ACu3
α-quartz	89.1 ± 0.4	84.2 ± 0.4	74.4 ± 0.4
Moganite	10.9 ± 0.3	13.5 ± 0.3	16.8 ± 0.3
Opal-CT	-	2.2 ± 0.1	6.2 ± 0.1
Calcite	-	-	2.6 ± 0.1

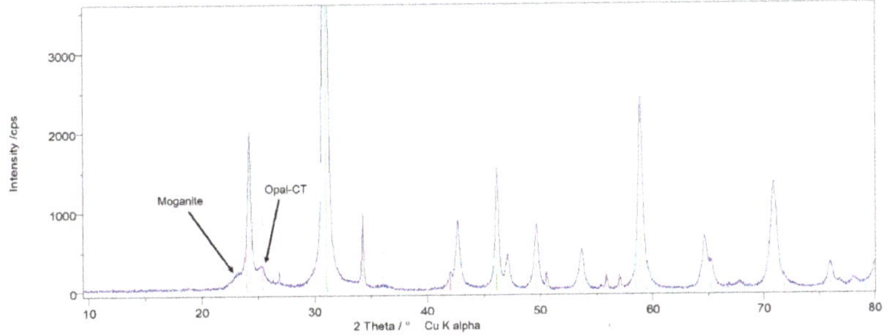

Figure 5. The X-ray diffraction pattern of sample ACub3 (powder sample, background corrected); search-match results in stick patterns: magenta, opal-CT (PDF 00-066-0177); light blue, moganite (PDF 01-073-2991); green, quartz (PDF 01-070-3755); red, calcite (PDF 01-083-0577). The arrows point to the characteristic broad maxima of moganite and opal-CT.

The microtexture of the agates is, in general, characterized by a sequence of granular, microcrystalline quartz, alternating layers of chalcedony with micro-granular quartz, fibrous chalcedony, and macrocrystalline quartz from the agate rim to the center (Figure 6a). In particular, the granular quartz layers are clearly distinguishable from chalcedony bands due to the varying porosity (Figure 6b). In contrast, the transition of fibrous chalcedony to macrocrystalline quartz in the center is smooth (Figure 6c).

The presence and spatial distribution of the silica polymorph moganite is difficult to detect by microscopic methods because of the similar optical properties and the narrow intergrowth with chalcedony. Although the moganite content in the agate samples from El Picado exceeds 10 wt% (compare Table 1), it was not possible to identify the monoclinic silica polymorph by polarizing or scanning electron microscopy. In contrast to moganite, opal-CT can be detected in the scanning electron microscope because of its specific morphological characteristics (Figure 7). Opal-CT forms typical lepispheres, spherical aggregates consisting of small crystal blades [8]. Local chemical analyses by SEM-EDX revealed elevated concentrations of Al (up to 0.55 wt%—see Figure 7b) in opal-CT. This contrasts with Al concentrations in chalcedony and macrocrystalline quartz (8–13 ppm) in the same agate samples [9].

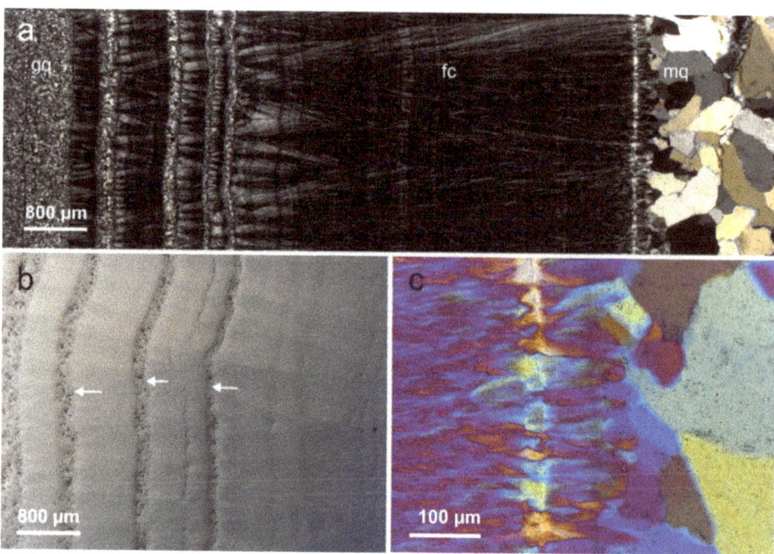

Figure 6. The microtexture of sample ACu1: (**a**) sequence of silica phases from the margin of the agate to the center with granular, microcrystalline quartz (gq), alternating layers of chalcedony with fine-grained quartz, fibrous chalcedony (fc) and macrocrystalline quartz (mq); (**b**) Nomarski DIC micrograph illustrating the differences in porosity between chalcedony and quartz layers; the surface relief imaged by Nomarski DIC microscopy reveals higher porosity in the micro-granular quartz layers (arrows); (**c**) micrograph in polarized light and additional nγ/λ-compensator showing the smooth transition from fibrous chalcedony to macrocrystalline quartz in the agate center.

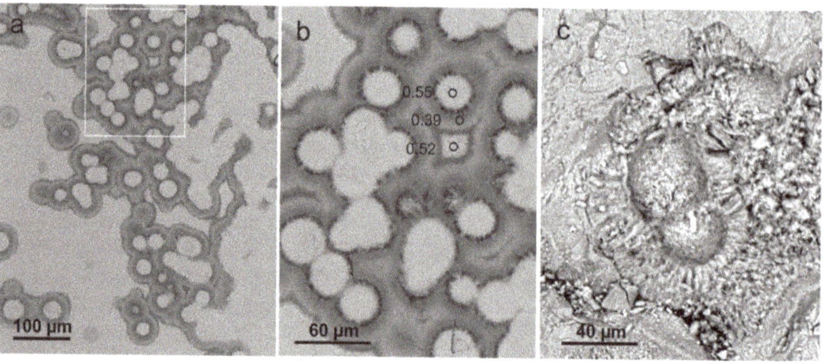

Figure 7. SEM-micrographs showing the appearance of opal-CT in agate sample ACu3: (**a**) a SEM-BSE image of the spherical opal-CT aggregates; (**b**) a close-up of (**a**) (see rectangle) with details and results of local chemical analyses of Al (in wt%) by SEM-EDX; (**c**) a SEM composite micrograph of opal-CT lepispheres revealing the characteristic spheres consisting of small crystal blades.

In addition to the different silica minerals, certain other phases were detected in the agates from El Picado. Calcite was detected in sample ACu3 in the contact area to the altered host rocks. The carbonate grains are partially corroded and replaced by silica minerals (Figure 8a); moreover, inclusions of dark material could be observed within the chalcedony matrix that is texturally arranged along the agate banding (Figure 8b). Local chemical analyses by SEM-EDX confirmed the presence of manganese oxides/hydroxides, which have probably been primarily intercalated into the SiO_2 matrix. According to local chemical analyses by SEM-EDX, the Mn concentrations in the dark areas varied between 2.52 and 5.41 wt%.

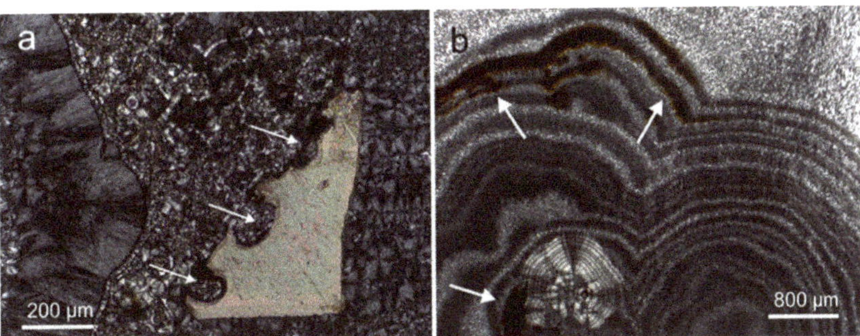

Figure 8. Micrographs in transmitted light (crossed polars) of non-silica phases in the agates from El Picado: (**a**) calcite grain (sample ACu3), which is partially corroded by SiO$_2$ (arrows); (**b**) intercalation of dark banded manganese oxides/hydroxides (sample ACu2) arranged along the chalcedony banding (see arrows) pointing to simultaneous accumulation and precipitation from mineralizing fluids.

4.2. Occurrence and Distribution of Moganite

Because of the limitations of microscopic techniques for the identification of moganite, Raman spectroscopy was used to provide information about the occurrence and spatial distribution of moganite in the agates from El Picado/Los Indios. Previous studies have shown that quartz and moganite can be distinguished due to their different spectral characteristics. The identification of the two silica polymorphs is possible by the characteristic main symmetric stretching–bending vibrations (A$_1$ modes) of alpha-quartz at 465 cm^{-1} and moganite at 502 cm^{-1}, respectively [10].

Point analyses in different areas of the El Picado agates revealed variations in the moganite-to-quartz ratio detectable by different intensities of the relevant peaks in the Raman spectra (Figure 9a). The analytical data indicate that the moganite content in the agates is not homogeneously distributed. There exist chalcedony bands with high contents of moganite as well as bands without—more or less—any moganite. This conclusion confirms previous results, which documented the variations in moganite content in different parts of agates and even within the chalcedony banding [11–13].

In continuation of the local Raman spectroscopic analyses, Raman mapping was performed to provide detailed information about the semi-quantitative spatial distribution of moganite in the agates. The integral ratio of the 502/465 cm^{-1} Raman bands of moganite and quartz, respectively, was used to depict the abundance of moganite in the silica matrix. Figure 8b illustrates a heterogeneous distribution of moganite in the agate, which strongly correlates with the structural banding; however, moganite content may even fluctuate within individual fibrous chalcedony layers. Considering the average moganite contents detected with XRD (Table 1), the local moganite content in moganite-rich areas must be significantly higher than 15 wt%.

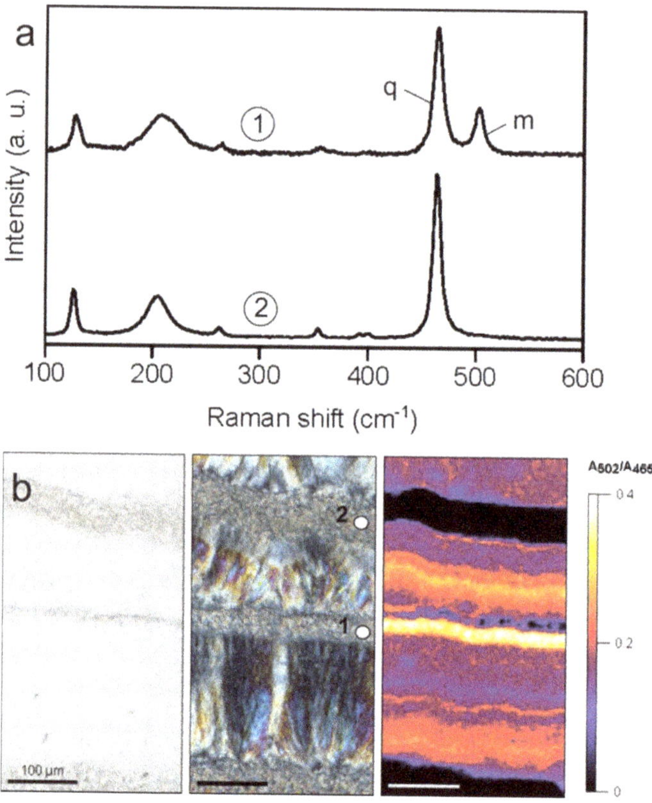

Figure 9. Results of Raman spectroscopy and Raman mapping of agate sample ACu1: (**a**) Raman spectra of two local analyses in different agate areas (compare b); spectrum 1 shows the characteristic main symmetric stretching-bending vibrations (A_1 modes) of alpha-quartz (q) at 465 cm^{-1} and moganite (m) at 502 cm^{-1}, respectively; whereas spectrum 2 shows only the band of alpha-quartz; (**b**) agate area in transmitted light with parallel and crossed polars, respectively, and the related Raman map based on the integral ratio of the 502/465 cm^{-1} bands; high ratios (yellow) point to areas with high moganite content.

4.3. Cathodoluminescence (CL) Properties of the Agates

CL microscopy and spectroscopy of the agates illustrate relatively uniform CL properties, exemplarily shown for ACu2 in Figure 10. The visible CL color is mostly a greenish-blue turning into a reddish-brown during electron irradiation (Figure 10a,b). Spectral CL measurements revealed a complex emission consisting of a broad composite blue band and a strong emission band centered at ca. 650 nm (Figure 10d). In addition, multiple emission peaks between 500 and 600 nm can be related to the luminescence of the UO_2^{2+} uranyl ion [14]. This greenish luminescence signal is characterized by a typical emission line at ~500 nm accompanied by several equidistant lines due to the harmonic vibrations of oxygen atoms in the uranyl complex (Figure 10d).

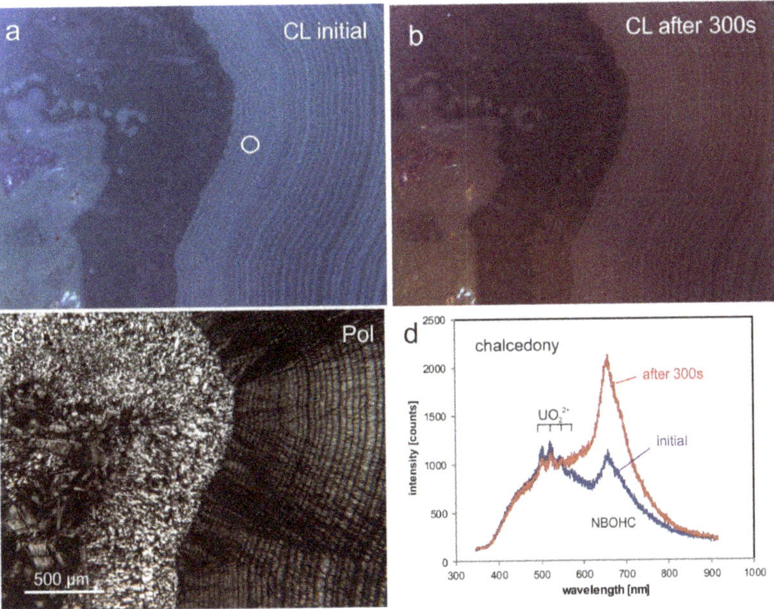

Figure 10. Micrographs of agate ACu2 showing the initial CL (**a**) compared with the CL after 300 s electron irradiation (**b**) and the sample area in transmitted light (crossed polars) (**c**) as well as the related CL spectra, which are dominated by a strong emission band centered at ca. 650 nm due to the non-bridging oxygen hole center (NBOHC) and multiple emission lines of the uranyl ion (**d**). The circle in (**a**) marks the position of the spectral CL measurement.

The 650 nm band (1.91 eV) is the most common CL emission in chalcedony and can be related to the non-bridging oxygen hole center (NBOHC) [15]. Whereas the broad blue emission is relatively stable under the electron beam, the 650 nm emission is very sensitive to electron irradiation, resulting in an increase in the band intensity during electron bombardment due to the conversion of different precursors (e.g., silanol groups: Si–O–H, Na impurities: Si–O–Na) into hole centers [16]. This is detectable by a change of the visible CL color from a greenish-blue to a reddish-brown (Figure 10a,b).

In contrast to the common CL behavior of most agates, the opal-CT rich parts of sample ACu3 exhibit a quite contrasting luminescence behavior (Figure 11). The visible luminescence color is bright blue and spectral measurements reveal a dominating broad band at ca. 500 nm, whereas the 650 nm emission band is subordinate. Such luminescence spectra were found in amorphous silica samples such as opal and opal-CT [17].

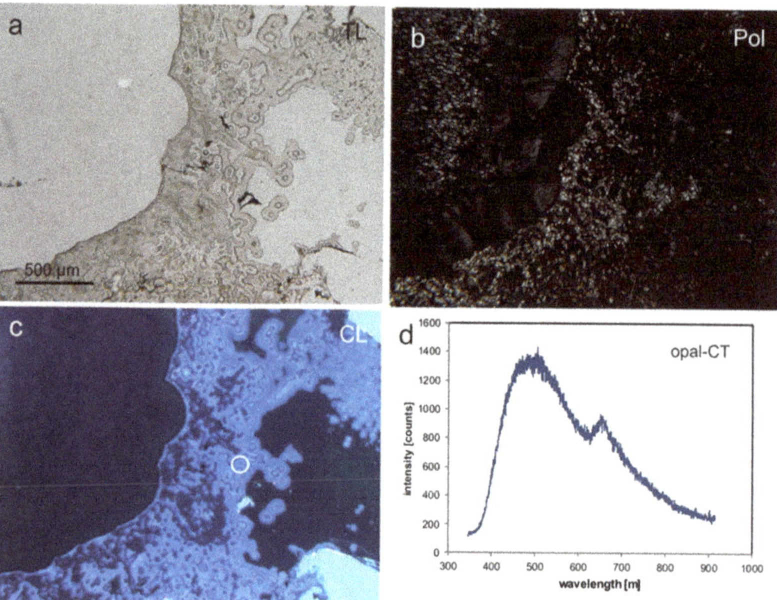

Figure 11. Micrographs of agate ACu3 in transmitted light (**a**) with crossed polars (**b**) and CL mode (**c**) showing the CL of an opal-CT rich area with a bright blue CL. (**d**) The related CL spectrum is dominated by a broad band centered around 500 nm; the circle in (**c**) marks the position of the spectral CL measurement.

5. Discussion

The discussion concerning the formation of agates in the Paleocene/Eocene tuffs from El Picado first focusses on the question about the origin of silica. Both the geological conditions of exposure of the volcanic rocks observed during field work and their mineralogical composition point to strong secondary alteration processes. The XRD results document high amounts of dioctahedral illite-smectite mixed-layer minerals and zeolites as typical alteration products of basic volcanic rocks. The primary mineral composition of the agate-bearing volcanic rocks is strongly modified, containing only residual plagioclase from the original inventory. These alteration processes must have caused the release of remarkable amounts of SiO_2 that have been available for silicification processes and the formation of massive agates. The SiO_2 supply is thereby associated with hydrothermal activities and secondary alteration of the predominantly SiO_2-poor, basic rocks.

Further indications concerning the formation processes were provided by previous investigations of fluid inclusions in macrocrystalline quartz crystals in the agates. Measured homogenization temperatures of aqueous inclusions scatter in a narrow range between 130 and 145 °C [7], emphasizing a source from low temperature hydrothermal solutions and/or heated meteoric water.

The detection of uranyl ions in the silica matrix by CL spectroscopy is another indication of intensive alteration processes. Uranium might be released from volcanic rocks during alteration processes and accumulated together with silica [18]. In result, uranium can be adsorbed from solution by natural silica compounds (silica colloids) and then trapped as a uranyl–silicate complex in a stable silica matrix [19]. A substitution of Si ions in the quartz lattice by U is unlikely due to the different crystal–chemical properties. The CL studies confirm the existence of uranyl compounds in quartz and chalcedony of agates, as it was already found in agates from other locations [14]. This binding mechanism of uranyl–silicate complexes in microcrystalline SiO_2 suggests a new mechanism of U deposition involving uranyl co-precipitation with silica during alteration processes.

Both field observations, concerning the appearance of agates in the host rocks and their mineralogical characteristics, illustrate that agates in El Picado can mostly be related to the crystallization from silica-rich solutions in fissures and cavities of the volcanic host rocks. The presence of opal-CT as an intermediate silica phase between amorphous and crystalline SiO_2 is a strong indication that the agate formation processes started from non-crystalline silica; particularly, in geologically young volcanic host rocks worldwide, opal-CT is detectable, since it has not been completely transformed to quartz [20,21]. Therefore, the preservation of opal-CT in the Cuban agates can be explained by the relatively young age (~55 Ma) of the hosting tuff layers.

The same applies for the unstable monoclinic SiO_2-modification moganite. A correlation between the formation age of the agates and the corresponding moganite content was first published by Moxon and Rios [22]. The amount of moganite in the agates strongly decreases with increasing age. For instance, less than 6 wt% moganite in agates from six 300 to 1100 Ma old host rocks was measured [23]; agates from a further three locations contained even less than 1 wt% moganite. In addition, Moxon [24,25] proved that the conversion of metastable silica phases into quartz is accompanied by loss of structural water and the coarsening of crystallite size.

The analyses of the present study emphasize that the agates from El Picado contain more than 15 wt% moganite. Considering the results of Moxon and Carpenter [23], these high moganite contents can be related to the geological age of the agates. However, it is noteworthy that the distribution of moganite within the silica matrix is not homogeneous. Raman mapping revealed extreme variations of the moganite content between different silica layers; particularly in zones with granular and macrocrystalline quartz, the moganite content approaches almost zero, whereas microcrystalline chalcedony contains elevated moganite contents (compare Figure 9).

The differences in moganite content are accompanied by variations in textural characteristics, including porosity (compare Figure 6). Therefore, the zoned distribution of moganite corresponds with the banded structure of the agates pointing to a multi-phase deposition of SiO_2. These structural and textural features strongly indicate variations in the physico-chemical conditions and a discontinuous silica supply.

6. Conclusions

The present study provides first detailed mineralogical data of agates from El Picado/Los Indios in the Eastern part of Cuba (Moa region). The investigations aimed to characterize the silica minerals in the agates and to provide data for the reconstruction of the agate forming processes.

The mineral composition of the agate-bearing Paleocene/Eocene tuffs is dominated by illite-smectite mixed-layer minerals, zeolites, opal-CT and calcite and has only minor amounts (<10 wt%) of residual primary plagioclase (andesine). These results document that the primary mineral composition of the predominantly SiO_2-poor, basic volcanic rocks is strongly modified during intensive secondary alteration. These surficial alteration processes must have caused the release of enormous amounts of aqueous silica that have been available for silicification processes and the formation of massive agates. Homogenization temperatures of fluid inclusions in macrocrystalline quartz (130–145 °C) point to the participation of hydrothermal fluids and/or heated meteoric water.

The mineral composition of the agates is dominated by alpha-quartz (microcrystalline chalcedony and macrocrystalline quartz), but also shows remarkable high abundances of moganite (>16 wt%) and opal-CT (>6 wt%). These elevated amounts of "immature" silica phases can probably be related to the geological age (~55 Ma) of the mineralization processes. The presence of opal-CT confirms the supposed formation of the agates via an amorphous silica precursor. Moreover, the heterogeneous distribution of moganite within the silica matrix, revealed by micro-Raman mapping as well as detected variations in the microtexture and porosity of the agates, indicate a multi-phase deposition of SiO_2 due to a discontinuous silica supply and probably varying physico-chemical conditions.

Author Contributions: K.S., G.O. and J.G. collected the studied samples and provided the geological data; J.G., T.M.-W. and M.L. conducted different analytical measurements, evaluated the mineralogical and spectroscopic data and provided appropriate parts of the manuscript; J.G. compiled and wrote the final version of the manuscript. All authors have read and agreed to the published version of the manuscript.

Funding: This research received no external funding.

Institutional Review Board Statement: Not applicable.

Informed Consent Statement: Not applicable.

Data Availability Statement: All data are contained within the article.

Acknowledgments: We would like to thank Reinhard Kleeberg (TU Bergakademie Freiberg, Germany) for assistance with XRD data acquisition and evaluation. Comments from Galina Palyanova (Novosibirsk, Russia) and the reviews of three anonymous reviewers improved the quality of the manuscript.

Conflicts of Interest: The authors declare no conflict of interest.

References

1. Barros de Oliveira, S.M.; De Moya Patiti, C.S.; Enzweiler, J. Ochreous laterite: A nickel ore from Punta Gorda, Cuba. *J. S. Am. Earth Sci.* **2001**, *14*, 307–317.
2. Gómez Narbona, L.J. Caracteristicas, usos, requisitos y metodos de evaluacion tecnologica de las manifestaciones de piedras preciosas Palmira, Loma de los Opalos, Pontezuela y Corralillo. In Proceedings of the IX Congreso Cubano de Geología (GEOLOGIA'2011), Influencia económica de la Microminería, La Habana, Cuba, 4–8 April 2011; pp. 1–11.
3. Moxon, T.; Palyanova, G. Agate genesis: A continuing enigma. *Minerals* **2020**, *10*, 953.
4. Götze, J.; Möckel, R.; Pan, Y. Mineralogy, geochemistry and genesis of agate—A review. *Minerals* **2020**, *10*, 1037.
5. Stanek, K.; Cobiella-Reguera, J.L.; Maresch, W.V.; Trujillo, G.M.; Grafe, F.; Grevel, C. Geological development of Cuba. *Z. Angew. Geol. Sonderh.* **2000**, *1*, 259–265.
6. Taut, T.; Kleeberg, R.; Bergmann, J. Seifert software: The new Seifert Rietveld program BGMN and its application to quantitative phase analysis. *Mater. Struct.* **1998**, *5*, 57–66.
7. Neuser, R.D.; Bruhn, F.; Götze, J.; Habermann, D.; Richter, D.K. Kathodolumineszenz: Methodik und Anwendung. *Zent. Geol. Paläontologie Teil I* **1995**, *H.1/2*, 287–306.
8. Graetsch, H. Structural characteristics of opaline and microcrystalline silica minerals. In *Silica—Physical Behaviour, Geochemistry and Materials Application*; Heaney, P.J., Prewitt, C.T., Gibbs, G., Eds.; Reviews in Mineralogy & Geochemistry; MSA: Washington, DC, USA, 1994; Volume 29, pp. 209–232.
9. Götze, J.; Blankenburg, H.-J. Zur Kathodolumineszenz von Achat—Erste Ergebnisse. *Der Aufschluß* **1994**, *45*, 305–312.
10. Kingma, K.J.; Hemley, R.J. Raman spectroscopic study of microcrystalline silica. *Am. Mineral.* **1994**, *79*, 269–273.
11. Götze, J.; Nasdala, L.; Kleeberg, R.; Wenzel, M. Occurrence and distribution of "moganite" in agate/chalcedony: A combined micro-Raman, Rietveld, and cathodoluminescence study. *Contrib. Mineral. Petrol.* **1998**, *133*, 96–105.
12. Natkaniec-Nowak, L.; Dumańska-Słowik, M.; Pršek, J.; Lankosz, M.; Wróbel, P.; Gaweł, A.; Kowalczyk, J.; Kocemba, J. Agates from Kerrouchen (The Atlas Mountains, Morocco). *Minerals* **2016**, *6*, 77.
13. Zhang, X.; Ji, L.; He, X. Gemological characteristics and origin of the Zhanguohong agate from Beipiao, Liaoning province, China: A combined microscopic, X-ray diffraction, and Raman spectroscopic study. *Minerals* **2020**, *10*, 401.
14. Götze, J.; Gaft, M.; Möckel, R. Uranium and uranyl luminescence in agate/chalcedony. *Mineral. Mag.* **2015**, *79*, 983–993.
15. Siegel, G.H.; Marrone, M.J. Photoluminescence in as-drawn and irradiated silica optical fibers: An assessment of the role of non-bridging oxygen defect centers. *J. Non-Cryst. Solids* **1981**, *45*, 235–247.
16. Stevens-Kalceff, M.A. Cathodoluminescence microcharacterization of point defects in α-quartz. *Mineral. Mag.* **2009**, *73*, 585–606.
17. Götze, J.; Plötze, M.; Habermann, D. Cathodoluminescence (CL) of quartz: Origin, spectral characteristics and practical applications. *Mineral. Petrol.* **2001**, *71*, 225–250.
18. Zielinski, R.A. Uranium mobility during interaction of rhyolitic obsidian, perlite and felsite with alkaline carbonate solution: T=120 °C, P=210 kg/cm². *Chem. Geol.* **1979**, *27*, 47–63.
19. Pan, Y.; Li, D.; Feng, R.; Wiens, E.; Chen, N.; Götze, J.; Lin, J. Uranyl binding mechanism in microcrystalline silicas: A potential missing link for uranium mineralization by direct uranyl co-precipitation and environmental implications. *Geochim. Cosmochim. Acta* **2021**, *292*, 518–531.
20. Dumańska-Słowik, M.; Natkaniec-Nowak, L.; Weselucha-Birczyńska; Gaweł, A.; Lankosz, M.; Wróbel, P. Agates from Sidi Rahal, in the Atlas Mountains of Morocco: Gemmological characteristics and proposed origin. *Gems Gemol.* **2013**, *49*, 148–159.
21. Götze, J.; Hofmann, B.; Machałowski, T.; Tsurkan, M.V.; Jesionowski, T.; Ehrlich, H.; Kleeberg, R.; Ottens, B. Biosignatures in subsurface filamentous fabrics (SFF) from the Deccan Volcanic Province, India. *Minerals* **2020**, *10*, 540.

22. Moxon, T.; Rios, S. Moganite and water content as a function of age in agate: An XRD and thermogravimetric study. *Eur. J. Mineral.* **2004**, *16*, 269–278.
23. Moxon, T.; Carpenter, M.A. Crystallite growth kinetics in nanocrystalline quartz (agate and chalcedony). *Mineral. Mag.* **2009**, *73*, 551–568.
24. Moxon, T. Agates: A study of ageing. *Eur. J. Mineral.* **2002**, *14*, 1109–1118.
25. Moxon, T. A re-examination of water in agate and its bearing on the agate genesis enigma. *Mineral. Mag.* **2017**, *81*, 1223–1244.

MDPI

St. Alban-Anlage 66

4052 Basel

Switzerland

Tel. +41 61 683 77 34

Fax +41 61 302 89 18

www.mdpi.com

Minerals Editorial Office

E-mail: minerals@mdpi.com

www.mdpi.com/journal/minerals